Springer Transactions in Civil
and Environmental Engineering

More information about this series at http://www.springer.com/series/13593

Natarajan Narayanan · Berlin Mohanadhas
Vasudevan Mangottiri
Editors

Flow and Transport in Subsurface Environment

Editors
Natarajan Narayanan
Department of Civil engineering
Dr. Mahalingam College of Engineering
  and Technology
Pollachi, Tamil Nadu
India

Vasudevan Mangottiri
Department of Civil Engineering
Bannari Amman Institute of Technology
Sathyamangalam, Tamil Nadu
India

Berlin Mohanadhas
Department of Civil Engineering
National Institute of Technology,
  Arunachal Pradesh
Yupia, Arunachal Pradesh
India

ISSN 2363-7633   ISSN 2363-7641 (electronic)
Springer Transactions in Civil and Environmental Engineering
ISBN 978-981-10-8772-1   ISBN 978-981-10-8773-8 (eBook)
https://doi.org/10.1007/978-981-10-8773-8

Library of Congress Control Number: 2018934863

© Springer Nature Singapore Pte Ltd. 2018
This work is subject to copyright. All rights are reserved by the Publisher, whether the whole or part of the material is concerned, specifically the rights of translation, reprinting, reuse of illustrations, recitation, broadcasting, reproduction on microfilms or in any other physical way, and transmission or information storage and retrieval, electronic adaptation, computer software, or by similar or dissimilar methodology now known or hereafter developed.
The use of general descriptive names, registered names, trademarks, service marks, etc. in this publication does not imply, even in the absence of a specific statement, that such names are exempt from the relevant protective laws and regulations and therefore free for general use.
The publisher, the authors and the editors are safe to assume that the advice and information in this book are believed to be true and accurate at the date of publication. Neither the publisher nor the authors or the editors give a warranty, express or implied, with respect to the material contained herein or for any errors or omissions that may have been made. The publisher remains neutral with regard to jurisdictional claims in published maps and institutional affiliations.

Printed on acid-free paper

This Springer imprint is published by the registered company Springer Nature Singapore Pte Ltd. part of Springer Nature
The registered company address is: 152 Beach Road, #21-01/04 Gateway East, Singapore 189721, Singapore

Foreword

The book entitled "Flow and Transport in Subsurface Environment" by Natarajan, Vasudevan, and Berlin is comprised of three parts. This book addresses subsurface flow in geological structures. A combination of experimental and numerical works in this area is presented. Coverage essentially encompasses the groundwater and petroleum applications. Part 1 of this book deals with contaminant migration and is composed of four chapters. Part 2 covers numerical modeling and covers a couple of chapters. The final Part 3 has the most extensive coverage of all three parts. It deals with transport through porous media and includes seven chapters. This book presents a collection of different aspects of heat and mass transfer in porous media. The mathematical formulation mostly adopted is based on Darcy's law and non-linear coupled partial differential equations for mass, momentum, induction, and energy transfer. The employed numerical solution is mainly based on implicit finite difference for combined heat and mass transfer under the influence of multiple physical parameters such as spatial inclination, non-uniform heating, viscous dissipation, and magnetic field. The experimental methodology adopted in few chapters deals with field applications such as immiscible foam displacement for enhanced oil recovery and straining in reservoirs as well as migration and entrapment of mercury in porous media. The covered topics are quite interesting and timely. This book will be of interest for groundwater researchers.

Kambiz Vafai
Distinguished Professor of Mechanical Engineering
Director of the Online Master of Science Program
in Engineering at UCR
Editor-in-Chief of JPM & STRPM
University of California, Riverside, USA
http://vafai.engr.ucr.edu

Preface

Subsurface fluid flow occurs in a complex and heterogeneous hydrogeologic environment whose structural, morphological, and petro-physical characteristics vary not only spatially but also with induced chemical transformations causing temporal changes in these parameters. Intellectual attempts to understand the complexities of flow and chemical transport in highly distinctive porous environment help us to critically evaluate some of the systemic limitations in classical computational methodology as well as challenges in advanced analytical tools. Being an important field of interest for many, such as hydrogeologists, petroleum and chemical engineers, researchers working in fluid and soil mechanics as well as scientists of many other disciplines, we have attempted to present the recent advances in experimental, analytical, and numerical studies in porous media.

Providing insight on the latest face of research in the porous system by collectively presenting an interconnected rather ensemble view of multifaceted applications forms the motivation for bringing out this book. Essentially, this book is a blend of experimental and numerical studies conducted in the porous domain by researchers from various backgrounds such as petroleum, groundwater, fluid mechanics. Experimental and modeling studies pertaining to multiphase flow and combined heat and mass transfer have been specifically addressed. A brief description of each of the chapters in this book is as follows.

Part I: Contaminant Migration in Complex Environment

The chapter entitled "Fines Migration in Aquifers and Oilfields: Laboratory and Mathematical Modelling" provides a detailed study on the mobilization, migration, and straining of fines in the natural reservoirs through experimental and mathematical modeling.

The chapter entitled "Migration and Capillary Entrapment of Mercury in Porous Media" describes the behavior of entrapped mercury in the subsurface porous media through pore-scale micro-model experiments.

The chapter entitled "New Insight into Immiscible Foam for Enhancing Oil Recovery" examines the effect of surfactant concentration on the strength and propagation of foams using X-ray CT scans.

Part II: Numerical Modeling of Fluid Flow Under Heterogeneous Conditions

The chapter entitled "Numerical Simulation of Flows in a Channel with Impermeable and Permeable Walls Using Finite Volume Methods" analyzes the laminar flows in a channel with permeable and impermeable walls using FV-CFD techniques.

The chapter entitled "A Comparative Analysis of Mixed Finite Element and Conventional Finite Element Methods for One-Dimensional Steady Heterogeneous Darcy Flow" uses mixed finite element method to obtain highly accurate flux distribution in groundwater flow applications.

The chapter entitled "Subsurface Acid Sulphate Pollution and Salinity Intrusion in Coastal Groundwater Environments" provides a brief review on the nature of groundwater modeling with the focus on acid sulfate soils and salinity pollution.

Part III: Synergetic Effects of Heat and Mass Transfer in Porous Media

The chapter entitled "Fully Developed Magnetoconvective Heat Transfer in Vertical Double-Passage Porous Annuli" examines the combined analytical and numerical investigation of fully developed mixed convective flow in a vertical double-passage porous annuli formed by three vertical concentric cylinders.

The chapter entitled "Effect of Nonuniform Heating on Natural Convection in a Vertical Porous Annulus" investigates the natural convection in a vertical annular cavity filled with saturated porous media numerically.

The chapter entitled "Natural Convection in an Inclined Parallelogrammic Porous Enclosure" investigates the natural convection in an inclined parallelogrammic porous enclosure numerically.

The chapter entitled "Natural Convection of Cold Water Near Its Density Maximum in a Porous Wavy Cavity" investigates the buoyant convective flow and heat transfer of cold water near its density maximum in a wavy porous square cavity.

The chapter entitled "Convective Mass and Heat Transfer of a Chemically Reacting Fluid in a Porous Medium with Cross Diffusion Effects and Convective Boundary" examines the effects of Soret, Dufour, and convective boundary condition on unsteady free convective flow, heat and mass transfer over a stretching surface subjected to first-order chemical reaction, heat generation/absorption and suction/injection effects.

The chapter entitled "Local Non-similar Solution of Induced Magnetic Boundary Layer Flow with Radiative Heat Flux" deals with the motion of steady 2D boundary layer viscous heat transfer flow past a stretching surface under the influence of aligned magnetic field with thermal radiation.

We deeply express our heartfelt and immense gratitude to the authors for their valuable scientific contribution. But for their tireless efforts, this book would not have been a success. Also, we appreciate the support rendered by the reviewers for their critical and unbiased comments. We express our sincere thanks to Springer for supporting our small but inspiring effort to publish this work. We believe that this book would be highly beneficial to the scientific community by enlightening them on the crucial elements of simulating flow through porous media under complex environmental conditions.

Pollachi, India	Natarajan Narayanan
Sathyamangalam, India	Vasudevan Mangottiri
Yupia, India	Berlin Mohanadhas

Contents

Part I Contaminant Migration in Complex Environment

Fines Migration in Aquifers and Oilfields: Laboratory and Mathematical Modelling 3
Y. Yang, F. D. Siqueira, A. Vaz, A. Badalyan, Z. You, A. Zeinijahromi, T. Carageorgos and P. Bedrikovetsky

Migration and Capillary Entrapment of Mercury in Porous Media 69
M. Devasena and Indumathi M. Nambi

New Insight into Immiscible Foam for Enhancing Oil Recovery 91
Mohammad Simjoo and Pacelli L. J. Zitha

Part II Numerical Modeling of Fluid Flow Under Heterogeneous Conditions

Numerical Simulation of Flows in a Channel with Impermeable and Permeable Walls Using Finite Volume Methods 119
Z. F. Tian, C. Xu and P. A. Dowd

A Comparative Analysis of Mixed Finite Element and Conventional Finite Element Methods for One-Dimensional Steady Heterogeneous Darcy Flow 141
Debasmita Misra and John L. Nieber

Subsurface Acid Sulphate Pollution and Salinity Intrusion in Coastal Groundwater Environments 189
Gurudeo Anand Tularam and Rajibur Reza

Part III Synergetic Effects of Heat and Mass Transfer in Porous Media

Fully Developed Magnetoconvective Heat Transfer in Vertical Double-Passage Porous Annuli 217
M. Sankar, N. Girish and Z. Siri

Effect of Nonuniform Heating on Natural Convection in a Vertical Porous Annulus 251
M. Sankar, S. Kiran and Younghae Do

Natural Convection in an Inclined Parallelogrammic Porous Enclosure 279
Bongsoo Jang, R. D. Jagadeesha, B. M. R. Prasanna and M. Sankar

Natural Convection of Cold Water Near Its Density Maximum in a Porous Wavy Cavity 305
S. Sivasankaran

Convective Mass and Heat Transfer of a Chemically Reacting Fluid in a Porous Medium with Cross Diffusion Effects and Convective Boundary 325
M. Bhuvaneswari and S. Sivasankaran

Local Non-similar Solution of Induced Magnetic Boundary Layer Flow with Radiative Heat Flux 343
M. Ferdows and Sakawat Hossain

Editors and Contributors

About the Editors

Dr. Natarajan Narayanan received his Ph.D. from the Indian Institute of Technology Madras, Chennai, India, and subsequently worked as a Postdoctoral Fellow at the University of Adelaide, South Australia. His research expertise includes numerical modeling of flow and transport in homogeneous and heterogeneous porous media, and modeling of coupled processes in fractured porous media. He has publications in international journals and conference proceedings.

Dr. Berlin Mohanadhas is an Assistant Professor in the Department of Civil Engineering, National Institute of Technology, Arunachal Pradesh, India. He secured his doctoral degree from the Indian Institute of Technology Madras, Chennai, India. His research interests include numerical modeling of nitrate transport and transformation through unsaturated porous media. He is also working on the transport of petroleum hydrocarbon and radionuclides in the vadose zone and has published in various international journals and conference proceedings.

Dr. Vasudevan Mangottiri received his B.Tech. in Agricultural Engineering from Kerala Agricultural University, Kerala, in 2007; his M.Tech. in Civil (Environmental) Engineering from Motilal Nehru National Institute of Technology, Allahabad, in 2009; and his Ph.D. from the Indian Institute of Technology Madras, Chennai, India, in 2015. He is currently working as an Assistant Professor in the Department of Civil Engineering, Bannari Amman Institute of Technology, Sathyamangalam, Tamil Nadu. His research areas include numerical modeling of contaminant transport in porous media and environmental remediation of hazardous chemicals. He has published in various international journals and conference proceedings.

Contributors

A. Badalyan Australian School of Petroleum, University of Adelaide, Adelaide, Australia

P. Bedrikovetsky Australian School of Petroleum, University of Adelaide, Adelaide, Australia

M. Bhuvaneswari Department of Mathematics, King Abdulaziz University, Jeddah, Saudi Arabia

T. Carageorgos Australian School of Petroleum, University of Adelaide, Adelaide, Australia

M. Devasena Department of Civil Engineering, Sri Krishna College of Technology, Coimbatore, Tamil Nadu, India

Younghae Do Department of Mathematics, KNU-Center for Nonlinear Dynamics, Kyungpook National University, Daegu, Republic of Korea

P. A. Dowd School of Civil, Environmental and Mining Engineering, The University of Adelaide, Adelaide, SA, Australia

M. Ferdows Research Group of Fluid Flow Modeling and Simulation, Department of Applied Mathematics, University of Dhaka, Dhaka, Bangladesh

N. Girish Department of Mathematics, JSS Academy of Technical Education, Bangalore, India

Sakawat Hossain Research Group of Fluid Flow Modeling and Simulation, Department of Applied Mathematics, University of Dhaka, Dhaka, Bangladesh

R. D. Jagadeesha Department of Mathematics, Government Science College, Hassan, India

Bongsoo Jang Department of Mathematical Sciences, Ulsan National Institute of Science and Technology (UNIST), Ulsan, Republic of Korea

S. Kiran Department of Mathematics, Sapthagiri College of Engineering, Bangalore, India

Debasmita Misra Department of Mining and Geological Engineering, College of Engineering and Mines, University of Alaska Fairbanks, Fairbanks, AK, USA

Indumathi M. Nambi EWRE Division, Department of Civil Engineering, Indian Institute of Technology-Madras, Chennai, Tamil Nadu, India

John L. Nieber Department of Biosystems and Agricultural Engineering, Biosystems and Agricultural Engineering Building, University of Minnesota, St. Paul, MN, USA

B. M. R. Prasanna Department of Mathematics, Siddaganga Institute of Technology, Tumkur, India

Rajibur Reza Department of Accounting, Finance and Economics, Environmental Futures Research Institute, Griffith Business School, Griffith University, Nathan, QLD, Australia

M. Sankar Department of Mathematics, School of Engineering, Presidency University, Bangalore, India; Department of Mathematics, KNU-Center for Nonlinear Dynamics, Kyungpook National University, Daegu, Republic of Korea

Mohammad Simjoo Sahand University of Technology, Tabriz, Iran

F. D. Siqueira North Fluminense State University of Rio de Janeiro, Lenep-UENF, Macaé-RJ, Brazil

Z. Siri Institute of Mathematical Sciences, University of Malaya, Kuala Lumpur, Malaysia

S. Sivasankaran Department of Mathematics, King Abdulaziz University, Jeddah, Saudi Arabia

Z. F. Tian School of Mechanical Engineering, The University of Adelaide, Adelaide, SA, Australia

Gurudeo Anand Tularam Mathematics and Statistics, Griffith Sciences [ENV], Environmental Futures Research Institute, Griffith University, Nathan, QLD, Australia

A. Vaz North Fluminense State University of Rio de Janeiro, Lenep-UENF, Macaé-RJ, Brazil

C. Xu School of Civil, Environmental and Mining Engineering, The University of Adelaide, Adelaide, SA, Australia

Y. Yang Australian School of Petroleum, University of Adelaide, Adelaide, Australia

Z. You Australian School of Petroleum, University of Adelaide, Adelaide, Australia

A. Zeinijahromi Australian School of Petroleum, University of Adelaide, Adelaide, Australia

Pacelli L. J. Zitha Delft University of Technology, Delft, The Netherlands

Part I
Contaminant Migration in Complex Environment

Fines Migration in Aquifers and Oilfields: Laboratory and Mathematical Modelling

Y. Yang, F. D. Siqueira, A. Vaz, A. Badalyan, Z. You, A. Zeinijahromi, T. Carageorgos and P. Bedrikovetsky

Nomenclature

A_{132}	Hamaker constant for interaction between materials 1 and 2 in medium 3, $ML^2\,T^{-2}$
c	Suspended particle concentration, L^{-3}
C	Dimensionless suspended particle concentration
C_{mi}	Molar concentration of i-th ion, L^{-3}
D	Dispersion coefficient
D_e	Dielectric constant
e	Electron charge, C
E	Young's modulus, $ML^{-1}\,T^{-2}$
F	Force, $ML\,T^{-2}$
h	Particle-surface separation distance, L
H	Half-width of the channel, L
J	Impedance (normalised reciprocal of mean permeability)
k	Permeability, L^2
k_{det}	Detachment coefficient
$\langle k \rangle$	Mean permeability, L^2
k_B	Boltzmann constant, $ML^2\,T^{-2}\,K^{-1}$
k_n	Number of data points in a given stage
K	Composite Young's modulus, $ML^{-1}\,T^{-2}$
l	Lever arm ratio
l_n	Normal lever, L
l_d	Tangential (drag) lever, L
L	Core length, L
p	Pressure, $MT^{-2}\,L^{-1}$

Y. Yang · A. Badalyan · Z. You (✉) · A. Zeinijahromi · T. Carageorgos · P. Bedrikovetsky
Australian School of Petroleum, University of Adelaide, Adelaide 5005, Australia
e-mail: zhenjiang.you@adelaide.edu.au

F. D. Siqueira · A. Vaz
North Fluminense State University of Rio de Janeiro, Lenep-UENF,
Rod. Amaral Peixoto km 163, Imboassica, Macaé-RJ 27925-310, Brazil

© Springer Nature Singapore Pte Ltd. 2018
N. Narayanan et al. (eds.), *Flow and Transport in Subsurface Environment*,
Springer Transactions in Civil and Environmental Engineering,
https://doi.org/10.1007/978-981-10-8773-8_1

P	Dimensionless pressure
n	Serial number of variant velocities in multi-rate test
N	Serial number of final velocity
r_s	Radius of a particle, L
r_{scr}	Critical radius of a particle that can be removed at certain velocity, L
S_a	Dimensionless attached particle concentration
S_s	Dimensionless strained particle concentration
ΔS_a	Dimensionless mobilised concentration of detached particles with velocity alteration
t	Time, T
T	Dimensionless time
$t_{st,n}$	Stabilisation time for n-th flow rate, T
$T_{st,n}$	Dimensionless stabilisation time for n-th flow rate
t_n	Initial time of n-th flow rate, T
T_n	Dimensionless initial time of n-th flow rate
$\bar{u}$	Average velocity through a slot
u_t	Tangential crossflow velocity of fluid in the centre of the particle
U	Darcy's velocity, LT^{-1}
U_s	Particle's seepage velocity, LT^{-1}
V	Potential energy, $ML^2 T^{-2}$
x	Linear coordinate, L
X	Dimensionless linear coordinate
z_i	Electrolyte valence of the i-th ion

Greek Symbols

α	Drift delay factor
β	Formation damage coefficient
Υ	Salinity
ε	Dimensionless delay time
ε_0	Free space permittivity, $C^{-2} J^{-1} m^{-1}$
η	Intersection of characteristic line and the T-axis
κ	Debye length, L^{-1}
λ_a	Filtration coefficient for attachment mechanism, L^{-1}
λ_s	Filtration coefficient for straining mechanism, L^{-1}
Λ_a	Dimensionless filtration coefficient for attachment mechanism
Λ_s	Dimensionless filtration coefficient for straining mechanism
μ	Dynamic viscosity, $ML^{-1} T^{-1}$
ν	Poisson's ratio
ρ	Fluid density, ML^{-3}
ρ_s	Particle density, ML^{-3}
σ_{cr}	Critical retention function, L^{-3}
$\Sigma_a(r_s)$	Size distribution of attached particles, L^{-3}

σ	Concentration of retained particles, L^{-3}
$\Delta\sigma_n$	Mobilised concentration of detached particles with velocity switch from U_{n-1} to U_n
σ_{LJ}	Atomic collision diameter, L
τ	Delay time of particle release, T
υ_i	Number of ions per unit volume
ω	Dimensionless coordinate of an immediate core point
χ	Lift factor
ϕ	Porosity
Ψ_{01}	Particle surface potential
Ψ_{02}	Collector surface potential
ω	Drag factor

Subscripts

a	Attached (for fine particles)
d	Drag (for force)
g	Gravitational (for force)
$i$ion	Injected ions
0ion	Initial ions
l	Lift (for force)
e	Electrostatic (for force)
max	Maximum
n	Normal (for force), flow rate number (for velocities, inherited retained concentrations, particle–fluid velocity ratios, inherited impedances)
BR	Born repulsion (for potential energy)
DLR	Electrostatic double layer (for potential energy)
LVA	London–van der Waal (for potential energy)
0	Initial value or condition (for permeability, retained concentrations)

1 Introduction

Fines migration with consequent permeability reduction has been widely recognised to cause formation damage in numerous petroleum, environmental and water resource processes (Noubactep 2008; Noubactep et al. 2012; Faber et al. 2016). Fines migration takes place during oil and gas production in conventional and unconventional reservoirs, significantly reducing well productivity (Sarkar and Sharma 1990; Byrne and Waggoner 2009; Byrne et al. 2014; Civan 2014). Natural and induced fines migration has occurred in the waterflooding of oilfields. It also causes drilling and completion fluids to invade the formation (Watson et al. 2008;

Fleming et al. 2007, 2010). Despite significant progress in the above-mentioned technologies, clogging of production and injection wells remains a major operational issue.

The distinguishing features of natural reservoir fines migration are mobilisation of the attached particles, their capture by straining in the rock, permeability reduction and consequent decline in well productivity and injectivity (Fig. 1). Several laboratory studies observed permeability decline during coreflooding with piecewise-constant increase in velocity in (Ochi and Vernoux 1998). Similar effects occur during piecewise-constant change in water salinity or pH during coreflooding (Lever and Dawe 1984). Numerous authors attribute the permeability reduction during velocity increase, salinity decrease and pH increase, to mobilisation of the attached fine particles and their migration into pore spaces until size exclusion in thin pore throats (Muecke 1979; Sarkar and Sharma 1990). Figure 1 shows a schematic for attached and size-excluded fine particles in the porous space, along with definitions of the concentrations of attached, suspended and strained particles. Detachment of fines from the grain surfaces yields an insignificant increase in permeability, whereas the straining in thin pore throats and consequent plugging of conducting paths causes significant permeability decline. The main sources of movable fine particles in natural reservoirs are kaolinite, chlorite and illite clays; quartz and silica particles can be mobilised in low-consolidated sandstones (Khilar and Fogler 1998). Usually, the kaolinite booklets of thin slices cover the grain surfaces (Fig. 2). Detachment of a thin, large slice from the booklet can result in plugging of a large pore.

Figure 3a, e show typical decreasing permeability curves during velocity increase.

The laboratory-based mathematical modelling supports planning and design of the above-mentioned processes. Classical filtration theory applied to particle detachment includes a mass balance equation for suspended, attached and strained particles:

$$\frac{\partial}{\partial t}[\phi c + \sigma_a + \sigma_s] + U\frac{\partial c}{\partial x} = 0 \qquad (1)$$

Fig. 1 Schematic for fines detachment, migration and straining with consequent permeability decline

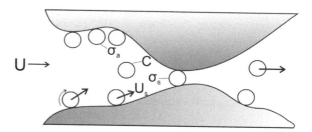

Fig. 2 Kaolinite particles attached to the grain surface (SEM image): **a** leaflet shape and **b** leaflets in the pore space

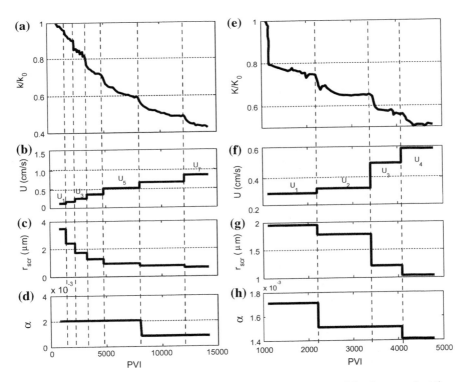

Fig. 3 Normalised permeability, flow rate, critical fine radius and drift delay factor against time, during coreflood with piecewise increasing velocity during test I (first column) and test II (second column): **a, e** experimentally determined permeability decline with time; **b, f** increasing velocity during the test; **c, g** decrease of the mobilised fines radius as velocity increases as calculated from torque balance; and **d, h** drift delay factor from the model adjustment

where c, σ_a and σ_s are the concentrations of suspended, attached and strained particles, respectively, and U is flow velocity of the carrier fluid, which coincides with particle speed.

The kinetics of simultaneous particle attachment and detachment is given by the relaxation equation (Bradford and Bettahar 2005; Tufenkji 2007; Bradford et al. 2012, 2013; Zheng et al. 2014; Bai et al. 2015b)

$$\frac{\partial \sigma_a}{\partial t} = \lambda_a c U - k_{\text{det}} \sigma_a \qquad (2)$$

where λ_a is the filtration coefficient for attachment and k_{det} is the detachment coefficient.

The irreversible fines straining rate in thin pore throats is expressed by the linear kinetics equation where the straining rate is proportional to the advective flux of suspended particles (Herzig et al. 1970; Yuan and Shapiro 2011a, b; You et al. 2013; Sacramento et al. 2015):

$$\frac{\partial \sigma_s}{\partial t} = \lambda_s c U \qquad (3)$$

Modified Darcy's law accounts for permeability damage due to both attachment and straining (Pang and Sharma 1997; Krauss and Mays 2014):

$$U = -\frac{k}{\mu(1 + \beta_s \sigma_s + \beta_a \sigma_a)} \frac{\partial p}{\partial x} \qquad (4)$$

Figure 1 illustrates the common assumption that the coating of grain by attached particles causes significantly lower permeability damage than does straining: $\beta_s \gg \beta_a$, i.e. the combination of particle detachment and straining is the primary cause of the decline in permeability. Therefore, the term $\beta_a \sigma_a$ in Eq. (4) that accounts for permeability increase due to detachment is negligible.

Civan (2010, 2014) presented numerous generalisations of the governing Eqs. (1)–(4), to account for non-Newtonian behaviour of suspension fluxes, non-equilibrium for deep-bed filtration of high-concentration suspensions and colloids, and particle bridging at thin pore throats.

Quasi-linear system of partial differential Eqs. (1)–(3) exhibits the delayed reaction to an abrupt injection rate alteration, whereas laboratory tests show an instant permeability and breakthrough concentration response (Ochi and Vernoux 1998; Bedrikovetsky et al. 2012a, b). This discrepancy between the modelling and laboratory data, and the corresponding shortcoming in the theory, has been addressed in the modified model for particle detachment, by introducing the maximum attached concentration as a velocity function $\sigma_a = \sigma_{cr}(U)$ (Bedrikovetsky et al. 2011a, b). If the attached concentration exceeds this maximum value, particle detachment occurs and the detached particles follow the classical filtration Eq. (3); otherwise, the maximum attached concentration holds. The dependency

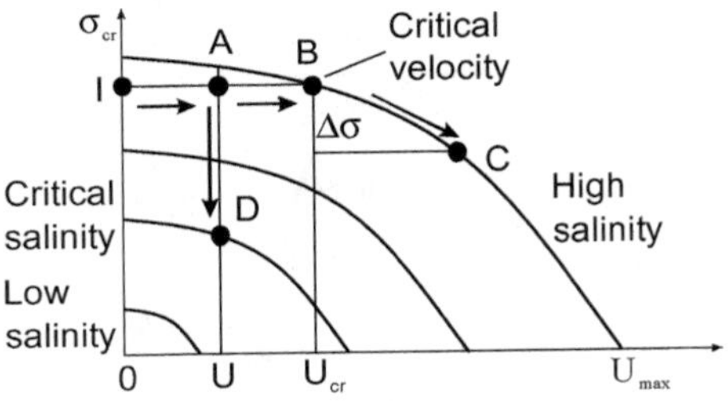

Fig. 4 Velocity and salinity dependencies of maximum retention function with introduction of critical velocity and salinity

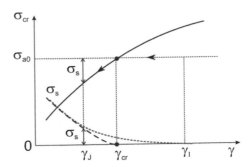

Fig. 5 Strained concentration σ_s in large-scale approximation is determined by the maximum retention function $\sigma_{cr}(\gamma)$. Here, concentrations σ_s and $\sigma_{cr}(\gamma)$ are extrapolated by the vanishing function into the domain $\sigma < \sigma_{cr}(\gamma)$, where no particles are mobilised

$\sigma_a = \sigma_{cr}(U)$ is called the maximum retention function. The following set of equations captures the above attachment–detachment scenario:

$$\begin{cases} \frac{\partial \sigma_a}{\partial t} = \lambda_a U c, & \sigma_a < \sigma_{cr}(U) \\ \sigma_a = \sigma_{cr}(U) \end{cases} \quad (5)$$

The maximum retention function decreases as the flow velocity increases. Therefore, the velocity increase causes instant release of the excess attached fine particles.

The maximum retention function $\sigma_{cr}(U)$ is an empirically based (material) function of the model and can be determined only by the inverse-problem approach applied to fines-migration tests (Figs. 4 and 5). However, it can be calculated theoretically for a simplified geometry of porous space, using the conditions of mechanical equilibrium of particles attached to the rock surface.

Freitas and Sharma (2001), Bergendahl and Grasso (2003), and Bradford et al. (2013) discussed the torque balance of attaching and detaching forces exerting on a particle situated at the rock or internal cake surface (Fig. 6):

$$F_d(U, r_{scr})l(r_{scr}) = F_e(r_{scr}) - F_l(U, r_{scr}) + F_g(r_{scr}), \quad l = l_d/l_n \quad (6)$$

Here, F_d, F_e, F_l and F_g represent drag, electrostatic, lift and gravitational forces, respectively; l_d and l_n are the lever arms for drag and normal forces, respectively. Substitution of the expressions for drag, electrostatic, lift and gravitational forces

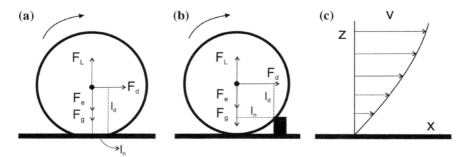

Fig. 6 Scenarios involving particle detachment in monolayer on the grain surface and forces exerting on the particles. Torque and force balance on a fine particle attached to the pore wall: **a** the lever arm is equal to the contact area radius, deformation due to attracting electrostatic force. **b** The lever arm is determined by the asperity size. **c** Velocity distribution in Hele–Shaw flow in a pore

into the torque balance equation (6) yields an expression of the maximum retention function (Bedrikovetsky et al. 2011a). The maximum retention function (5) for the case of poly-layer attachment of single-radius particles in rock having mono-sized cylindrical capillaries is a quadratic polynomial with flow velocity as the variable. The maximum retention function for a monolayer of polydispersed particles is expressed via size distribution for fine particles (You et al. 2015, 2016).

Expression (5) substitutes the equation for simultaneous attachment and detachment (2) in the mathematical model for colloidal-suspension transport (1)–(3). The modified model consists of three Eqs. (1), (3) and (5) for three unknown concentrations c, σ_a and σ_s. Equation (4) for pressure is separated from system (1), (3) and (5). Let us discuss the case of low straining concentration, where size exclusion does not affect the probability of particle capture. In this case, the filtration coefficient λ_s is constant, whereas it should be a σ_s-function in the general case. We also discuss the case of injection with timely decreasing rate, where the attached concentration is equal to the maximum retained concentration [second line in Eq. (5)]. The one-dimensional flow problem with attachment and detachment allows for exact solution, yielding suspended, attached and strained concentrations and pressure drop across the core. The laboratory- and theoretically determined maximum retention functions are in high agreement, which validates the maximum retention function as a mathematical model for particle detachment (Zeinijahromi et al. 2012a, b; Nguyen et al. 2013).

In the case of large stained concentration, $\lambda_s = \lambda_s(\sigma_s)$. Suspended concentration can be expressed from Eq. (4) as a time derivative of a σ_s-dependent potential. Its substitution into Eq. (1) and integration in t reduces the system to one non-linear first-order partial differential equation, which is solved analytically using the method of characteristics (Alvarez et al. 2006, 2007).

Usually, fines-migration tests are performed under piecewise-constant decreasing velocity (Ochi and Vervoux 1998; Bedrikovetsky et al. 2011a; Oliveira et al. 2014, 2016). The amount of released particles during abrupt velocity alteration forms the initial suspended concentration for system (1), (3) and (4) with unknowns c, σ_s and p. Thus, the basic governing equations for deep-bed filtration and fines migration are the same (Herzig et al. 1970). The initial suspended concentration for deep-bed filtration in clean beds is zero, whereas for fines migration the initial suspended concentration is defined by the velocity alteration. Inlet boundary condition for deep-bed filtration is equal to concentration of the injected suspension, whereas it equals zero for fines migration. Therefore, the methods of exact integration of direct problems and regularisation of inverse problems, developed by Alvarez et al. (2006, 2007) for deep-bed filtration, can be applied for fines migration also.

The axisymmetric analogue of Eqs. (1), (3)–(5) describes the near-well flows, allowing estimating the well inflow performance accounting for fines migration. Zeinijahromi et al. (2012a, b) derive an analytical model for intermediate times with steady-state suspension concentration. Bedrikovetsky et al. (2012b) present an analytical steady-state model for late times, when all fines are either produced or strained. Marques et al. (2014) derive the analytical transient model for the overall well inflow period.

However, the exact solution of system (1), (3), (4) and (5) shows stabilisation of the pressure drop after injection of one pore volume (Bedrikovetsky et al. 2011a, b), whereas numerous laboratory studies have exhibited periods of stabilisation within 30–500 PVI (here PVI stands for pore volume injected) (Ochi and Vernoux 1998; Oliveira et al. 2014). Figure 3a shows the permeability stabilisation within 70–3000 PVI, for various injection velocities (Ochi and Vernoux 1998). The stabilisation times for flow exhibited in Fig. 3e vary from 300 to 1200 PVI. Therefore, the modified model for colloidal-suspension transport in porous media (1), (3) and (5) approximates well the stabilised permeability but fails to predict the long stabilisation period.

Several works have observed slow surface motion of the mobilised particles and simultaneous fast particle transport in the bulk of the aqueous suspension. Li et al. (2006) attributed the slow surface motion to particles in the secondary energy minimum. Yuan and Shapiro (2011a) and Bradford et al. (2012) introduced slow particle velocity into the classical suspension flow model, resulting in a two-speed model that matched their laboratory data on breakthrough concentration. Navier–Stokes-based simulation of colloids' behaviour at the pore scale, performed by Sefrioui et al. (2013), also exhibited particle transport speeds significantly lower than the water velocity. However, classical filtration theory along with the modified particle detachment model assumes that particle transport is at carrier fluid velocity (Tufenkji 2007; Civan 2014).

Oliveira et al. (2014) attributed long stabilisation periods to slow drift of fine particles near the rock surface in the porous space. However, a mathematical model that depicts slow-particle migration and accurately reflects the stabilisation periods is unavailable in the literature.

Application of nanoparticles (NPs) can significantly decrease migration of the reservoir fines and the consequent permeability impairment (Habibi et al. 2012). Under certain salinity and pH, NPs attract both fines and grain. Low size of NPs causes high mobility and diffusion, spreading them over the grain surfaces. NPs 'glue' the fines and significantly increase the electrostatic fine-grain attraction (Ahmadi et al. 2013; Sourani et al. 2014a, b). The basic system of equations includes two mass balance equations for NPs and salt. Yuan et al. (2016) solved the system by the method of characteristics (Qiao et al. 2016). Combination of low-salinity and NP waterfloods in oilfields adds the above-mentioned mass balance and deep-bed filtration equations to Buckley–Leverett equation for two-phase flow in porous media (Bedrikovetsky 1993; Arab and Pourafshary 2013; Assef et al. 2014; Huang and Clark 2015; Dang et al. 2016).

Mahani et al. (2015a, b) observed delay between salinity alteration and corresponding surface change. This delay was attributed to saline water diffusion from the contact area between the deformed particle and rock surface. The Nernst–Planck diffusion in the thin slot between two plates subject to molecular-force action is significantly slower than the Brownian diffusion, so the Nernst–Planck diffusion can bring significant delay. The diffusive delay in particle mobilisation due to water salinity decrease can serve as another explanation for the long stabilisation period. Yet, a mathematical model that accounts for delay in particle mobilisation due to salinity alteration also seems absent from the literature.

In the current work, the long times for permeability stabilisation are attributed to slow surface motion of mobilised fine particles. The governing system (1), (3) and (5) is modified further by replacing the water flow velocity U by the particle velocity $U_s < U$ (Fig. 1). We also introduce a maximum retention function with delay, which corresponds to the Nernst–Planck diffusion from the grain–particle contact area into the bulk of the fluid. We derive the maximum retention function for a monolayer of size-distributed fines, which accounts for its non-convex form. We found that during continuous velocity/pH/temperature increase or salinity decrease, the largest particles were released first. The obtained system with slow fines migration and delayed maximum retention function allows for exact solution for cases of piecewise-constant velocity/pH/temperature increase or salinity decrease. High agreement between the laboratory and modelling data validates the proposed model for slow surface motion of released fine particles in porous media.

The structure of the text is as follows. Section 2 presents the laboratory study of fines migration due to high velocities and presents the mathematical model for slow-particle migration that explains the long stabilisation periods. The derived analytical model provides explicit formulae for concentration profiles, histories and the pressure drop. Section 3 presents the laboratory study of fines migration due to low salinities, and it derives the mathematical model that accounts for slow-particle migration and for delayed fines mobilisation. Here, we also derive the analytical model. Section 4 presents the analytical model for fines mobilisation at high temperature. The recalculation method for varying salinity, temperature, pH, or velocity is developed. Section 5 presents fines migration in gas and coal-bed-methane reservoirs. Section 6 presents the conclusions.

2 Fine Particles Mobilisation, Migration and Straining Under High Velocities

This section presents the modelling and laboratory study of fine particles that migrate after having been detached by drag and lift forces at increased velocities. Section 2.1 presents a brief physical description of fines detachment in porous media and introduces the maximum retention function for a monolayer of size-distributed particles. A qualitative analysis of the laboratory results on long-term stabilisation gives rise to a slow-particle modification of the mathematical model for fines migration in porous media. Section 2.2 presents those basic equations accounting for slow-particle transport. Section 2.3 derives the analytical model for one-dimensional flow under piecewise increasing flow velocity with consequent fines release and permeability impairment. Section 2.4 describes the laboratory coreflood tests with fines mobilisation and examines how closely the analytical model matches the experimental data. Section 2.5 discusses the model's validity, following the results of the laboratory and analytical modelling.

2.1 Physics of Fines Detachment, Transport and Straining in Porous Media

In this section, we discuss the physics of fines detachment/mobilisation on a micro-scale. In the presence of low ionic strength or high flow velocity, reservoir fines are detached from the rock surface, mobilise and flow through the porous media as shown in Fig. 1. Four forces act on a fine particle attached to the surface of the grain: drag, lift, electrostatic and gravity (Fig. 6). For calculation of drag, we use the expression proposed by Bergendahl and Grasso (2003) and Bradford et al. (2013); lift is calculated using the formula of Akhatov et al. (2008); and the electrostatic forces are calculated using DLVO (Derjaguin–Landau–Verwey–Overbeek) theory (Khilar and Fogler 1998; Israelachvili 2011; Elimelech et al. 2013).

Elastic particles located on the grain surface undergo deformation due to gravitational, lift and electrostatic forces acting normal to the grain surface. The right side of Eq. (6) contains the resultant of these forces (normal force). We assume that at the mobilisation instant, a particle rotates around the rotation-touching point in the boundary of the particle–grain contact area (Fig. 6a). Also assumed is that the lever arm is equal to the radius of the contact area of particle deformation, which is subject to the normal force (Freitas and Sharma 2001; Schechter 1992; Bradford et al. 2013). The contact area radius is equal to the lever arm l_n and is calculated using Hertz's theory of mutual grain–particle deformation:

$$l_n^3 = \frac{F_n r_s}{4K} \quad l_d = \sqrt{r_s^2 - l_n^2} \quad K \equiv \frac{4}{3\left(\frac{1-v_1^2}{E_1} + \frac{1-v_2^2}{E_2}\right)} \tag{7}$$

Here, K is the composite Young modulus that depends on Poisson's ratio v and Young's elasticity modulus E of the particle and of the surface. Indices 1 and 2 refer to the particle and solid matrix surfaces, respectively.

Figure 6b depicts the scenario in which a particle revolves around the contacting roughness (asperity) on the surface of a grain. The elastic properties of rock and particle determine the value of l_n in the first scenario (Fig. 6a). In the second scenario (Fig. 6b), the value of l_n is determined by the surface roughness.

Two coreflood tests (I and II) at piecewise-constant increasing fluid velocity on Berea sandstone cores were carried out by Ochi and Vernoux (1998) and resulted in mobilisation of kaolinite particles (Fig. 3). We used the following electrostatic constants and parameters for quartz and kaolinite in order to calculate F_e in Eq. (6): surface potentials ψ_{01} and ψ_{02} (−55, −50 mv) for test I and (−70, −80 mv) for test II (Ochi and Vernoux 1998); the Hamaker constant 2.6×10^{-20} J (Welzen et al. 1981); atomic collision diameter 0.4×10^{-9} m (Das et al. 1994); and salinity 0.1 mol/L for test I and 0.01 mol/L for test II. The Hamaker constant was calculated using dielectric constant for water $D = 78.0$ and permittivity of free space (vacuum) $\varepsilon_0 = 8.854 \times 10^{-12}$ C^{-2} J^{-1} m^{-1} (Israelachvili 2011; Khilar and Fogler 1998). Electron charge was $e = 1.6 \times 10^{-19}$ C, Boltzmann's constant was $k_B = 1.3806504 \times 10^{-23}$ J/K and temperature was $T = 25$ °C. Young's modulus for kaolinite was 6.2 GPa and for quartz was 12 GPa (Prasad et al. 2002), and Poisson's ratios were 0.281 and 0.241 (Gercek 2007) and were used to evaluate the lever arm ratio according to Eq. (7). The above parameters were used to construct graphs for electrostatic potential and force versus separation particle–grain distance (Fig. 7a, b).

The total potential of interaction V determines electrostatic force F_e. Zero values for F_e correspond to energy extremes V_{max} and V_{min}, and the minimum value of electrostatic force is obtained from the inflection point of the total potential of interaction curve. The first test of Ochi and Vernoux (Fig. 3a–d) was favourable for attachment of kaolinite particles to the grain surface, which resulted in the absence of the secondary minimum on the total potential of interaction curve. For the second test (Fig. 3e–h), the values of the primary and secondary energy minima equalled 550 and 19 kT, respectively. The energy barrier was 87 kT, exceeding the values of the secondary minimum and allowing a particle to jump from secondary minimum to primary minimum (Elimelech et al. 2013). Therefore, the second test was unfavourable for kaolinite particle attachment to the grain surface.

Under the condition of mechanical equilibrium given by the torque balance Eq. (6), fluid flow velocity affects lift and drag forces, whereas particle size determines the magnitudes of all forces in the equation. Therefore, the critical radius of the particle mobilised by fluid flow with velocity U can be determined as follows: $r_{scr} = r_{scr}(U)$. This is an implicit function from Eq. (6), i.e. Equation (6) is a transcendent equation for implicit dependency $r_{scr} = r_{scr}(U)$. The stationary iterative numerical procedure can be used to solve Eq. (6) (Varga 2009). The graph of

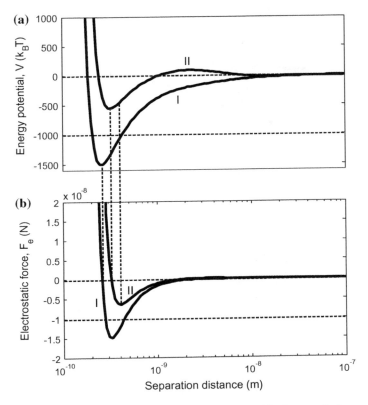

Fig. 7 Measured values for tests I and II: **a** energy potential and **b** electrostatic force

the function $r_{scr} = r_{scr}(U)$ obtained using the above parameters shows that the size of each mobilised particle $r_{scr}(U)$ decreases monotonically as fluid velocity increases. Therefore, those particles that remained immobilised on the grain surface at fluid velocity U have sizes $r < r_{scr}(U)$. The magnitude of F_e increases as the Hamaker constant increases (see Fig. 8), resulting in the right-shift of the $r_{scr}(U)$-curve.

Assume that the attached particles form a monolayer on the rock surface. The initial concentration distribution of attached particle sizes is denoted as $\Sigma_a(r_s)$. Particles are mobilised by descending size, as mentioned above. Thus, the critical retention concentration in Eq. (5) includes all particles with radii smaller than $r_{scr}(U)$:

$$\sigma_{cr}(U) = \sigma_{a0} \int_0^{r_{scr}(U)} \sum_a (r_s) dr_s \qquad (8)$$

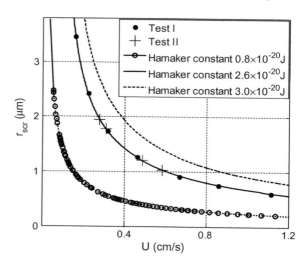

Fig. 8 Critical particle size (minimum size of the fine particles lifted by the flux with velocity U)

Now assume that the attached particles are size-distributed according to the breakage algorithm (i.e. log-normal distribution for attached particle sizes $\Sigma_a(r_s)$ holds (Jensen 2000). The forms of the maximum retention function as calculated by Eq. (8) for different size distributions of the attached particles, using the above values for electrostatic and elastic constants, are shown in Fig. 9.

Bedrikovetsky et al. (2011a, b) found that $\sigma_{cr}(U)$ for mono-sized particles that form the poly-layer coating on the surface of cylindrical pores is a quadratic polynomial. The corresponding curves $\sigma_{cr}(U)$ are convex. The model for the poly-layer coating can be modified by the introduction of size distributions for spherical particles and cylindrical pores; the resulting maximum retention curves can contain the concave parts (Fig. 9).

Figure 9 indicates that the maximum retention function for monolayer fines is not convex. The calculated $\sigma_{cr}(U)$-curves for three particle-size distributions characterised by equal variance coefficients support the above observation that the larger the particle, the higher the drag on the particle and the fewer the remaining particles (Fig. 9a). Similar calculations for log-normal distributions with the same average particle size and different variance coefficients result in the $\sigma_{cr}(U)$-curves shown in Fig. 9b. The higher fraction of mobilised large particles corresponds to larger coefficient of variation C_v. For the low-velocity range, $\sigma_{cr}(U)$-curves having high standard deviation lie lower; whereas with the increase in fluid velocity, $\sigma_{cr}(U)$-curves shift to higher σ_{cr}-values. Equation (8) shows that the $\sigma_{cr}(U)$-curve has a step shape for mono-sized particles ($C_v \to 0$), meaning that the maximum retention function is a step function. The wider the attached particle-size distribution, the wider the transitional spread of the $\sigma_{cr}(U)$-curves. The phenomenological model for fines detachment in porous media (1), (3)–(5) assumes the existence of a maximum retention function whose form is unconstrained.

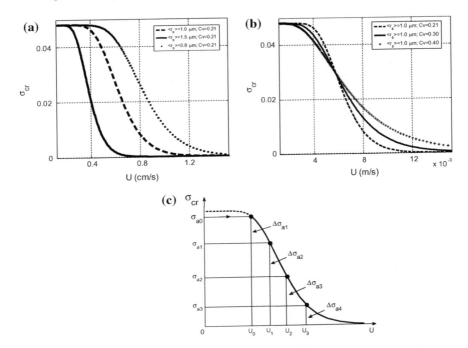

Fig. 9 Maximum retention function for the attached fines forming a monolayer on the pore surface: **a** for log-normal particle-size distributions with varying mean particle size. **b** For log-normal particle-size distributions with varying variance coefficient. **c** Determining the maximum retention function from the number of particles released at each abrupt velocity alteration

Now consider particle-free water being injected with increasing piecewise-constant velocity into a core. The movable attached fines concentration is σ_{a0}. There is no particle mobilisation at low fluid velocities (see Fig. 9c), because the attaching torque from Eq. (6) exceeds the detaching torque for all size particles: points $(U < U_0, \sigma_{a0})$ are located below the maximum retention curve. Concentration of attached particles remains constant along the horizontal arrow from the point $U = 0$ to critical velocity $U = U_0$. Value U_0 corresponds to the minimum velocity that results in mobilisation of particles and the consequent first fine appearance in the core effluent.

The initial concentration of attached fines σ_{a0} determines the critical velocity U_0 (Miranda and Underdown 1993; Hassani et al. 2014) as follows:

$$\sigma_{a0} = \sigma_{cr}(U_0) \tag{9}$$

Movement along the $\sigma_{cr}(U)$-curve corresponds to velocity increase above the critical value $U > U_0$. An instant rate change from U_1 to U_2 is accompanied by instant particle mobilisation with concentration $\Delta\sigma_{a1} = \sigma_{cr}(U_1) - \sigma_{cr}(U_2)$ and

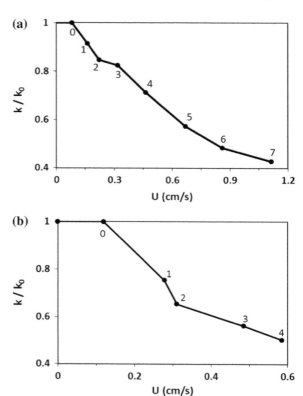

Fig. 10 Stabilised normalized permeability versus velocity: **a** test I and **b** test II

increase in suspended concentration by $[\sigma_{cr}(U_1) - \sigma_{cr}(U_2)]/\phi$. The mobilised particle moves along the rock surface with velocity $U_s < U$ until it is strained by a pore throat smaller than the particle size. This results in rock permeability decline due to plugging of the pores. The increased strained particle concentration yields the permeability decline according to Eq. (4). The stabilised damaged permeability values at various fluid velocities, due to increasing strained particle concentration for tests I and II, is presented in Fig. 10a, b.

Let us introduce the non-dimensional pressure drop across the core, normalised by the initial pressure drop. This is denoted as the impedance J:

$$J(T) = \frac{\Delta P(T) U(0)}{U(T) \Delta P(0)} = \frac{k_0}{\langle k \rangle (T)} \tag{10}$$

where $\langle k \rangle(T)$ is the average core permeability and T is dimensionless time (pore volume injected). The permeability decline curves in Fig. 3a, e are recalculated to yield the impedance growth curves in Fig. 11a, b, respectively.

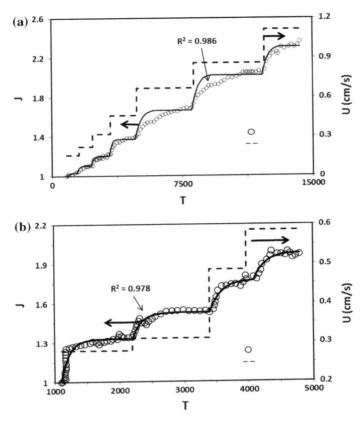

Fig. 11 Comparing the pressure drop across the core obtained from coreflood against the mathematical model: **a** test I and **b** test II

The increase in pressure drop across the core from Δp_{n-1} to Δp_n, or permeability decline from k_{n-1} to k_n, is caused by increasing fluid velocity U_n, $n = 1, 2, 3...$, which leads to particle mobilisation.

Permeability decline during the increase in fluid velocity is shown in Fig. 3. According to the data from tests I (Fig. 3a) and II (Fig. 3b), rock permeabilities stabilise after injection of many pore volumes. Classical filtration theory implies that for a mobilised particle to appear at the end of the core, it must traverse the overall core length. Each fine particle is transported by the carrier fluid; it is strained in the core or must arrive at the outlet after injection of at most one pore volume. According to Fig. 3a, e, permeability stabilisation times are significantly higher than 1 PVI. This can be explained by slow-particle drift along the rock surface: the mobilised particles move along the rock with velocity U_s that is significantly lower than the carrier fluid velocity U.

The next section introduces the basic governing equations for the transport of suspended colloids in porous media. The basic system includes the maximum retention function σ_{cr} which models particle detachment and its slow drift along the porous medium with low velocity $U_s < U$.

2.2 Governing System for Suspension-Colloidal Transport and Detachment in Porous Media

The following assumptions are introduced for the development of the mathematical model for detachment/mobilisation of particles and transport of suspended colloids in porous media (Yuan and Shapiro 2011a; Yuan et al. 2012, 2013):

- the mobilised fine particles cannot reattach to the rock surface;
- the mobilised particles do not diffuse in long micro-homogeneous cores;
- the carrier fluid is incompressible;
- small concentrations of suspension in flowing fluid do not change the density or viscosity of the carrier fluid, which equal those of injected water;
- there exists a phenomenological maximum retention function for particles attached to the rock surface;
- volume balance of the incompressible carrier fluid is not affected by the presence of small concentrations of suspended, attached or strained particles; and
- the mobilised particles move with velocity U_s which is smaller than fluid velocity U.

Mobilised particles move along the surface of grains with velocity $U_s < U$, meaning that the drifted particle concentration is significantly higher than the suspended concentration of fine particles carried by water stream (see Fig. 1). The drift speed U_s is a phenomenological constant of the model.

The slow-fines-drift assumption $U_s < U$ determines the difference between the above formulated assumptions and those for modified model (1), (3)–(5). Thus, the system of governing equations includes a mass balance equation for suspended, attached and strained fines, where the suspended particles are transported by water flux with reduced velocity U_s:

$$\frac{\partial(\phi c + \sigma_s + \sigma_a)}{\partial t} + U_s \frac{\partial c}{\partial x} = 0 \qquad (11)$$

The straining rate is assumed to be proportional to particle advection flux, cU_s (Herzig et al. 1970; Xu 2016):

$$\frac{\partial \sigma_s}{\partial t} = \lambda(\sigma_s) U_s c \qquad (12)$$

If the maximum retention concentration is greater than attached concentration, the particle attachment rate of Eq. (5) is also assumed to be proportional to the particle advection flux cU_s:

$$\frac{\partial \sigma_a}{\partial t} = \lambda_a U_s c, \quad \sigma_a < \sigma_{cr}(U) \qquad (13)$$

$$\sigma_a = \sigma_{cr}(U)$$

Otherwise, the attached particle concentration is expressed by the maximum retention function given by Eq. (8).

Four Eqs. (4), (11), (12) and (13) with four unknowns c, σ_a, σ_s and p constitute a closed system and a mathematical model for fine-particle migration in porous media.

Now we introduce the following dimensionless parameters:

$$S_a = \frac{\sigma_a}{\sigma_{a0}}, \quad S_s = \frac{\sigma_s}{\sigma_{a0}}, \quad C = \frac{c\phi}{\sigma_{a0}}, \quad \Lambda_a = \lambda_a L, \quad \Lambda_s = \lambda_s L,$$

$$T = \frac{\int_0^t U(y)dy}{\phi L}, \quad X = \frac{x}{L}, \quad \alpha_n = \frac{U_{sn}}{U_n}, \quad P = \frac{kp}{\mu L U} \qquad (14)$$

Here, the particle drift velocities U_{sn} and delay factors α_n, $n = 1, 2, 3\ldots$ correspond to flow velocities U_n; T is the accumulated non-dimensional volume of injected water. For the case of piecewise-constant flow velocity $U(t)$, the dimensionless accumulated injected volume $T(t)$ is piecewise linear.

Substitution of dimensionless parameters (14) into governing Eqs. (4), (11)–(13) yields the following dimensionless system, which consists of the particle balance:

$$\frac{\partial(C + S_s + S_a)}{\partial T} + \alpha_n \frac{\partial C}{\partial X} = 0 \qquad (15)$$

particle straining kinetics (Xu 2016),

$$\frac{\partial S_s}{\partial t} = \Lambda_s \alpha_n C \qquad (16)$$

particle attachment–detachment kinetics,

$$\frac{\partial S_a}{\partial T} = \Lambda_a \alpha_n C, \quad S_a < S_{cr}(U) \qquad (17)$$
$$S_a = S_{cr}(U)$$

and the modified Darcy's law that accounts for permeability damage due to fines retention

$$1 = -\frac{1}{1 + \beta_s \sigma_{a0} S_s} \frac{\partial P}{\partial X} \tag{18}$$

In the next section, we solve non-dimensional governing system (15)–(18) for the conditions of laboratory tests with piecewise-constant increasing velocity.

2.3 Analytical Model for One-Dimensional Suspension-Colloidal Flow with Fines Mobilisation and Straining

During coreflood when velocity U_1 is higher than critical velocity U_0, i.e. $\sigma_{a0} > \sigma_{cr}(U_1)$, the excess of the attached concentration is instantly released into the colloidal suspension. Particle straining in the proposed model is irreversible; therefore, it is assumed that initial porosity and permeability already account for the strained particle initial concentration. Coreflood with constant fluid velocity results in constant attached concentration S_a given by the maximum retention function. Thus, the initial conditions are

$$T = 0: \quad C = \Delta S_{a1} = S_{a0} - S_{cr}(U_1), \quad S_s = 0, \quad S_a = S_{cr}(U_1) \tag{19}$$

The inlet boundary condition corresponds to injection of water without particles:

$$X = 0: \quad c = 0 \tag{20}$$

Substituting the expression for straining rate (16) into mass balance Eq. (15) and accounting for steady-state distribution of attached particles S_a yields the linear first-order hyperbolic equation

$$\frac{\partial C}{\partial T} + \alpha_1 \frac{\partial C}{\partial X} = -\Lambda_s \alpha_1 C \tag{21}$$

The next section uses the method of characteristics to solve Eq. (21).

2.3.1 Exact Analytical Solution for Injection at Constant Rate

The characteristic velocity in Eq. (21) is equal to α_1. The solution $C(X, T)$ is presented in Table 1, and integration of Eq. (16) over T determines the strained concentration profile $S_s(X, T)$.

As illustrated in Fig. 12a, the concentration front of the injected particle-free fluid moves along the path $X = \alpha_1 T$. Behind the concentration front, suspended particle concentration is zero and the 'last' mobilised particle arrives at the core outlet at $T = 1/\alpha_1$. The mobilised particles are assumed to be uniformly distributed:

Table 1 Exact analytical solution for 1-D fines migration during the increase in piecewise-constant velocity

Term	Explicit formulae for 1-D solution	(X, T)-domain
Suspension concentration during stage 1	$C = 0$	$X \leq \alpha_1 T$
	$C = \Delta S_{a1} e^{-\alpha_1 \Lambda_s T}$	$X > \alpha_1 T$
Retention concentration during stage 1	$S_s = \Delta S_{a1}\left(1 - e^{-\Lambda_s X}\right)$	$X \leq \alpha_1 T$
	$S_s = \Delta S_{a1}\left(1 - e^{-\alpha_1 \Lambda_s T}\right)$	$X > \alpha_1 T$
Impedance during stage 1	$J(T) = 1 + \beta_s \sigma_{a0} \Delta S_{a1}\left[1 - \frac{1}{\Lambda_s} - \left(1 - \frac{1}{\Lambda_s} - \alpha_1 T\right)e^{-\alpha_1 \Lambda_s T}\right]$	$T < \alpha_1^{-1}$
	$J(T) = 1 + \beta_s \sigma_{a0} \Delta S_{a1}\left[1 - \frac{1}{\Lambda_s} - \frac{e^{-\Lambda_s}}{\Lambda_s}\right]$	$T \geq \alpha_1^{-1}$
Suspension concentration during stage n	$C = 0$	$X \leq \alpha_1 T$
	$C = \Delta S_{an} e^{-\alpha_n \Lambda_s (T-T_n)}$	$X > \alpha_1 T$
Retention concentration during stage n	$S_s - S_{sn} = \Delta S_{an}\left(1 - e^{-\Lambda_s X}\right)$	$X < \alpha_n(T - T_n)$
	$S_s - S_{sn} = \Delta S_{an}\left[1 - e^{-\alpha_n \Lambda_s (T-T_n)}\right]$	$X \geq \alpha_n(T - T_n)$
Impedance during stage n	$J(T) = J_{0n} + \beta_s \sigma_{a0} \Delta S_{an}\left[1 - \frac{1}{\Lambda_s} - \left(1 - \frac{1}{\Lambda_s} - \alpha_1(T - T_n)\right)e^{-\alpha_n \Lambda_s (T-T_n)}\right]$	$T < T_n + \alpha_1^{-1}$
	$J(T) = J_{0n} + \beta_s \sigma_{a0} \Delta S_{an}\left[1 - \frac{1}{\Lambda_s} - \frac{e^{-\Lambda_s}}{\Lambda_s}\right]$	$T \geq T_n + \alpha_1^{-1}$

they move with the same velocity and have the same probability of capture by pore throats. Therefore, the profile of suspended particle concentration is uniform during fluid flow, and concentration of suspension at $X > \alpha_1 T$ is independent of X (as indicated by the second line in Table 1). This leads to the conclusion that the concentration of particles strained in thin pores is independent of X ahead of the concentration front. Particle straining occurs for non-zero concentration of suspended particles. Therefore, the strained particles accumulate at a reservoir point X until the arrival of the concentration front, after which the concentration of suspended particles remains unchanged. Therefore, the concentration of strained particles behind the concentration front is steady-state.

The profiles of suspended particle concentration at $T = 0$, T_a (before the arrival of the concentration front at the outlet of the core) and T_b (after the arrival of the concentration front) are shown in Fig. 12b. We denote ΔS_{a1} as the initial concentration of the released suspended particles. The profile of the concentration of suspended particles equals zero behind the concentration front and is constantly ahead of the front. After the front's arrival, the breakthrough concentration of

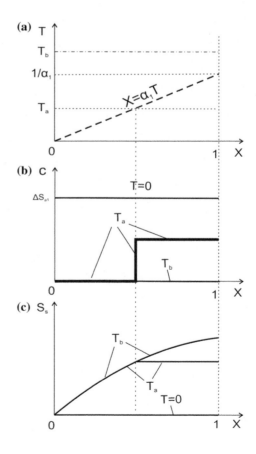

Fig. 12 Schematic for the analytical solution of 1-D fines migration under piecewise increasing velocity at times before and after the breakthrough (moments T_a and T_b, respectively): **a** trajectory of fronts and characteristic lines in (X, T) plane. **b** Suspended concentration profiles in three moments $T = 0$, T_a, and T_b. **c** Strained concentration profiles at three instants

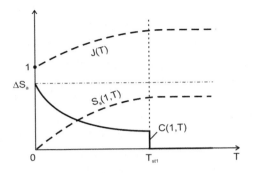

Fig. 13 Histories for dimensionless pressure drop across the core J and breakthrough concentration C, with strained concentration S_s at the outlet

suspended particles equals zero because all mobilised particles either are strained in thin pore throats or emerge at the rock effluent.

Three profiles of concentration of strained particles, for times 0, T_a and T_b, are shown in Fig. 12c. No strained particles are present in the rock before particle mobilisation. The concentration of strained particles continues to grow until the front's arrival, after which it remains constant. The duration of particle straining during the flow becomes longer with X. Therefore, the profiles of strained particle concentration grow as X increases. The probability of particle capture ahead of the front remains constant. Thus, the particle advective flux is uniform, and the strained profile is uniform.

Figure 13 shows the history of particle breakthrough concentration. The later the arrival, the higher the particle capture probability. According to Herzig et al. (1970), the coefficient of filtration λ_s equals the probability of particle straining per unit length of the particle trajectory. Therefore, the number of particles captured by thin pore throats increases with time, and breakthrough concentration $C(1, T)$ decreases with time. All mobilised fine particles either are strained or exit the core at time $T_{st,1}$, i.e. concentration of suspended particles becomes zero and the rock permeability stabilises at time $T_{st,1}$.

$$T_{st,1} = \frac{1}{\alpha_1} \tag{22}$$

The impedance can be calculated directly from Eq. (10). Impedance for the time interval having the constant fluid velocity from Eq. (18) equals

$$J(T) = \int_0^1 \left(-\frac{\partial P}{\partial X}\right) dX = 1 + \beta_s \sigma_{a0} \int_0^1 S_s(X, T) dX \tag{23}$$

Substituting the solution for the concentration of the retained particles (rows 4 and 5 in Table 1) into Eq. (23) and integrating over X results in the following explicit formula for impedance increase with time:

$$J(T) = 1 + \beta_s \sigma_{a0} \Delta S_{a1} \left[1 - \frac{1}{\Lambda_s} - \left(1 - \frac{1}{\Lambda_s} - \alpha_1 T\right) e^{-\alpha_1 \Lambda_s T} \right], \quad T < \alpha_1^{-1} \quad (24)$$

Substituting the expression for stabilisation time (22) into Eq. (24) yields the stabilised value of the impedance

$$J(T_{st1}) = 1 + \beta_s \sigma_{a0} \Delta S_{a1} \left(1 - \frac{1}{\Lambda_s} + \frac{e^{-\Lambda_s}}{\Lambda_s} \right), \quad T \geq \alpha_1^{-1} \quad (25)$$

Monotonic increase of dimensionless pressure drop from one to the maximum stabilised value is achieved at time equal to $1/\alpha_1$, which coincides with the arrival of the 'last' mobilised particle at the core outlet.

2.3.2 Analytical Solution for Multiple Injections

The solution of problem (15)–(18) where fluid velocity has changed from U_{n-1} to U_n is similar to that of the first stage under U_1. The only difference is that the concentration of strained particles for $T > T_n$ equals the total of the concentration of strained particles before the fluid velocity alteration at time $T = T_n - 0$ and the concentration of particles that have been strained during time $T > T_n$. We discuss the case where the change in fluid velocity occurs after permeability stabilisation.

When the fluid velocity changes from U_{n-1} to U_n at instant $T = T_n$, the attached particles are immediately mobilised. The mobilised particle concentration is $\Delta S_{an} = S_{cr}(U_{n-1}) - S_{cr}(U_n)$. The strained concentration at $T = T_n$ equals that before velocity alteration at $T = T_n - 0$:

$$T = T_n: \quad C = \Delta S_{an} = S_{cr}(U_{n-1}) - S_{cr}(U_n), \quad S_s = S_s(X, T_n - 0), \quad S_a = S_{cr}(U_n) \quad (26)$$

Substituting the strained concentration (see Table 1) into Darcy's law (18) and integrating for pressure gradient over X along the core yields the impedance for $T > T_n$:

$$T < T_n + \alpha_n^{-1}:$$
$$J(T) = J_{0n} + \beta_s \sigma_{a0} \Delta S_{an} \left[1 - \frac{1}{\Lambda_s} - \left(1 - \frac{1}{\Lambda_s} - \alpha_n(T - T_n)\right) e^{-\alpha_n \Lambda_s (T - T_n)} \right] \quad (27)$$

Substituting stabilised time $T = T_n + 1/\alpha_n$ into Eq. (27) gives the stabilised impedance after the n-th injection with velocity U_n.

The analytical model-based formulae for impedance (rows 12 and 13 in Table 1) will be validated against the laboratory tests in the next section.

2.4 Using the Laboratory Results to Adjust the Analytical Model

In order to replicate water injection in a well, Ochi and Vernoux (1998) performed two laboratory corefloods using Berea cores at conditions similar to bottom-hole pressures and temperatures and at various fluid velocities. Permeability and flow velocity for test I are shown in Fig. 3a, b, respectively, and for test II in Fig. 3e, f, respectively. Initial and boundary conditions (19) and (20) correspond to injection of particle-free water with piecewise-constant increasing velocity. Pressure drop along the core was measured. Both tests used Berea sandstone cores prepared from the same block, so that the rock properties for both cores would be similar. As fluid velocity increased, kaolinite particles detached from the grain surface. The mobilised fines migrated and were strained by the rock. Pressure drop predicted by the analytical model proposed in Sect. 2.3 was compared to the actual pressure drop across the cores during the tests. Minimisation of the difference between the modelled and measured pressure drop was used to adjust the phenomenological constants of the model: α, $\Delta\sigma$, λL and β.

2.4.1 Tuning the Rheological Model Parameters from Laboratory Coreflooding Data

Formation damage and filtration coefficients were assumed to remain constant for the duration of the experiment. Therefore, these parameters would be independent of fluid velocity and concentration of the retained particles. The drift delay factor was assumed to vary, i.e. the alteration of rock surface during detachment/mobilisation of particles affects drift velocity.

For stabilised permeability, according to Eq. (4), the permeability values k_n fulfil the following relationship:

$$\beta[\sigma_{cr}(U_{n-1})]\Delta\sigma_{an} = \frac{k_{n-1}}{k_n} - 1 \tag{28}$$

Pressure drops along the core, which define the permeabilities k_i, were measured during coreflood tests with varying fluid velocities U_i. The least-squares method was used to tune the above experimental pressure drop data and obtained filtration coefficient λ_s, the products $\beta_s \Delta\sigma_{cr}(U_n)$, $n = 1, 2\ldots$, and the drift delay factors α_n for different fluid velocities U_n. The optimisation problem (Coleman and Li 1996) was solved using the reflective trust region algorithm in Matlab (Mathworks 2016).

The average core permeabilities (Fig. 3a,e) were used to calculate the impedances in Fig. 11. For Berea sandstone, we assumed typical porosity of 0.2 and typical concentration of kaolinite particles of 0.06 (Khilar and Fogler 1998). The attached volumetric concentration is equal to $\sigma_{a0} = 0.06 \times 0.8 = 0.048$, which is equivalent to σ_{cr} for $U_0 < U_1$ (see Eq. (9)). We calculated the formation damage coefficient for the condition of total removal of all attached particles at the maximum fluid velocity, during the last fluid injection:

$$\beta_s = \frac{\sum_{n=1}^{N}(\beta_s \Delta \sigma_{an})}{\sigma_{a0}} \quad (29)$$

Substituting formation damage coefficient (29) into the products $\beta_s \Delta \sigma_{an}$ results in the values of released concentrations $\Delta \sigma_{an}$. The maximum retention concentrations at different fluid velocities can be calculated as follows:

$$\sigma_{cr}(U_n) = \sigma_{a0} - \sum_{n=1}^{N} \Delta \sigma_{an} \quad (30)$$

2.4.2 Results

Table 2 and Fig. 11 show results for history matching of impedance for the two coreflood tests by Ochi and Vernoux (1998). Tuning the model parameters resulted

Table 2 Tuned values of the model parameters

Model parameter	Test I	Test II
α_1	0.0020	0.0017
α_2	0.0020	0.0015
α_3	0.0020	0.0015
α_4	0.0020	0.0014
α_5	0.0020	–
α_6	0.0008	–
α_7	0.0008	–
$\Delta\sigma_{a1}$	0.0017	0.0178
$\Delta\sigma_{a2}$	0.0039	0.0087
$\Delta\sigma_{a3}$	0.0045	0.0087
$\Delta\sigma_{a4}$	0.0076	0.0066
$\Delta\sigma_{a5}$	0.0114	–
$\Delta\sigma_{a6}$	0.0114	–
$\Delta\sigma_{a7}$	0.0076	–
λL	2.2869	3.2842
β	30.9328	22.327

in monotonically decreasing dependency of the drift delay factor $\alpha = \alpha(S)$ for corefloods in cores I and II (Fig. 3d, h, respectively). The higher the strained concentration, the higher the rock tortuosity, which decelerates particle drift. Also, the higher the strained concentration, the smaller the mobilised particles, which drifted at lower velocity.

The experimental data closely matched the model (R^2 values of 0.99 and 0.98 for cores I and II, respectively). Fixing the drift delay factor for the overall period of fluid injection and then comparing the impedance data resulted in significantly lower R^2 values during adjustment of the proposed model. Using Eq. (30) to tune the model for two cores yielded the maximum retention function shown in Fig. 14. The increase in fluid velocity resulted in the increase in drag and lift forces, which detached the kaolinite particles from the surface of the rock grains and reduced the concentration of kaolinite particles remaining immobilised on the rock grain surface. If a monolayer of poly-sized kaolinite particles is attached to the surface of rock grains, mechanical equilibrium model (6) and (8) indicates that the obtained σ_{cr}-curves would not be convex.

The proposed model can be used to calculate the size distribution of attached fine particles $\Sigma_a(r_s)$ from the maximum retention function $\sigma_{cr}(U)$: the minimum mobilised size $r_{scr}(U)$ is determined from Eq. (6), and size distribution function $\Sigma_a(r_s)$

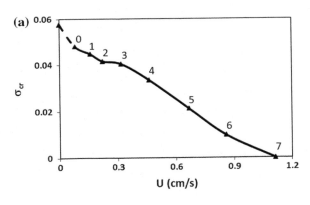
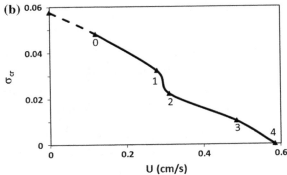

Fig. 14 Maximum retention curves $\sigma_{cr}(U)$ obtained from **a** test I and **b** test II

is calculated from Eq. (8) by regularised numerical differentiation (Coleman and Li 1996).

Because the fluid velocity was changing stepwise during coreflood tests (seven velocities for test I, and four velocities for test II), the calculated kaolinite particle distributions are given in the form of a histogram (Fig. 15). As follows from Fig. 3c, g, the minimum radius of detached particles decreases as fluid velocity increases. This observation agrees with the shape of the velocity dependency of the critical radius exhibited in Fig. 8.

Figure 16a, b compare the various forces acting on a particle at the critical instant of its mobilisation. According to Fig. 15, the ranges of particle radii cover the ranges of size distributions for particles attached to the surface of rock grains. The drag force was two orders of magnitude smaller than the electrostatic force. The drag force was significantly larger than the gravitational or lift force. Because lever arm ratio l significantly exceeded one, the small drag torque exceeded the torque developed by electrostatic force.

The maximum retention function can be parameterised by the critical particle radius $\sigma_{cr}(U) = \sigma_{cr}(r_{scr}(U))$. Considering the value $\sigma_{cr}(r_{scr})$ as an accumulation function of retained concentration for all particles with radius smaller than r_{scr} yields the corresponding histogram (Fig. 15), representing the concentration

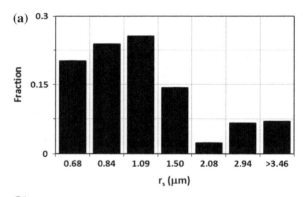
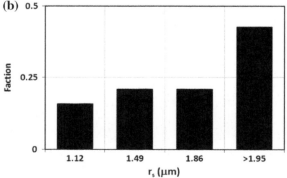

Fig. 15 Size distributions of movable fine particles on the matrix surface: **a** test I and **b** test II

Fig. 16 Forces exerting on the attached particles: **a** test I and **b** test II

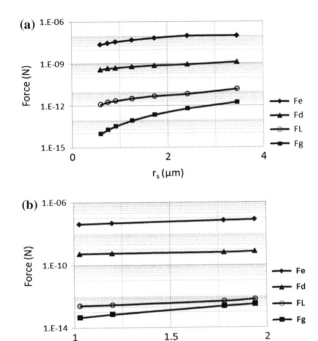

distribution of initial reservoir fines for various radii. Thus, the maximum retention function, σ_{cr}, for various sized particles attached to the grain surface can be explained: if particles attached to the grain surface cannot be mobilised by fluid flowing with velocity U, then their radii are smaller than $r_{scr}(U)$ and their concentration is expressed as $\sigma_{cr}(U)$. Increasing fluid velocity U results in the decrease of minimum radius of particles detached by flow with velocity U.

The calculated values of filtration and formation damage coefficients (Table 2) fall within the common ranges of these coefficients reported by Pang and Sharma (1997). The orders of magnitude of drift delay factor, which vary between 10^{-3} and 10^{-4} in the present work, are the same as those reported by Oliveira et al. (2014).

2.5 Summary and Discussion

According to mathematical model (1), (3) and (5), which accounts for the maximum retention function for detachment and migration of particles at velocity equal to carrier fluid velocity, rock permeability should stabilise after 1 PVI. However, the experimental data showed that the permeability stabilisation periods are significantly greater than 1 PVI. Such behaviour can be explained only if a mobilised particle moves significantly slower than the flowing fluid. This behaviour could be described by a two-speed particle-transport model (Yuan and Shapiro 2011a, b; Bradford et al. 2012).

The model contains six constants of mass exchange between particles moving slow and fast, detachment coefficient and filtration coefficient for fast particles, corresponding coefficients for slow particles and velocity of slow particles. The model tuning for the experimental breakthrough curves is not unique. Complete characterisation of the two-speed model would require complex experiments in which pressure drops along a core and along the particle breakthrough curve are measured and the retained particle concentrations are calculated. Yet, most core-flood studies have reported data for pressure drop along the core only. For this reason, the present study considers a rapid exchange between populations of particles migrating with fast and slow velocities along each rock pore, yielding a unique particle drift velocity. Also, this exchange is assumed to occur significantly faster than the capture of particles by the rock after a free run in numerous pores, resulting in equal concentrations of particles moving with fast and slow velocities. The above assumptions translate to a single-velocity model (You et al. 2015, 2016).

Proposed model (15)–(18) is applied to the data treatment of laboratory tests. The modelling results show that the migrating particles move significantly slower than the carrier fluid. Hence, there is a delay in the permeability stabilisation due to fines migration. The delay time is 500–1250 PVI, which corresponds to the drift delay factor α_n varying within 0.0008–0.002.

Migrating particles can be divided into two groups according to the velocity. One group of particles travel with the same velocity as the carrier fluid; while the other group drifts along the grain surfaces with significantly lower velocity (slow particles). The percentage of slow particles depends on particle-size distribution, velocity of the carrier fluid and electrostatic forces between the particles and grains (both magnitude of the electrical forces and whether they are repulsive or attractive). The slow particles can slide or roll on the grain surfaces, or temporarily move away from grain surface to the bulk of the fluid, before colliding with grain surface asperities again (Li et al. 2006; Yuan and Shapiro 2011b; Sefrioui et al. 2013).

The largest size of particles that can stay attached to the grain surface at each velocity can be calculated using torque balance Eq. (6), i.e. there exists a critical particle radius for each velocity such that all particles with larger radii will be mobilised by the carrier fluid: $r_{scr} = r_{scr}(U)$ (see Fig. 6).

The maximum retention function, $\sigma_{cr}(U)$, can be defined for a monolayer of attached particles as the concentration of attached particles with $r < r_{scr}$. The minimum size of mobilised particles as a function of fluid velocity follows from torque balance given by Eq. (6). It allows calculation of σ_{cr} using size distribution of the particles that can be mobilised at each fluid velocity. The function σ_{cr} depends on particle-rock electrostatic constants, Young's moduli, Poisson's ratio and size distribution of the attached particles.

Maximum retention is a monotonically decreasing function of mean particle size (Fig. 9a). The higher the variance coefficient of particle-size distribution, the lower the σ_{cr} at low fluid velocities, and the higher the σ_{cr} at high velocities (Fig. 9b).

The σ_{cr} curve for a monolayer of multi-sized particles has a convex shape at low fluid velocities and a concave shape at high velocities (Fig. 9c). However, for a multilayer attachment of mono-sized particles, that curve has a convex shape for all velocities.

For 1-D (one-dimensional) suspension flow with piecewise increase of the fluid velocity, the exact analytical solution can be obtained. The process includes particle migration and subsequent capture (straining) at the pore throats. Changing the fluid velocity creates a particle concentration front that starts moving from the core inlet. The concentration front coincides with the trajectory of the drifting particles and separates the particle-free region (behind the front) from particle-migration region (ahead of the front). The concentration profiles of the suspended and strained particles are uniform ahead of the front.

The drift delay factor α_n is a function of the particle size and the geometry of the porous media, which undergoes a continuous change during straining of migrating particles at the pore throats. Small particles move along the grain surface more slowly than do large particles. Hence, the drift delay factor is smaller for small particles than for large particles. Because larger particles are detached at lower fluid velocities, the size of the released particles decreases during the coreflood test with piecewise increase of fluid velocity. This explains the decrease in drift delay factor during the experiment (Fig. 3d, h). Hence, the introduction of a phenomenological function of the form $\alpha_n = \alpha_n(r_s, \sigma_s)$ can further improve the proposed model for colloidal-suspension transport in porous media with instant particle release and slow drift (Eqs. 15–18).

The proposed mathematical model has been found to yield pressure data that closely approximate those from coreflood tests with piecewise increase of velocity. To completely validate the model for where velocity of the migrating particles differs significantly from that of the carrier fluid, parameters such as particle retention profiles, breakthrough concentration of particles and size distribution of produced particles should also be measured. Then, these measured data would be compared against the analytical solutions presented in Table 1. Such a test with measurement of all required parameters is not available in the literature.

3 Fines Detachment and Migration at Low Salinity

Salinity alteration affects the electrostatic forces between fines and the rock surface, thereby influencing fines detachment. Decreasing salinity of the flowing brine increases the repulsive component of the electrical forces (double-layer electrical force). This weakens the total attraction between the attached fines and rock surface, which may cause fines to be mobilised by the viscous forces from flowing brine (Eq. 6). The detached fine particles migrate with the carrier fluid and plug pore throats smaller than they are, leading to a significant permeability reduction.

Section 3.1 describes the methodology and experimental setup for coreflood tests with piecewise salinity decrease. Section 3.2 presents the experimental results.

Section 3.3 derives a mathematical model for fines detachment and migration in porous media, accounting for slow fines migration and delayed fines release during salinity alteration. Section 3.4 compares the experimental data and the mathematical model's prediction.

3.1 Laboratory Study

This section presents the experimental setup, properties of the core and fluids, and the experimental methodology.

3.1.1 Experimental Setup

A special experimental setup was developed to conduct colloidal-suspension flow tests in natural reservoir rocks. The core permeability and produced fines concentration were measured. In addition, an extra pressure measurement was taken at the midpoint of the core, which complements the routine core inlet and outlet pressure measurements. The schematic drawing and the photograph of the apparatus are shown in Figs. 17 and 18, respectively. The system consisted of a Hassler type core

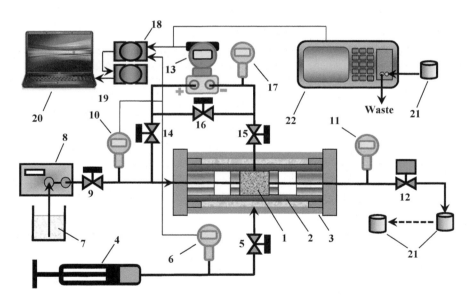

Fig. 17 Schematic of laboratory setup for fines migration in porous media: (1) core plug. (2) Viton sleeve. (3) Core holder. (4) Pressure generator. (5, 9, 14, 15, 16) Manual valves. (6, 10, 11, 17) Pressure transmitters. (7) Suspension. (8) HPLC pump. (12) Back-pressure regulator. (13) Differential pressure transmitter. (18) Data acquisition module. (19) Signal converter. (20) Computer. (21) Beakers. (22) PAMAS particle computer/sizer

Fig. 18 Photo of laboratory setup for fines migration in porous media

holder, a high-pressure liquid chromatography pump (HPLC) and a dome back-pressure regulator to maintain a constant pressure at the core outlet. Three Yokogawa pressure transmitters were used to record the pressure data at the core inlet, outlet and the intermediate point, and a fraction collector was used to collect samples of the produced fluid. The overall volumetric concentration of solid particles at the effluent was measured using a PAMAS SVSS particle counter with a particle-size range of 1–200 μm. Prior to the tests, the concentration of solid particles in the solution was measured and used as the background particle concentration in the calculations.

3.1.2 Materials

A sandstone core plug was used to perform a coreflood test with piecewise salinity decrease. The core was taken from the Birkhead Formation (Eromanga Basin, Australia) and had a permeability of 34.64 mD. A water-cooled diamond saw was used to fashion several core plugs, each having a diameter of 37.82 mm and a length of 49.21 mm. These were subsequently dried for 24 h.

Table 3 Mineralogical composition of the rock

Mineral	A1 core % (w/w)
Quartz	59.9
K-feldspar	2.3
Plagioclase	1.0
Kaolinite	9.2
Illite/mica	18.6
Illite/smectite	2.0
Chlorite	5.1
Siderite	1.9
Total	100.00

The XRD test showed that the core sample contains a considerable amount of movable clay, including 9.2 w/w% of kaolinite and 18.6 w/w% of illite (see Table 3 for the full mineral composition).

The ionic composition of the formation fluid (FF) is listed in Table 4a (supplied by Amdel Laboratories, Adelaide, Australia). The ionic composition was expressed as salt concentration in order to prepare an artificial formation fluid (AFF) with similar ionic strength (0.23 mol/L) to the formation water (Table 4b). The AFF was prepared by dissolving the calculated salt concentrations in deionized ultrapure water (Millipore Corporation, USA; later in the text it is called the DI water). NaCl was then added to the AFF, in order to increase the ionic strength of the solution to 0.6 mol/L, equivalent to the ionic strength of the completion fluid. The composition of high-salinity AFF is listed in Table 4b (AFF (NaCL)). In order to decrease salinity of the injected fluid during the experiment, the AFF was diluted using DI water to obtain the desired ionic strength for each injection step (maintaining the salinity 0.6, 0.4, 0.2, 0.1, 0.05, 0.025, 0.01, 0.005, 0.001 mol/L).

3.2 Methodology

Prior to the experiment, the air was displaced from the core by saturating the core sample with 0.6 M AFF under a high vacuum. Then, the core plug was installed inside the core holder, and the overburden pressure was gradually increased to 1000 psi and maintained during the experiment. Afterwards, the 0.6 M AFF was injected into the core sample with a constant volumetric flow rate of 1.0 mL/min (superficial velocity: 1.483×10^{-5} m/s). The pressure drop was recorded at three points: inlet, outlet and the intermediate point. The intermediate pressure point was placed 25.10 mm from the core inlet.

Fluid injection continued until permeability stabilisation was achieved with uncertainty of 3.2% or less (Badalyan et al. 2012). The test then proceeded by stepwise decreasing the ionic strength of injected brine in nine consecutive steps: 0.4, 0.2, 0.1, 0.05, 0.025, 0.01, 0.005, 0.001 mol/L and DI water.

Table 4 **a** Ionic compositions for formation fluid (FF), and **b** ionic compositions for artificial formation fluid (AFF)

(a)

Parameter or ion	Unit	FF
Electrical conductivity	µS/cm	24,000
pH	N/A	7.6
Total dissolved solids (TDS)	mg/L	15,000
Ionic strength	mol/L	0.231
Chloride	mg/L	7,300
Sulphate as SO_4^{2-}	mg/L	350
Bicarbonate as HCO_3^-	mg/L	450
Calcium	mg/L	260
Magnesium	mg/L	18
Sodium	mg/L	1,600
Potassium	mg/L	5,400

(b)

Parameter or ion	Units	AFF	AFF (NaCl)
Electrical conductivity	µS/cm	25,257	49,200
pH	N/A	7.9	8.1
Total dissolved solids (TDS)	mg/L	15,275	36,851
Ionic strength	mol/L	0.230	0.601
NaCl	mg/L	3,118	24,693
$MgCl_2$	mg/L	70.5	70.5
Na_2SO_4	mg/L	517.5	517.5
$CaCl_2$	mg/L	720.0	720.0
$NaHCO_3$	mg/L	553.4	553.4
KCl	mg/L	10,296	10,296

Permeability stabilisation was achieved at each step. The produced fluid was sampled using an automatic fraction collector. The sampling size was 0.17 PV at the beginning and then increased to 0.86 PV after 2 PVI.

The overall solid particle concentration in the produced samples was measured using a PAMAS particle counter, which uses laser scattering in a flow-through cell to measure the number and size distribution of the solid particles (from 0.5 to 5.0 µm) in the suspension, assuming spherical particles. Multiplying the size distribution function by the sphere volume and integrating with respect to radius yields the overall volumetric particle concentration. The electrolytic conductivity of the produced samples was also measured, to calculate breakthrough ionic strength.

3.3 Experimental Results

The initial core permeability and porosity were 34.64 mD and 0.13, respectively. This allowed calculation of mean pore radius (r_p = 6.62 μm, $r_p^2 = k/(4.48\phi^2)$) (Katz and Thompson 1986). The analysis of effluent samples shows that the mean size of produced particles was 1.47 μm. The so-called 1/7–1/3 rule of filtration was introduced by Van Oort et al. (1993). They suggested that particles larger than 1/3 of the pore size cannot enter the porous media and form an external filter cake, but that particles smaller than 1/7 of the pore size can travel through porous media without being captured. Particles between 1/3 and 1/7 of the pore size can enter the porous media; however, they can be retained at the pore throats, which impairs permeability. In the current experiment, the particle-to-pore size ratio (jamming ratio) was 0.22 (between 1/3 and 1/7), implying that particles can be captured at pore throats after being released by reduction of fluid salinity. This explains the observed impedance growth (Fig. 19a) during reduction of injection fluid salinity (ionic strength).

If all mobilised particles were released instantly and moved at the same velocity as the carrier fluid, permeability would be expected to stabilise in 1 PVI after each salinity alteration. However, the measured pressure data show that permeability stabilisation takes much longer (Fig. 19a). This delay in permeability stabilisation could be attributed to delay in particle detachment after salinity alteration or to the slow migration of released particles drifting along the rock surface (Yuan and Shapiro 2011a; You et al. 2015).

Figure 20 shows the cumulative produced particles and effluent ionic strength. The salinity and produced particle fronts coincide after each salinity alteration. The salinity alteration is accompanied by fine particle production and permeability reduction. This confirms that the fines mobilisation during salinity alteration is the mechanism for the permeability impairment. Similar to permeability behaviour, the fines were produced for a much longer period than 1 PVI.

3.4 Analytical Model for Slow Fines Migration and Delayed Particle Release

The impedance (reciprocal of permeability) growth curve presented in Fig. 19a indicates that after each salinity alteration, permeability stabilisation was achieved in tens to hundreds of PVI rather than the expected 1 PVI. As mentioned previously, one possible mechanism for the slow permeability stabilisation is the delay in particle detachment after salinity alternation. This phenomenon can be explained by electrokinetic ion-transport theory (Nernst–Planck model). It describes the diffusion of ions between the bulk fluid and the particle–grain area (Mahani et al. 2015a, b, 2016), which is not considered in slow-migration model (15)–(18). In this section,

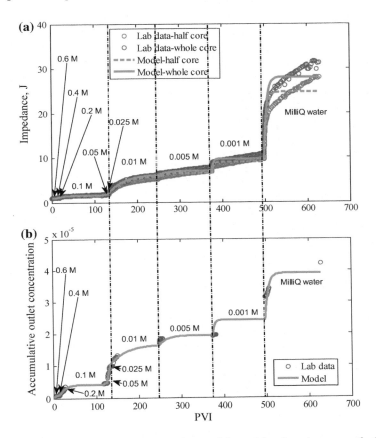

Fig. 19 Matching the coreflood data by the slow-particle model: **a** impedance growth along the whole core and the first core section. **b** Accumulated fine particle production

the slow-migration model is modified to account for both slow fines migration and delayed particle release.

Introducing a delay τ into the maximum retention function results in an expression for delayed fines detachment: $\sigma_a(x, t + \tau) = \sigma_{cr}(\gamma(x,t))$, where $\gamma(x, t)$ is the fluid salinity at time t. Retaining the first two terms of Taylor's expansion for a small value of τ results in

$$\tau \frac{\partial \sigma_a}{\partial t} = \sigma_{cr}(\gamma_1) - \sigma_a \tag{31}$$

The equation for maximum retention function (13) can now be replaced by kinetic Eq. (31). The system of three Eqs. (11), (12) and (31) describes suspension transport in porous media and accounts for delayed release of the reservoir fines

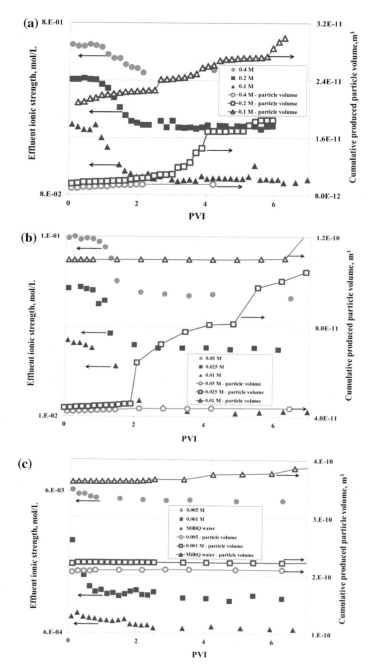

Fig. 20 Variation in effluent ionic strength and cumulative produced particle volume versus PVI during injection of water with piecewise-constant decreasing salinity: **a** for salinities 0.4, 0.2 and 0.1 M; **b** for salinities 0.05, 0.025 and 0.01 M; and **c** for salinities 0.005, 0.001 M and MilliQ water

during salinity reduction. This system can be solved for the unknown values c, σ_a and σ_s.

Replacing σ_{a0} with released particles concentration $\Delta\sigma_{cr}$ in dimensionless group (14) yields a dimensionless system of equations for suspension transport in porous media that accounts for delayed release of reservoir fines:

$$\frac{\partial(C + S_s + S_a)}{\partial T} + \alpha \frac{\partial C}{\partial X} = 0 \tag{32}$$

$$\varepsilon \frac{\partial S_a}{\partial T} = S_{cr}(\gamma_1) - S_a \tag{33}$$

$$\frac{\partial S_s}{\partial T} = \alpha \Lambda C \tag{34}$$

where the delay factor ε is defined as $\varepsilon = U\tau/\phi L$.

Because it takes significantly longer than 1 PVI for the mobilised particles to reach the core outlet ($1/\alpha \gg 1$), i.e. $\alpha \ll 1$, the initial and boundary conditions for injection of particle-free fluid with salinity γ_1 are

$$\begin{gathered} C(X,0) = 0, \quad \gamma(X,0) = \gamma_1, \quad S_s(X,0) = 0, \\ S_a(X,0) = S_{a0} = \frac{\sigma_{a0}}{\Delta\sigma_a} \\ C(0,T) = 0, \quad \gamma(0,T) = \gamma_1 \end{gathered} \tag{35}$$

where γ_0 and γ_1 are initial and injected salinities, respectively. The initial concentration of attached particles is $\sigma_{a0} = \sigma_{cr}(\gamma_0)$ for $\gamma_0 > \gamma_1$.

The solution to linear ordinary differential Eq. (33) with initial condition (35) is

$$S_a = S_{acr}(\gamma_1) - [S_{acr}(\gamma_1) - S_{a0}] \exp\left(-\frac{T}{\varepsilon}\right) \tag{36}$$

It yields the following equation for the detaching rate:

$$\frac{\partial S_a}{\partial T} = \frac{1}{\varepsilon}[S_{acr}(\gamma_1) - S_{a0}] \exp\left(-\frac{T}{\varepsilon}\right) \tag{37}$$

Substituting the equations for straining rate (34) and detaching rate (37) into overall particle balance equation (32) results in the following equation for suspended concentration:

$$\frac{\partial C}{\partial T} + \alpha \frac{\partial C}{\partial X} = -\alpha \Lambda C + \frac{1}{\varepsilon}[S_{a0} - S_{acr}(\gamma_1)] \exp\left(-\frac{T}{\varepsilon}\right) \tag{38}$$

Introduction of the following constants:

$$y = \frac{1}{\varepsilon}[S_{a0} - S_{acr}(\gamma_1)], \quad b = 1/\varepsilon \tag{39}$$

simplifies Eq. (38) as

$$\frac{\partial C}{\partial T} + \alpha \frac{\partial C}{\partial X} = -\alpha \Lambda C + y \exp(-bT) \tag{40}$$

Ahead of the front of the mobilised fines ($T \leq X/\alpha$), the characteristic form of the linear hyperbolic Eq. (40) is

$$\frac{dx}{dT} = \alpha \tag{41}$$

$$\frac{dC}{dT} = -\alpha \Lambda C + y \exp(-bT), \quad C(X, 0) = 0 \tag{42}$$

If $\alpha \Lambda \neq b$, the solution to the linear ordinary differential Eq. (42) is

$$C(T) = \frac{y}{\alpha \Lambda - b}\left(e^{-bT} - e^{-\alpha \Lambda T}\right) \tag{43}$$

If $\alpha \Lambda = b$, the solution to (42) is

$$C(T) = yTe^{-bT} \tag{44}$$

Similar to fines migration due to abrupt velocity increase (row 3 in Table 1), the initial uniform profile of the suspended particles moves with an equal capture probability for all the suspended particles. Thus, the profile of the suspended particles remains uniform, and the suspended concentration is time-dependent only.

Behind the front of mobilised fines ($T > X/\alpha$), the characteristic form of Eq. (38) with a zero boundary condition is

$$\frac{dT}{dX} = \frac{1}{\alpha}, \quad T(0) = \eta \tag{45}$$

$$\frac{dC}{dX} = -\Lambda C + \frac{y}{\alpha}e^{-bT}, \quad C(0) = 0 \tag{46}$$

If $\alpha \Lambda \neq b$, the solution to the linear ordinary differential Eq. (46) is

$$C(X) = \frac{m}{\Lambda - \frac{b}{\alpha}}\left(e^{-\frac{b}{\alpha}X} - e^{-\Lambda X}\right), \quad m = \frac{y}{\alpha}e^{(-b\eta)} \tag{47}$$

Substituting the constant η along the characteristic line

$$\eta = T - \frac{X}{\alpha} \tag{48}$$

into solution (48) yields the following expression for the suspended concentration behind the front:

$$C(X,T) = \frac{y}{\alpha\Lambda - b} e^{-bT} \left(1 - e^{\left(\frac{b}{\alpha} - \Lambda\right)X}\right) \tag{49}$$

If $\alpha\Lambda = b$, the solution to Eq. (46) is

$$C(X) = mXe^{-\Lambda X} \tag{50}$$

Substituting the constant η along the characteristic line (48) into solution (50) yields the following expression for the suspended concentration behind the front:

$$C(X,T) = \frac{y}{\alpha} e^{-bT} X e^{\left(\frac{b}{\alpha} - \Lambda\right)X} \tag{51}$$

Formulae for strained concentration S_s are obtained by substituting suspended concentration from Eqs. (43), (44), (49) and (51) into the equation for straining rate (34) and then integrating with respect to T. This solution is listed in Table 5 for $\alpha\Lambda \neq b$ and in Table 6 for $\alpha\Lambda = b$. The profiles of suspended and strained concentrations are shown in Fig. 21.

3.5 Treatment of Experimental Data

The result of the coreflood test with piecewise salinity decrease (presented in Sect. 3.1) is modelled in this section. The three models that have been presented in previous sections are applied to the experimental data treatment: the slow-particle migration model (Table 1), the delayed-particle-release model (Tables 5 and 6 with $\alpha = 1$), and the general model that accounts for both effects (Tables 5 and 6). The pressure drop along the core (whole core) and between inlet and midpoint (half-core), and the accumulated produced fines are matched separately by all three models. The reflective trust region algorithm (Coleman and Li 1996) is applied for optimisation using Matlab (Mathworks 2016).

Figure 19 presents the treatment of impedance and cumulative produced fines data using the slow-particle migration model. The coefficient of determination R^2 is 0.9902. The model tuning parameters are formation damage coefficient β, filtration coefficient λ, released concentration $\Delta\sigma$, delay factor ε, and drift delay factor α. The pore space geometry and values of the tuning parameters change during coreflood, as a result of pore throat plugging by mobilised fines. Hence, the tuning parameters

Table 5 Analytical model for fines mobilisation, migration and suspension ($\alpha\Lambda \neq b$)

Term	Notation	Zones	Expression
Suspended concentration	$C(X,T)$	$T \leq X/\alpha$	$\frac{y}{\alpha\Lambda - b}\left(e^{-bT} - e^{-\alpha\Lambda T}\right)$
		$T > X/\alpha$	$\frac{y}{\alpha\Lambda - b} e^{-bT}\left(1 - e^{\left(\frac{b}{\alpha} - \Lambda\right)X}\right)$
Strained concentration	$S_s(X,T)$	$T \leq X/\alpha$	$\alpha\Lambda\left[\frac{y}{\alpha\Lambda(\alpha\Lambda - b)} e^{-\alpha\Lambda T} - \frac{y}{b(\alpha\Lambda - b)} e^{-bT}\right] - \alpha\Lambda\left[\frac{y}{\alpha\Lambda(\alpha\Lambda - b)} - \frac{y}{b(\alpha\Lambda - b)}\right]$
		$T > X/\alpha$	$\alpha\Lambda\left[-\frac{y}{b(\alpha\Lambda - b)} e^{-\Lambda X} - \frac{y}{b(\alpha\Lambda - b)}\left(1 - e^{\left(\frac{b}{\alpha} - \Lambda\right)X}\right) e^{-bT}\right] - \alpha\Lambda\left[\frac{y}{\alpha\Lambda(\alpha\Lambda - b)} - \frac{y}{b(\alpha\Lambda - b)}\right] \cdots$ $+ \alpha\Lambda\left[\frac{y}{b(\alpha\Lambda - b)}\left(1 - e^{\left(\frac{b}{\alpha} - \Lambda\right)X}\right) e^{-b\frac{X}{\alpha}}\right]$
Permeability	$k(t_D)$		$k_0\left(1 + \frac{\beta\phi}{\omega}\int_0^\infty S_s(x_D, t_D)\mathrm{d}x_D\right)^{-1}$

Table 6 Analytical model for fines mobilisation, migration and suspension ($\alpha\Lambda = b$)

Term	Notation	Zones	Expression
Suspended concentration	$C(X,T)$	$T \leq X/\alpha$	yTe^{-bT}
		$T > X/\alpha$	$\frac{y}{\alpha}e^{-bT}X$
Strained concentration	$S_s(X,T)$	$T \leq X/\alpha$	$-ye^{-bT}T + \frac{y}{b}\left(1 - e^{-bT}\right)$
		$T > X/\alpha$	$\frac{y}{b}\left(1 - e^{-b\frac{X}{\alpha}}\right) - \frac{y}{\alpha}Xe^{-bT}$
Permeability	$k(t_D)$		$k_0\left(1 + \frac{\beta\phi}{\omega}\int_0^\omega S_s(x_D, t_D)\mathrm{d}x_D\right)^{-1}$

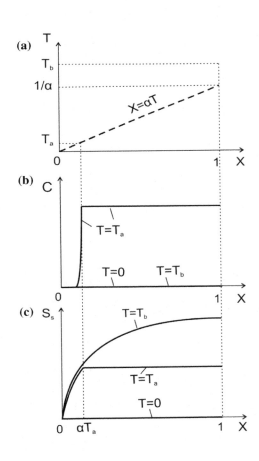

Fig. 21 Analytical slow-fines delay-release model: **a** trajectory of fronts and characteristic lines in (X, T) plane. **b** Suspended concentration profiles at three instants. **c** Strained concentration profiles at three instants

vary with change in injected fluid salinity, which causes fines mobilisation and permeability impairment. Table 7 and Fig. 22 show the values of the tuning parameters at each salinity.

The drift delay factor α decreases as salinity decreases (Fig. 22a). This can be attributed to the decreasing size of released particles during the salinity decrease.

Table 7 Fitted parameters for the slow-particle model

Salinity (M)	β	$\Delta\sigma$	α	λ	ε
0.6	32,803	3.17E−06	0.0138	12.06	0
0.4	93,310	4.35E−07	0.0439	17.78	0
0.2	916,406	3.72E−07	0.0628	1.14	0
0.1	89,189	4.63E−06	0.0094	8.28	0
0.05	100,070	1.04E−06	0.0500	9.99	0
0.025	41,985	1.30E−05	0.0128	2.81	0
0.01	99,875	2.15E−05	0.0019	15.78	0
0.005	2,928	6.27E−05	0.0010	99.87	0
0.001	4,451	8.87E−05	0.0030	84.44	0
0	8,879	2.10E−04	0.0022	69.37	0

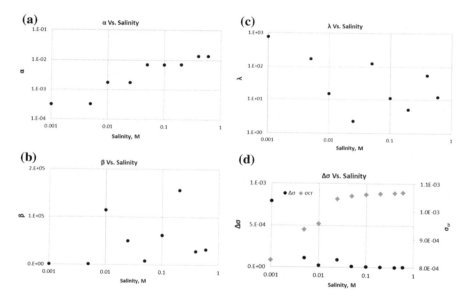

Fig. 22 Tuned values of the slow-particle model parameters: **a** drift delay factor, **b** formation damage coefficient, **c** filtration coefficient and **d** maximum retention function

Smaller particles are subject to lower drag force and therefore move at lower velocity. Also, the rock tortuosity increases due to straining, so that the fines move at lower velocity. Regarding formation damage coefficient β, there are two competing factors during the salinity decrease. The first is permeability, which inversely affects the formation damage coefficient. The other is particle size, with which the formation damage coefficient varies. The retained-concentration dependency for the formation damage coefficient, shown in Fig. 22b, is attributed to particle size's

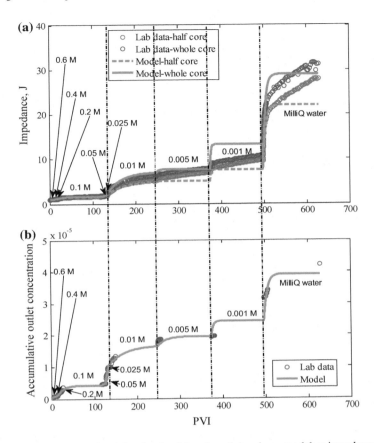

Fig. 23 Matching the coreflood data by the delayed-particle-release model: **a** impedance growth along the whole core and the first core section; **b** accumulated fine particle production

dominating the permeability effect. The filtration coefficient for straining, λ, increases with rock tortuosity increase during salinity reduction (Fig. 22c). Yet, it should decrease due to decrease in released particle size. Figure 22d presents the salinity dependency of the maximum retention function, which exhibits a typical form (Bedrikovetsky et al. 2012a, b; Zeinijahromi et al. 2012a, b).

The results of comparison with the delayed-particle-release model are presented in Fig. 23a for impedance, and in Fig. 23b for accumulated particle concentration. Table 8 shows the tuning parameter values. The values of tuned parameters β, λ, ε and $\Delta\sigma$ versus the salinity injected are also presented in Fig. 24a–d, respectively. The coefficient of determination is equal to $R^2 = 0.9874$ and is slightly lower than that for the slow-particle model.

Figure 24a shows that formation damage coefficient β increases as salinity declines, which contradicts the above conclusion for the slow-particle model. We attribute this to the dominant role of permeability decline on the formation damage

Table 8 Fitted parameters for the delayed-particle-release model

Salinity (M)	β	$\Delta\sigma$	α	λ	ε
0.6	12,876	1.07E−05	1.0000	31.40	8
0.4	99,876	6.09E−07	1.0000	21.07	4
0.2	99,976	5.86E−05	1.0000	13.27	338
0.1	99,973	3.97E−06	1.0000	6.65	11
0.05	199,985	5.60E−07	1.0000	6.65	1
0.025	99,990	3.56E−05	1.0000	2.23	84
0.01	1,100,000	3.51E−06	1.0000	2.23	33
0.005	100,000	1.77E−06	1.0000	2.23	5
0.001	200,000	6.02E−06	1.0000	2.23	2
0	900,000	5.14E−06	1.0000	2.34	6

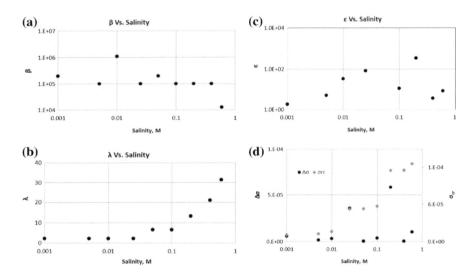

Fig. 24 Tuned values of the delayed-particle-release model parameters for 10 different-salinity stages: **a** formation damage coefficient, **b** filtration coefficient, **c** delay factor and **d** concentration of detached fines

coefficient: the lower the permeability, the higher the formation damage. Yet, this explanation contradicts the observation made above for the slow-particle model, where the formation damage coefficient decreases with the deposit increase.

The filtration coefficient λ decreases during salinity reduction (Fig. 24b), which also contradicts the above observation for the slow-particle model. We attribute this to the blocking filtration function, where λ is proportional to the vacancy concentration: the filtration function approaches zero as the number of pores smaller than particles approaches zero. Also, the released particle size decreases during salinity decrease, so the probability of particle straining declines, which is another explanation why the filtration coefficient declines during the injection.

The delay factor ε decreases with increase in strained concentration (Fig. 24c), which we attribute to more confined porous space and smaller diffusive path. Also, the interstitial velocity increases during straining, resulting in higher effective (Taylor's) diffusion in each pore and yielding the delay decline.

The maximum retention function (Fig. 24d) is unlike the usual release of fine particles at very low salinity, which is close to freshwater. The effect of fines release decrease during salinity decrease might be explained by low concentration of small particles on the rock surface. This hypothesis could be tested by measuring particle-size distributions in the breakthrough fluid.

Mahani (2015a, b) measured the delay period, which is 10–20 times longer than that expected by diffusion alone and is explained by slow electrokinetic Nernst–Planck ion-diffusion in the field of electrostatic DLVO forces. The delay time t_τ varies from 10,800 to 363,600 s, which corresponds to dimensionless time ($\varepsilon = Ut_\tau/\phi L$) varying from 15.49 to 521.62 PVI for conditions of the test presented in Sect. 3.1 ($\phi = 0.13$, $U = 1.48 \times 10^{-5}$ m/s). These values have the same order of magnitude as those obtained by tuning the parameter ε from laboratory tests and are presented in Tables 8 and 9.

Figure 25a presents the results of impedance matching by the general model that accounts for both phenomena of slow-particle migration and delayed particle release. Figure 25b presents the results for accumulated particle concentration. Table 9 shows the tuning parameter values. The coefficient of determination is equal to $R^2 = 0.9899$. The modelling data are in close agreement with the laboratory results, with deviation observed only for freshwater injection. The strained concentration dependencies for α, β, λ and $\Delta\sigma$ (Fig. 26a–c, e, respectively) follow the same tendencies as those exhibited by the slow-particle model. The delay factor ε (Fig. 26d) has the same form as that for the delay-detachment model. Thus, the tendencies for tuned values as obtained by the general model agree with the results of both particular models.

Table 9 Fitted parameters for the slow-particle delayed-release model

Salinity	β	$\Delta\sigma$	α	λ	ε
0.6 M	99,998	1.14E−06	0.4835	5.14	3.99
0.4 M	99,998	8.06E−07	0.2682	10.00	4.59
0.2 M	64,851	8.21E−06	0.0503	14.85	23.64
0.1 M	80,078	4.63E−06	0.0103	8.54	0.91
0.05 M	11,084	9.38E−06	0.0055	101.73	0.01
0.025 M	2,370	8.02E−05	0.0067	52.28	2.94
0.01 M	208,613	9.90E−06	0.0043	7.30	0.63
0.005 M	9,963	2.91E−05	0.0033	50.65	0.17
0.001 M	10,070	5.44E−05	0.0014	29.74	0.17
0 M	30,211	6.52E−05	0.0084	29.74	0.17

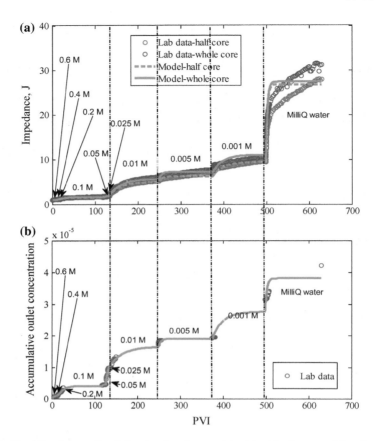

Fig. 25 Matching the coreflood data by the slow-particle delayed-release model: **a** impedance growth along the whole core and the first core section. **b** Accumulated fine particle production

3.6 Summary and Discussion

The laboratory study of fines migration due to decreasing brine salinity provided three measurement histories during each injection step with constant salinity: impedance across the half-core, impedance across the whole core and the outlet concentration of fine particles. Each pressure curve and each concentration curve has at least two degrees of freedom. Thus, the three measured curves have six degrees of freedom for constant-salinity periods.

The slow-particle migration model has four independent coefficients; therefore, a six-dimensional dataset was compared to the model with four tuned coefficients, and the latter was found to be highly accurate. We explained the strained saturation dependencies of the tuned parameters by the well-known dependencies for formation damage parameters of particle and pore sizes. Under the conditions where

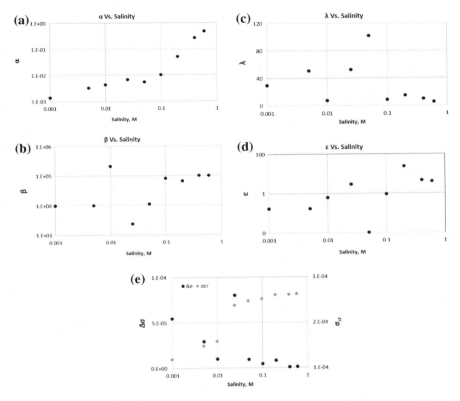

Fig. 26 Tuned values of the slow-particle delayed-release model parameters for 10 different-salinity stages: **a** drift delay factor, **b** formation damage coefficient, **c** filtration coefficient, **d** delay factor and **e** concentration of detached fines

the degree of freedom for the experimental data are higher than the number of tuned constants, close agreement between the experimental data and the model allows concluding the validity of the slow-particle migration model. The delay-release model has five independent coefficients. This is higher than that for the slow-particle migration model, but still lower than the six degrees of freedom of the laboratory dataset. The agreement coefficient is also very high between the laboratory data and the model-predicted data. The obtained delay periods have the same order of magnitude as do those observed in laboratory tests by Mahani et al. (2015a, b). However, the model does not exhibit a common form of the maximum retention function after the laboratory-data adjustment.

Thus, the advantages of the slow-particle model over the delay-release model are the smaller number of tuned parameters and common form of the revealed maximum retention function.

The general model has five independent coefficients, which is lower than the six degrees of freedom of the laboratory dataset. The agreement coefficient is also very high. The obtained delay periods have the same order of magnitude as do those observed in laboratory tests. The model exhibits a common form of the maximum retention function.

The slow-fines-migration model with four free parameters already exhibits very close agreement with laboratory data. Adding the delay factor into the slow-fines model does not change its accuracy.

The proposed interpretation of the model-parameter variations with the salinity decrease includes several competitive factors. It is impossible to declare a priori which factor dominates. Therefore, the proposed explanations must be verified by micro-scale modelling.

4 Fines Detachment and Migration at High Temperature

This section discusses the temperature dependency of fine particle detachment and migration in geothermal reservoirs (Rosenbrand et al. 2015).

As previously discussed, the DLVO theory is used for calculating the electrostatic forces. Because the electrostatic forces are temperature-dependent, fines mobilisation is also a function of temperature. The temperature-dependent parameters of the DLVO forces are listed in Table 10.

Figure 27 presents the effect of temperature on critical particle size according to Eq. (6). Because electrical attraction decreases with temperature increase, particles can be mobilised at a lower velocity if the temperature is increased (illustrated by the curve for 25 °C being above the curve for 80 °C).

Table 10 Temperature effects on the parameters in DLVO interaction energy model

Parameter	Temperature effect	Reference
λ	N/A	Gregory (1981)
ε_1	Table 11	Leluk et al. (2010)
ε_2	Negligible if $T < 170$ °C (Fig. 1 in Ref.)	Stuart (1955)
ε_3	Table 12	Marshall (2008)
n_1	N/A	Egan and Hilgeman (1979)
n_2	Interpolation from Fig. 1 in Ref.	Leviton and Frey (2006)
n_3	Equation (8) in Ref.	Aly and Esmail (1993)
ζ_s	Equation (9) in Ref.	Schembre and Kovscek (2005)
ζ_{pm}	Equation (9) in Ref.	Schembre and Kovscek (2005)
σ_c	N/A	Elimelech et al. (2013)

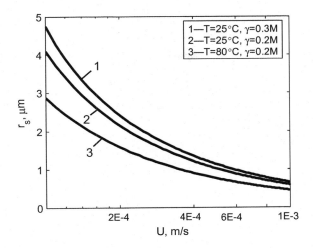

Fig. 27 Critical particle size for monolayer of size-distributed particles mobilised by the fluid flow at various velocities

4.1 Experimental Results and Model Prediction

The experimental study was undertaken to help analyse formation damage due to fines migration in the Salamander-1 geothermal well (Pretty Hill Formation, Otway Basin, South Australia). However, there is no core from the Salamander-1 well, and only drilling cuttings are available. Thus, a core with analogous mineral characteristics from the same formation and basin (Ladbroke Grove-1 well) was used for this study. This core was taken from depth 2557.12 m and has porosity of 17.2%. The core is 6.33 cm long and has a diameter of 3.92 cm.

In order to characterise the mineralogy of fines present in the core sample, the produced fines were collected by filtering the effluent fluid through a 0.45 μm Nylon filter. The collected volume of produced fines was insufficient for performing an XRD analysis. Thus, SEM-EDX analyses were performed on the produced fines, the results of which are shown in Fig. 28. The plate-like 'booklets' on the SEM image show the typical characteristic of kaolinite fines (Fig. 28a). The 'peak height ratio' (ratio of relative molar proportions) for Al and Si are shown in the EDX spectrum (Fig. 28b). The similar ratios between Al and Si indicate that the observed booklets on the SEM image are kaolinite, where the compound $Al_2(Si_2O_5)(OH)_4$ is typical.

A coreflood test with piecewise salinity (ionic strength) decrease was performed using the Ladbroke Grove-1 core, using the methodology given in Sect. 3.

Figure 29a presents the normalised permeability of the core for each salinity versus time (black circles). Figure 29b shows that reduction in injection fluid salinity resulted in decrease of the core permeability and production of fine particles. As expected, the graph shows a similar trend as those presented in Figs. 20, 23 and 25. The core permeability drops instantly after each salinity reduction, implying

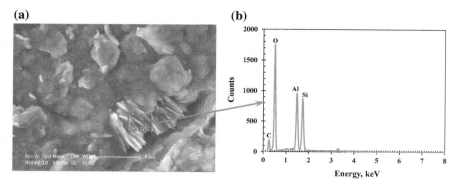

Fig. 28 SEM-EDX results for the core sample (Ladbroke Grove-1 well): **a** SEM image and **b** EDX spectra for kaolinite

that a significant fraction of attached fine particles are mobilised when salinity is reduced. The mobilised particles then plug pore throats gradually.

The analytical model presented in Sect. 2.3 (rows 12 and 13 in Table 1) was used to treat the experimental data from the Ladbroke Grove-1 core. A typical log-normal distribution of particle sizes is assumed.

Because quartz and kaolinite are the most abundant minerals in the studied sandstone core, the following DLVO parameters were used in the calculations: refractive index of kaolinite $n_1 = 1.502$ (Egan and Hilgeman 1979), refractive index of quartz n_2 and brine n_3 as functions of temperature (Leviton and Frey 2006; Aly and Esmail 1993), dielectric constant of quartz $\varepsilon_2 = 4.65$ (Stuart 1955), dielectric constant of kaolinite $\varepsilon_1 = 6.65$ at $T = 25\,°C$ and $\varepsilon_1 = 6.35$ at 80 °C (Leluk et al. 2010; see Table 10), and dielectric constant of brine ε_3 (Leluk et al. 2010; see Table 11). The zeta potentials for fines and grains were calculated using the correlation presented by Schembre and Kovscek (2005), and the water viscosity as a function of temperature was calculated using $\mu(T) = 2.414 \times 10^{-5} \times 10^{247.8/(T-140)}$ (Al-Shemmeri 2012).

No available data were found for temperature dependency of the characteristic wavelength of interaction λ, refractive index of clay n_1, or collision diameter σ_c. Therefore, it was assumed that the parameters mentioned above are constant with temperature (Schembre and Kovscek 2005; Schembre et al. 2006; Lagasca and Kovscek 2014). These constant values were taken from Egan and Hilgeman (1979) and Elimelech et al. (2013).

The model tuning parameters are as follows: filtration coefficient λ, formation damage coefficient β, drift delay factor α, variance coefficient for particle-size distribution C_v, and mean particle size $\langle r_s \rangle$. As discussed previously, the sizes of particles that are mobilised depend on the injected fluid velocity and ionic strength (salinity). The DLVO calculation showed that particles with smaller sizes can be mobilised during the reduction of injected fluid salinity (Fig. 27). We assume that particles of only one size can be mobilised during each salinity step. This allows the

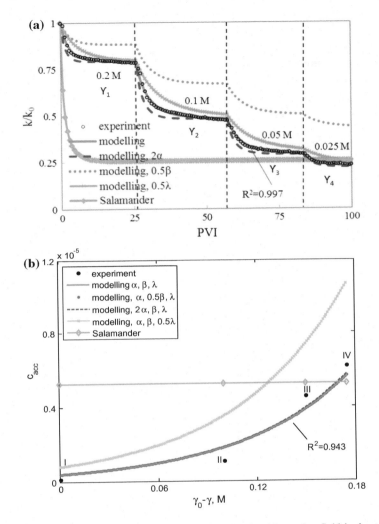

Fig. 29 Results of tuning the laboratory data from corefloods with varying fluid ionic strength at $T = 25$ °C, using the analytical model and prediction for Salamander geothermal field ($T = 129$ °C): **a** decrease of core permeability during tests with piecewise-decreasing ionic strength. **b** Cumulative breakthrough concentration at different fluid ionic strengths

Table 11 Dielectric constant of kaolinite ε_1 [interpolated from data in Leluk et al. (2010)]

T (°C)	ε_1
25	6.65
80	6.35
130	6.11
180	5.89

following parameters to remain unchanged during each salinity step: filtration coefficient, formation damage coefficient and drift delay factor.

A least-square goal function was used to tune the analytical model. The minimisation of the difference between model prediction and the experimental data was carried out using the Levenberg–Marquardt minimisation algorithm (Matlab 2016). The obtained values of the tuning parameters are presented in Table 12. Figure 29 presents the results of modelling for permeability and effluent concentration. It shows high agreement between modelling results and experimental data for both core permeability ($R^2 = 0.997$) and produced particle concentration ($R^2 = 0.943$).

We now calculate the number of degrees of freedom for the experimental dataset. Assuming that pressure drop is exponential with time during each constant-salinity injection, we obtain three degrees of freedom for each time interval in Fig. 29a. Yet, the initial permeability for each time interval is equal to the final permeability from the previous interval, which results in $2 \times 4 = 8$ degrees of freedom of the pressure drop measurements. Four independent breakthrough particle concentrations, averaged over the constant-salinity injection periods, add four degrees of freedom, giving a total of 12. Table 13 shows 15 independent constants tuned from the experimental data, which exceeds the number of degrees of freedom of the laboratory dataset. There is close agreement between the experimental and modelling data. We conclude that the model matches the laboratory results with high accuracy.

However, validating the proposed model would require significantly more laboratory data. Such data could include online measurements of breakthrough particle concentration against time. Another possibility is the three-point pressure measurement discussed in Sect. 3. Pressure measurements in an intermediate core port will double the number of degrees of freedom of the pressure drop information. It would allow increasing the number of degrees of freedom for the experimental dataset and validating the mathematical model.

The sensitivity analysis was performed using the following tuning parameters: drift delay factor, formation damage coefficient and filtration coefficient. Figure 29a, b presents the results of the sensitivity calculations. The blue curve in Fig. 29a indicates that α-increase yields stabilisation-time decrease, because the particles move faster. The light-blue curve indicates that lower probability of particle capture by thin pores translates to lower values of λ, leading to higher values for permeability stabilisation period. The green curve shows that smaller values of β correspond to decrease in permeability damage with time. The only modelling parameter that has a significant effect on breakthrough concentration of mobilised particles is the filtration coefficient λ, because it reflects the probability of mobilised particle capture by thin pore throats (Fig. 29b). The light-blue curve in Fig. 29b shows that the lower values of λ correspond to higher breakthrough particle concentrations. The blue curve indicates that breakthrough particle concentration is insensitive to drift delay factor. The green curve shows that the breakthrough concentration is insensitive to β. The obtained values of λ and β presented in Table 13 along with the drift delay factor fall within their common intervals given by Nabzar and Chauveteau (1996), Pang and Sharma (1997), and Civan (2010, 2014).

Table 12 Dielectric constant of brine ε_3[a]

T (°C)	0.6 M	0.4 M	0.2 M	0.1 M	0.05 M	0.025 M	0.01 M	0.005 M	0.001 M	0.00013 M
25	72.767	74.995	77.222	78.336	78.893	79.172	79.339	79.395	79.439	79.449
100	50.711	52.263	53.816	54.592	54.980	55.174	55.291	55.329	55.361	55.367
129	44.191	45.544	46.897	47.574	47.912	48.081	48.183	48.216	48.243	48.249
200	31.589	32.556	33.524	34.007	34.249	34.370	34.442	34.466	34.486	34.490
300	18.744	19.318	19.892	20.179	20.323	20.394	20.437	20.452	20.463	20.466

[a]Values are calculated from formula $\varepsilon_3(T)$ given by Marshall (2008) based on laboratory-measured ε_3 values with different ionic strengths at ambient condition

Table 13 Values of the model tuning parameters in the coreflood test

Parameter	Value
r_s, μm	1.80
C_v	0.66
σ_0	3.04e−4
α_1	4.10e−3
α_2	2.96e−3
α_3	2.81e−3
α_4	2.74e−3
β_1	9,793
β_2	7,631
β_3	7,391
β_4	7,158
λ_{D1}	67.14
λ_{D2}	53.79
λ_{D3}	51.11
λ_{D4}	50.13

The tuned data allow predicting the maximum retention function using Eq. (8). Maximum retention functions for a monolayer of multi-sized particles versus fluid velocity and fluid ionic strength are shown in Fig. 30a, b, respectively. Point I in Fig. 30a corresponds to initial attached particle concentration. While fluid velocity increases from 0 to U_A under the ambient temperature, the state point moves along the path $I \rightarrow A$ without particle mobilisation. Increasing fluid velocity from U_A to U_B (represented by migrating from point A to point B) leads to initiation of particle mobilisation at critical fluid velocity $U = U_B$. If fluid velocity further increases from U_B to U_C, the state point moves down along the maximum retention curve from point B to point C. Figure 30a, b depict the typical shape of the maximum retention function with gradual decrease in fluid ionic strength corresponding to increase in fluid velocity. This results in similarity between the critical ionic strength γ_B and the critical fluid velocity U_B. Miranda and Underdown (1993) reported that critical velocity corresponds to the first-particle release when fluid velocity increases. Khilar and Fogler (1998) reported that critical ionic strength corresponds to the first-particle release when fluid ionic strength decreases. The maximum retention function allows predicting the amount of the released fines as a result of altering the critical parameters. The amount of mobilised particles during alteration of fluid velocity or salinity is denoted as $\Delta\sigma$ in Fig. 30a, b.

Electrostatic attractive force decreases with increasing temperature, thereby somewhat decreasing the maximum retention function. On the contrary, reduction in water viscosity with temperature results in decrease in (detaching) lift and drag forces, yielding an increase in the maximum retention function. These two competing effects determine whether the maximum retention function increases or decreases.

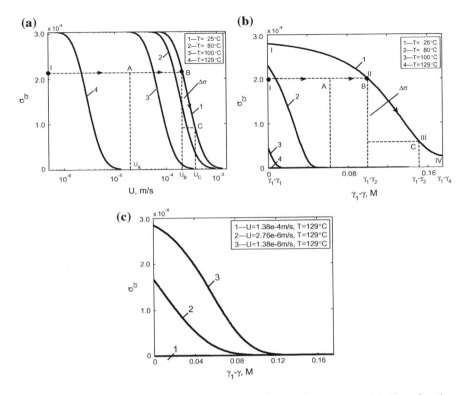

Fig. 30 Temperature, velocity, and salinity dependence of maximum retention function: **a** maximum retention concentration versus velocity at different temperatures. **b** Ionic strength dependency of maximum retention concentration ($\gamma_1 = 0.2$ M NaCl) at different temperatures. **c** Ionic strength dependency at geothermal reservoir temperature and different velocities

Figure 30a, b show that the dominance of temperature influence on electrostatic attaching forces leads to decrease in the maximum retention function and consequent permeability decline with temperature rise. σ_{cr}-curves were calculated for the following temperatures: room temperature (25 °C), 80, 100, and 129 °C (curves 1–4, respectively, in Fig. 30a, b). Comparing these curves shows decline in the maximum retention function with temperature increase. The geothermal well from the Salamander-1 field is characterised by a moderate temperature of $T = 129$ °C and ionic strength equivalent to 0.2 M NaCl. At these field conditions, according to curve 4 (Fig. 30b), almost all fines were mobilised when the maximum retention function has decreased to zero.

Fluid velocity alteration also affected particle mobilisation within a porous medium. Figure 30c shows σ_{cr}-curves for the following fluid velocities: wellbore velocity r_w, 50 r_w, and 100 r_w (curves 1–3, respectively). A decrease in fluid velocity yielded the maximum retention concentration increase, due to reduced detaching drag force acting on the attached particles.

4.2 Using Sensitivity of Ionic Strength to Characterise Fines Mobilisation

Torque balance equation (6) shows that the critical concentration of the attached particle σ_{cr} is a function of drag and lift forces. These detaching forces highly depend on the fluid velocity and viscosity. In a geothermal reservoir, both fluid velocity and viscosity change with time and position.

Consider inflow performance in a geothermal production well. The production rate per unit of the reservoir thickness is $q = 2\pi r U(r)$, where $U(r)$ is fluid velocity and r is radius of the drainage contour.

During production, the fluid velocity decreases significantly with radius (two or more orders of magnitude) in the direction from the wellbore toward the drainage contour. Therefore, the rheological dependence of σ_{cr} on temperature should be studied in a wide range of fluid velocities. Having studied this relationship, one can reliably estimate well fines migration and consequently the productivity index.

Experimental study of fines migration at very high velocities is limited by injection pump capacity. However, coreflood tests can be performed at a wide range of fluid ionic strength. Therefore, in the laboratory, it is more practical to perform fines mobilisation tests by varying fluid ionic strength than by varying velocity alteration. Torque balance Eq. (6) has solutions $\sigma_{cr} = \sigma_{cr}(\gamma)$ and $\sigma_{cr} = \sigma_{cr}(U)$. This allows the effect of fluid ionic strength on fines mobilisation ($\sigma_{cr} = \sigma_{cr}(\gamma)$) to be translated into the effect of fluid velocity on fines mobilisation ($\sigma_{cr} = \sigma_{cr}(U)$, and vice versa. From $\sigma_{cr}(\gamma_0) = \sigma_{cr}(U_0)$, the translation formula is

$$\sigma_{cr}(U, \gamma_0) = \sigma_{cr}(U_0, \gamma) \tag{52}$$

The methodology of this translation is described below for a monolayer of multi-sized particles (see Eq. (8)). Performing a coreflood test at constant velocity U_0 with varying fluid ionic strength γ yields critical concentration of attached particles $\sigma_{cr}(U_0, \gamma)$. Equation (6) determines critical attached particle radius at each salinity $r_{scr}(U, \gamma)$. The critical radius for varying velocity at a constant ionic strength can be recalculated using Eq. (6): $r_{scr}(U_0, \gamma) = r_{scr}(U, \gamma_0)$. This allows translating the experimentally obtained $\sigma_{cr}(U_0, \gamma)$ into $\sigma_{cr}(U, \gamma_0)$. Figure 30a, b show such a translation. Curve 1 in Fig. 30a presents the critical particle concentration at constant ionic strength $\sigma_{cr}(U, \gamma_0)$ that is equivalent to curve 1 in Fig. 30b, the critical particle concentration at constant velocity $\sigma_{cr}(U_0, \gamma)$.

Figure 30b presents the maximum retention function (critical concentration of the attached fine particles) that corresponds to the permeability decline curve presented in Fig. 29a. The cumulative produced particle concentration is presented in Fig. 29b. Point I in Fig. 30b corresponds to the initial condition (injection of high-salinity fluid γ_1). A negligible volume of fines was produced at this salinity (point I in Fig. 29b). The path for reduction of injection fluid salinity (ionic strength) from point I to III on the maximum retention function curve goes through point II (horizontally from I → II and then along the curve II → III, indicated by arrows in Fig. 30b). Reduction in fluid salinity (ionic strength) from point I to point III results in a significant mobilisation of fines and a sharp increase in produced fines (points II and III in Fig. 29b). This sharp increase in fines mobilisation can be explained by the increase of slope of the maximum retention function curve from point II to point III. Further decrease of fluid ionic strength to point IV is accompanied by a smaller increase of released and produced fines. This corresponds to the inflection of the maximum retention function curve.

4.3 Summary and Discussion

This laboratory study on fines migration at high temperatures and micro-scale modelling of fines mobilisation allows drawing the following conclusions:

- For geothermal reservoir conditions, the lifting and gravity forces are two to four orders of magnitude weaker than the drag and electrostatic forces. Mechanical equilibrium of attached fines and the maximum retention function is determined by drag and electrostatic forces; the lift and gravity forces are negligible.
- The fines release capacity—maximum retention function—for a monolayer deposit of multi-sized particles, as well as for a poly-layer of single-size particles, can be expressed by explicit formula.
- Experiment-based model predictions for high-temperature geothermal conditions showed that the electrostatic attraction weakens with temperature increase, and the detaching drag force reduces with water viscosity decrease. The former effect dominates, resulting in the decrease of the maximum retention function with temperature. Therefore, geothermal reservoirs are more susceptible to fines migration than conventional aquifers or oilfields.
- The laboratory 'temperature-ionic strength' transformation procedure along with mechanical equilibrium modelling allows determining temperature dependency of the maximum retention concentration from the tests with varying ionic strength; it allows predicting particle detachment at high temperatures based on laboratory tests with salinity variation.
- Laboratory-measured permeability history is consistent with the model prediction.
- The prediction from laboratory-based mathematical modelling closely approximates geothermal-well index history from field data.
- Kaolinite and illite/chlorite, as main clay minerals presented in released fines from coreflood in the present study, are responsible for formation damage.

5 Fines Migration in Gas and CBM Reservoirs

Byrne et al. (2009, 2014) reported the intensive fines production and associated formation damage in high-rate gas wells. Possible explanation of this phenomenon is drying-up the rock in well vicinity. Despite negligible equilibrium vapour concentration in methane, numerous gas volumes pass via the near-well zone, resulting in evaporation of the connate water. The adhesion grain–grain attraction (bridging between two particles by capillary menisci) consolidates the rock. As water saturation decreases below its connate value, the capillary water bridge between some grains disappeared, the capillary component of stress decreases, yielding rock dis-consolidation. It causes the release of some particles by the drag force, exerting on the particle by the flowing gas.

The maximum retention function can model the phenomenon of fines release during water saturation decreasing below the connate value. Lazouskaya et al. (2013) account for two-phase fines mobilisation in torque balance Eq. (6). The adhesive force attracts water-wet particles and repulses hydrophobic fines (Muecke 1979). Accounting for capillary forces makes maximum retention function saturation-dependent (Yuan and Shapiro 2011b). However, matching the laboratory or field data on fines migration in two-phase environment by mathematical model is not available.

Besides salinity decreasing and increase in velocity, pH and temperature, stress increasing can also be a cause of fines mobilisation (Bai et al. 2015a; Han et al. 2015). Fines migration has been reported during well fracturing. Another area of fines migration due to increase in the reservoir stress is methane production from coal beds (Yao et al. 2016). The mathematical model for fines lifting is a kinetic equation for detachment rate, where the kinetics coefficient (relaxation time) is stress-dependent (Civan 2010; Guo et al. 2015; Mitchell and Leonardi 2016). Validation of the fines detachment rate under stress-increase by laboratory testing and theoretical derivation of kinetics rate equation from entropy production (Onsager principle) are still not available in the literature.

6 Conclusions

This analytical modelling and laboratory study of fines migration due to velocity, salinity, temperature and pH alteration during coreflooding allows drawing the following conclusions:

1. Mechanical equilibrium of attached fines is determined mainly by drag and electrostatic forces. Neglecting lift and gravitational forces eliminates half of the tuning parameters in the torque balance equation.
2. Low-velocity fines' drifting along the rock surface (rolling, sliding) explains the long permeability stabilisation periods. Stabilisation time greatly exceeds the expected one pore volume injected, suggesting that the fine particles migrate at

a velocity that is two to three orders of magnitude lower than the carrier water velocity.

3. Another explanation of long permeability stabilisation periods might be the delay in particle release due to slow diffusion of salt from the grain–particle deformed contact area into the bulk of the fluid.
4. One-dimensional problems for slow fines migration with delayed particle release after velocity, salinity or pH alteration allow for exact solution. The analytical model contains explicit formulae for breakthrough and retained concentrations and pressure drop history.
5. Matching the measured permeability and cumulative outlet particle concentration by the analytical model, accounting for both slow fines migration and delayed release, shows strong agreement between the measured data and modelling results.
6. The slow-particle model matches the experimental data with higher accuracy than does the delay-release model. The straining concentration and salinity dependencies for model parameters obtained from tuning the laboratory data by the slow-particle model have typical forms observed in other studies. Using the delay-release model for laboratory-data tuning reveals a non-typical form of the maximum retention function.
7. Fine particles mobilisation occurs in order of decreasing particle size during velocity, temperature and pH increase, or salinity decrease.
8. The maximum retention function for size-distributed fine particles attached to pore walls as a monolayer is expressed by an explicit formula that includes the size distribution of attached particles and the critical detached-size curve. This function is equal to accumulated concentration of particles smaller than those mobilised by the flux with a given flow velocity U.
9. Size distribution of the attached movable particles can be determined from the maximum retention function and the critical-size curve.
10. The laboratory 'velocity-ionic strength' and 'temperature-ionic strength' translation procedures along with mechanical equilibrium modelling allow determining velocity- and temperature dependencies of the maximum retention concentration from tests that vary ionic strength.
11. Temperature effect on reduction of electrostatic attractive forces exceeds the effect of (detaching) drag-force reduction induced by viscosity decrease. This results in the maximum retention function decrease with temperature increase. Therefore, geothermal reservoirs are more susceptible to fines migration than are conventional aquifers or oil and gas fields.

Acknowledgements The authors are grateful to numerous researchers with whom they worked on colloidal-suspension transport in porous media: Prof. A. Shapiro and Dr. H. Yuan (Denmark Technical University), Dr. R. Farajzadeh and Profs. P. Zitha and H. Bruining (Delft University of Technology), Prof. A. Polyanin (Russian Academy of Sciences), Prof. Y. Osipov (Moscow University of Civil Engineering), and L. Kuzmina (National Research University, Russia).

References

Ahmadi M, Habibi A, Pourafshary P et al (2013) Zeta-potential investigation and experimental study of nanoparticles deposited on rock surface to reduce fines migration. SPE J 18:534–544

Akhatov IS, Hoey JM, Swenson OF et al (2008) Aerosol focusing in micro-capillaries: Theory and experiment. J Aerosol Sci 39:691–709

Al-Shemmeri T (2012) Engineering fluid mechanics. Bookboon, London

Alvarez AC, Bedrikovetsky P, Hime G et al (2006) A fast inverse solver for the filtration function for flow of water with particles in porous media. J Inverse Probl 22:69–88

Alvarez AC, Hime G, Marchesin D et al (2007) The inverse problem of determining the filtration function and permeability reduction in flow of water with particles in porous media. Transp Porous Media 70:43–62

Aly KM, Esmail E (1993) Refractive index of salt water: effect of temperature. Opt Mater 2:195–199

Arab D, Pourafshary P (2013) Nanoparticles-assisted surface charge modification of the porous medium to treat colloidal particles migration induced by low salinity water flooding. Colloids Surf A 436:803–814

Assef Y, Arab D, Pourafshary P (2014) Application of nanofluid to control fines migration to improve the performance of low salinity water flooding and alkaline flooding. J Petrol Sci Eng 124:331–340

Badalyan A, Carageorgos T, Bedrikovetsky P et al (2012) Critical analysis of uncertainties during particle filtration. Rev Sci Instrum 83:095106

Bai T, Chen Z, Aminossadati SM et al (2015a) Characterization of coal fines generation: A micro-scale investigation. J Nat Gas Sci Eng 27:862–875

Bai B, Li H, Xu T, Chen X (2015b) Analytical solutions for contaminant transport in a semi-infinite porous medium using the source function method. Comput Geotech 69:114–123

Bedrikovetsky PG (1993) Mathematical theory of oil & gas recovery (with applications to ex-USSR oil & gas condensate fields). Kluwer Academic Publishers, London

Bedrikovetsky P, Siqueira FD, Furtado CA et al (2011a) Modified particle detachment model for colloidal transport in porous media. Transp Porous Media 86:353–383

Bedrikovetsky P, Vaz AS, Furtado CJ et al (2011b) Formation-damage evaluation from nonlinear skin growth during coreflooding. SPE Reserv Eval Eng 14:193–203

Bedrikovetsky P, Zeinijahromi A, Siqueira FD et al (2012a) Particle detachment under velocity alternation during suspension transport in porous media. Transp Porous Media 91:173–197

Bedrikovetsky P, Vaz A, Machado F et al (2012b) Skin due to fines mobilisation, migration and straining during steady state oil production. J Petrol Sci Tech 30:1539–1547

Bergendahl JA, Grasso D (2003) Mechanistic basis for particle detachment from granular media. Environ Sci Technol 37:2317–2322

Bradford SA, Bettahar M (2005) Straining, attachment, and detachment of Cryptosporidium oocyst in saturated porous media. J Environ Qual 34:469–478

Bradford SA, Torkzaban S, Kim H et al (2012) Modeling colloid and microorganism transport and release with transients in solution ionic strength. Water Resour Res 48:W09509

Bradford SA, Torkzaban S, Shapiro A (2013) A theoretical analysis of colloid attachment and straining in chemically heterogeneous porous media. Langmuir 29:6944–6952

Byrne MT, Waggoner SM (2009) Fines migration in a high temperature gas reservoir-laboratory simulation and implications for completion design. Paper presented at the International Symposium and Exhibition on Formation Damage Control, Lafayette, 26–28 Feb 2014

Byrne M, Rojas E, Kandasamy R et al (2014) Fines migration in oil and gas reservoirs: quantification and qualification through detailed study. Paper presented at the International Symposium and Exhibition on Formation Damage Control, Lafayette, 26–28 Feb 2014

Civan F (2010) Non-isothermal permeability impairment by fines migration and deposition in porous media including dispersive transport. Transp Porous Media 85:233–258

Civan F (2014) Reservoir formation damage, 3rd edn. Gulf Professional Publishing, Burlington

Coleman TF, Li Y (1996) An interior trust region approach for nonlinear minimization subject to bounds. SIAM J Optimiz 6:418–445

Dang C, Nghiem L, Nguyen N et al (2016) Mechanistic modeling of low salinity water flooding. J Petrol Sci Eng 146:191–209

Das SK, Schechter RS, Sharma MM (1994) The role of surface roughness and contact deformation on the hydrodynamic detachment of particles from surfaces. J Colloid Interface Sci 164:63–77

Egan WG, Hilgeman TW (1979) Optical properties of inhomogeneous materials: applications to geology, astronomy chemistry, and engineering. Academic Press, New York

Elimelech M, Gregory J, Jia X (2013) Particle deposition and aggregation: measurement, modelling and simulation. Butterworth-Heinemann, Oxford

Faber S, Al-Maktoumi A, Kacimov A et al (2016) Migration and deposition of fine particles in a porous filter and alluvial deposit: laboratory experiments. Arab J Geosci 9:1–13

Fleming N, Mathisen AM, Eriksen SH et al (2007) Productivity impairment due to kaolinite mobilization: laboratory & field experience, Oseberg Sor. Paper presented at the European Formation Damage Conference, Scheveningen, 30 May–1 June 2007

Fleming N, Ramstad K, Mathisen AM et al (2010) Squeeze related well productivity impairment mechanisms & preventative/remedial measures utilised. Paper presented at the SPE International Conference on Oilfield Scale, Aberdeen, 26–27 May 2010

Freitas AM, Sharma MM (2001) Detachment of particles from surfaces: an AFM study. J Colloid Interface Sci 233:73–82

Gercek H (2007) Poisson's ratio values for rocks. Int J Rock Mech Min Sci 44:1–13

Gregory J (1981) Approximate expressions for retarded van der Waals interaction. J Colloid Interface Sci 83:138–145

Guo Z, Hussain F, Cinar Y (2015) Permeability variation associated with fines production from anthracite coal during water injection. Int J Coal Geol 147:46–57

Habibi A, Ahmadi M, Pourafshary P et al (2012) Reduction of fines migration by nanofluids injection: an experimental study. SPE J 18:309–318

Han G, Ling K, Wu H et al (2015) An experimental study of coal-fines migration in coalbed-methane production wells. J Nat Gas Sci Eng 26:1542–1548

Hassani A, Mortazavi SA, Gholinezhad J (2014) A new practical method for determination of critical flow rate in Fahliyan carbonate reservoir. J Petrol Sci Eng 115:50–56

Herzig JP, Leclerc DM, Goff PL (1970) Flow of suspensions through porous media—application to deep filtration. Ind Eng Chem 62:8–35

Huang TT, Clark DE (2015) Enhancing oil recovery with specialized nanoparticles by controlling formation-fines migration at their sources in waterflooding reservoirs. SPE J 20:743–746

Israelachvili JN (2011) Intermolecular and surface forces, 3rd edn. Academic press, Amsterdam

Jensen JL (2000) Statistics for petroleum engineers and geoscientists. Gulf Professional Publishing, Burlington

Katz AJ, Thompson AH (1986) Quantitative prediction of permeability in porous rock. Phys Rev B 34:8179

Khilar KC, Fogler HS (1998) Migrations of fines in porous media. Kluwer Academic Publishers, Dordrecht

Krauss ED, Mays DC (2014) Modification of the Kozeny-Carman equation to quantify formation damage by fines in clean, unconsolidated porous media. SPE Reserv Eval Eng 17:466–472

Lagasca JRP, Kovscek AR (2014) Fines migration and compaction in diatomaceous rocks. J Petrol Sci Eng 122:108–118

Lazouskaya V, Wang LP, Or D et al (2013) Colloid mobilization by fluid displacement fronts in channels. J Colloid Interface Sci 406:44–50

Leluk K, Orzechowski K, Jerie K (2010) Dielectric permittivity of kaolinite heated to high temperatures. J Phys Chem Solids 71:827–831

Lever A, Dawe RA (1984) Water-sensitivity and migration of fines in the hopeman sandstone. J Petrol Geol 7:97–107

Leviton DB, Frey BJ (2006) Temperature-dependent absolute refractive index measurements of synthetic fused silica. In: SPIE astronomical telescopes and instrumentation. International Society for Optics and Photonics

Li X, Lin CL, Miller JD et al (2006) Role of grain-to-grain contacts on profiles of retained colloids in porous media in the presence of an energy barrier to deposition. Environ Sci Technol 40:3769–3774

Mahani H, Berg S, Ilic D et al (2015a) Kinetics of low-salinity-flooding effect. SPE J 20:8–20

Mahani H, Keya AL, Berg S et al (2015b) Insights into the mechanism of wettability alteration by low-salinity flooding (LSF) in carbonates. Energ Fuel 29:1352–1367

Marques M, Williams W, Knobles M et al (2014) Fines migration in fractured wells: integrating modeling, field and laboratory data. SPE Prod Oper 29:309–322

Marshall WL (2008) Dielectric constant of water discovered to be simple function of density over extreme ranges from −35 to +600 °C and to 1200 MP_a (12000 Atm.), believed universal. Nature Preced

MATLAB and Optimization Toolbox (2016) The MathWorks Inc. Natick, Massachusetts

Miranda R, Underdown D (1993) Laboratory measurement of critical rate: a novel approach for quantifying fines migration problems. Paper presented at the SPE Production Operations Symposium, Oklahoma, 21–23 Mar 1993

Mitchell TR, Leonardi CR (2016) Micromechanical investigation of fines liberation and transport during coal seam dewatering. J Nat Gas Sci Eng 35:1101–1120

Muecke TW (1979) Formation fines and factors controlling their movement in porous media. J Pet Technol 31:144–150

Nabzar L, Chauveteau G, Roque C (1996) A new model for formation damage by particle retention. Paper presented at the SPE Formation Damage Control Symposium, Lafayette, 14–15 Feb 1996

Nguyen TKP, Zeinijahromi A, Bedrikovetsky P (2013) Fines-migration-assisted improved gas recovery during gas field depletion. J Petrol Sci Eng 109:26–37

Noubactep C (2008) A critical review on the process of contaminant removal in F_eO–H_2O systems. Environ Technol 29:909–920

Noubactep C, Caré S, Crane R (2012) Nanoscale metallic iron for environmental remediation: prospects and limitations. Water Air Soil Poll 223:1363–1382

Ochi J, Vernoux JF (1998) Permeability decrease in sandstone reservoirs by fluid injection: hydrodynamic and chemical effects. J Hydrol 208:237–248

Oliveira M, Vaz A, Siqueira F et al (2014) Slow migration of mobilised fines during flow in reservoir rocks: laboratory study. J Petrol Sci Eng 122:534–541

Pang S, Sharma MM (1997) A model for predicting injectivity decline in water-injection wells. SPEFE 12:194–201

Prasad M, Kopycinska M, Rabe U et al (2002) Measurement of Young's modulus of clay minerals using atomic force acoustic microscopy. Geophys Res Lett 29: 13-1–13-4

Qiao C, Han J, Huang TT (2016) Compositional modeling of nanoparticle-reduced-fine-migration. J Nat Gas Sci Eng 35:1–10

Rosenbrand E, Kjoller C, Riis JF et al (2015) Different effects of temperature and salinity on permeability reduction by fines migration in Berea sandstone. Geothermics 53:225–235

Sacramento RN, Yang Y et al (2015) Deep bed and cake filtration of two-size particle suspension in porous media. J Petrol Sci Eng 126:201–210

Sarkar A, Sharma M (1990) Fines migration in two-phase flow. J Pet Technol 42:646–652

Schechter RS (1992) Oil well stimulation. Prentice Hall, NJ, USA

Schembre JM, Kovscek AR (2005) Mechanism of formation damage at elevated temperature. J Energy Resour Technol 127:171–180

Schembre JM, Tang GQ, Kovscek AR (2006) Wettability alteration and oil recovery by water imbibition at elevated temperatures. J Petrol Sci Eng 52:131–148

Sefrioui N, Ahmadi A, Omari A et al (2013) Numerical simulation of retention and release of colloids in porous media at the pore scale. Colloid Surface A 427:33–40

Sourani S, Afkhami M, Kazemzadeh Y et al (2014a) Importance of double layer force between a plat and a nano-particle in restricting fines migration in porous media. Adv Nanopart 3:49153

Sourani S, Afkhami M, Kazemzadeh Y et al (2014b) Effect of fluid flow characteristics on migration of nano-particles in porous media. Geomaterials 4:47299

Stuart MR (1955) Dielectric constant of quartz as a function of frequency and temperature. J Appl Phys 26:1399–1404

Tufenkji N (2007) Colloid and microbe migration in granular environments: a discussion of modelling methods. In: Colloidal transport in porous media. Springer, Berlin, pp 119–142

Van Oort E, Van Velzen JFG, Leerlooijer K (1993) Impairment by suspended solids invasion: testing and prediction. SPE Prod Fac 8:178–184

Varga RS (2009) Matrix iterative analysis. Springer, Berlin

Watson RB, Viste P, Kageson-Loe NM et al (2008) Paper presented at the Smart mud filtrate: an engineered solution to minimize near-wellbore formation damage due to kaolinite mobilization: laboratory and field experience, Oseberg Sør. In: SPE International Symposium and Exhibition on Formation Damage Control, Lafayette, 13–15 Feb 2008

Welzen JTAM, Stein HN, Stevels JM et al (1981) The influence of surface-active agents on kaolinite. J Colloid Interface Sci 81:455–467

Xu J (2016) Propagation behavior of permeability reduction in heterogeneous porous media due to particulate transport. EPL 114:14001

Yao Z, Cao D, Wei Y et al (2016) Experimental analysis on the effect of tectonically deformed coal types on fines generation characteristics. J Petrol Sci Eng 146:350–359

You Z, Badalyan A, Bedrikovetsky P (2013) Size-exclusion colloidal transport in porous media—stochastic modeling and experimental study. SPE J. https://doi.org/10.2118/162941-PA

You Z, Bedrikovetsky P, Badalyan A (2015) Particle mobilization in porous media: temperature effects on competing electrostatic and drag forces. Geophys Res Lett 42:2852–2860

You Z, Yang Y, Badalyan A et al (2016) Mathematical modelling of fines migration in geothermal reservoirs. Geothermics 59:123–133

Yuan H, Shapiro AA (2011a) A mathematical model for non-monotonic deposition profiles in deep bed filtration systems. Chem Eng J 166:105–115

Yuan H, Shapiro AA (2011b) Induced migration of fines during waterflooding in communicating layer-cake reservoirs. J Petrol Sci Eng 78:618–626

Yuan H, Shapiro A, You Z et al (2012) Estimating filtration coefficients for straining from percolation and random walk theories. Chem Eng J 210:63–73

Yuan H, You Z, Shapiro A et al (2013) Improved population balance model for straining-dominant deep bed filtration using network calculations. Chem Eng J 226:227–237

Yuan B, Moghanloo RG, Zheng D (2016) Analytical evaluation of nanoparticle application to mitigate fines migration in porous media. SPE J. https://doi.org/10.2118/174192-PA

Zeinijahromi A, Vaz A, Bedrikovetsky P et al (2012a) Effects of fines migration on well productivity during steady state production. J Porous Med 15:665–679

Zeinijahromi A, Vaz A, Bedrikovetsky P (2012b) Well impairment by fines production in gas fields. J Petrol Sci Eng 88–89:125–135

Zheng X, Shan B, Chen L et al (2014) Attachment–detachment dynamics of suspended particle in porous media: experiment and modelling. J Hydrol 511:199–204

Migration and Capillary Entrapment of Mercury in Porous Media

M. Devasena and Indumathi M. Nambi

1 Introduction

Elemental mercury (Hg^0) is considered to be a widespread contaminant in the environment. It is highly toxic, persistent and difficult to handle. As Hg^0 is a liquid under ambient conditions, it stands unique among all metals. In addition, its high specific gravity and electrical conductivity have brought about extensive usage of Hg^0 in various scientific and domestic equipment such as thermometers, barometers and compact fluorescent lights (CFLs). Much of the Hg^0 in the broken and discarded CFLs and thermometers are generally not recovered and become an environmental and human health concern. Improper disposal of Hg^0 from chlor-alkali plants, thermometer manufacturing units, metal smelting facilities and pharma industries has dramatically increased the spread and distribution of Hg^0 around the globe. Due to low aqueous solubility and high density, it penetrates into the subsurface and migrates as a typical dense non-aqueous phase liquid (DNAPL) and gets trapped in the void spaces as residual Hg^0. Any non-aqueous phase liquid (NAPL) migrates under the influence of groundwater forces such as capillary, gravity, viscous and buoyancy forces. The bulk of the NAPL migrates downwards and in the direction of groundwater flow and displaces water from the saturated zone. NAPL finally becomes discontinuous and immobile and gets trapped within the pore space. The entrapped NAPL serves as a long-term source of groundwater contamination. Oil, petroleum compounds, creosote and chlorinated organic solvents are few common examples of NAPLs. The entrapped liquid is termed as

M. Devasena (✉)
Department of Civil Engineering, Sri Krishna College of Technology,
Coimbatore 641042, Tamil Nadu, India
e-mail: nanduewreiitm@gmail.com

I. M. Nambi
EWRE Division, Department of Civil Engineering, Indian Institute
of Technology-Madras, Chennai 600036, Tamil Nadu, India

© Springer Nature Singapore Pte Ltd. 2018
N. Narayanan et al. (eds.), *Flow and Transport in Subsurface Environment*,
Springer Transactions in Civil and Environmental Engineering,
https://doi.org/10.1007/978-981-10-8773-8_2

residual oil saturation in petroleum reservoir engineering (Morrow 1979; Chatzis and Morrow 1984; Anderson 1988) whereas all chlorinated organic solvents entrapped in the subsurface are referred to as residual NAPL saturation or residual organic saturation. A residual liquid can serve as a long-term source of groundwater contamination both by direct dissolution and by volatilization into the soil gas (Mercer and Cohen 1990; Kueper and Frind 1991a, b; Chevalier and Fonte 2000). In order to determine the rate at which a DNAPL such as mercury gets entrapped depends on the aquifer characteristics, groundwater flow and properties of Hg^0.

Mercury entrapment is an important phenomenon and is expected to be governed by the same principles of commonly referred DNAPLs such as trichloroethylene (TCE) and perchloroethylene (PCE). However, investigations on Hg^0 as a DNAPL and its migration into soils and aquifers are limited. The fundamental properties of Hg^0 have been well documented in numerous publications and are summarized in Table 1. Properties of a comparable DNAPL (PCE) are also shown in Table 1.

Hg^0, being a highly dense and viscous liquid, with high interfacial tension with water, behaves as a typical DNAPL and gets trapped in the interparticle void spaces of the subsurface Entrapped Hg^0, termed as residual Hg^0 saturation, brings widespread groundwater contamination by its biochemical transformations to ionic and methylated forms of mercury. High density and high surface tension control the immediate behaviour of Hg^0 spills. Being an element, Hg^0 never breaks down but it persists in the environment, continuously cycling through the interconnected terrestrial, atmospheric, aquatic and biotic compartments. However, land spills and disposals are more serious since Hg^0 can percolate the subsurface. Depending upon the heterogeneity, it can travel longer distances and to deeper depths without boundaries. Several questions need to be answered in order to fully understand mercury entrapment process in porous media: What are the predominant forces influencing its migration in the subsurface? Do elemental mercury's high density

Table 1 Properties of Hg^0 and PCE

Parameter	Units	Mercury	PCE
Chemical formula		Hg^0	C_2Cl_4
Molecular weight		200	166
Solubility	mg/l	0.056[a]	200[b]
Vapour pressure	kPa	0.2×10^{-3}[a]	2.462[c]
Density	kg/m^3	13,500[a]	1630[b]
Viscosity	kg/m s	0.001554[a]	0.00089[d]
Surface tension	dynes/cm	480[a]	72.8[e]
Interfacial tension with water	dynes/cm	375[a]	47.8[f]

[a]U.S.DOE (2001)
[b]Gilham and Rao (1990)
[c]Watts (1996)
[d]Li et al. (2007)
[e]Demond and Linder (1993)
[f]Pennell et al. (1994)

and interfacial tension play a significant role in Hg^0 entrapment and residual Hg^0 saturation values? To what extent does Hg^0 behave according to the well-established DNAPL theories?

Mercury porosimetry technique is based on forcing mercury into a porous structure, observing Hg^0 intrusion and extrusion curves and analysing the structure of porous media. Capillary pressure–water saturation (P_c–S_w) curves have been used widely by soil scientists and petroleum engineers to find the pore size distribution and pore connectivity for pairs of wetting and non-wetting fluids. Ioannidis et al. (1991) found the relation between mercury saturation and capillary pressure for mercury–air system using glass-etched micromodels as porous media. Rigby et al. (2003) investigated the same using sol-gel silica powder fragments and pellets as porous media They demonstrated that no mercury entrapment was observed with silica gel powders of size $\sim$30–40 μm whereas pellets of size $\sim$3 mm showed apparent hysteresis and mercury entrapment. The reason attributed for entrapment was the specific arrangement of porous media where the larger pore sizes were surrounded by smaller pore sizes creating non-random heterogeneity. The samples were first subjected to a pressure of 6.7 Pa and then gradually up to 412 MPa during mercury intrusion. During mercury retraction, the pressure was reduced to 0.131 MPa. Wardlaw and McKellar (1981) also attributed the cause of mercury entrapment to non-random structural heterogeneity while conducting mercury porosimetry experiments with glass-etched micromodels. Their study revealed that only a non-random model, which had clusters of larger pores within a continuous network of smaller pores, showed mercury entrapment. Grids of uniform pore size and grids with different and randomly distributed pore size did not result in entrapment of mercury. In the above studies, neither residual saturation quantification nor capillary entrapment parameters have been reported. Nevertheless, the Young–Laplace equation given in Eq. 1 is the basis of capillary trapping of such immiscible liquids. The fundamentals of two-phase flow phenomena, capillary entrapment, two-phase P_c–S_w experiments conducted for Hg^0–water systems in homogenous porous media and investigation of pore-scale distribution of entrapped Hg^0 and identification of pore-scale entrapment parameters in glass bead-flow through micromodel experiments are discussed in this chapter.

2 Two-Phase Flow Phenomena

Migration and entrapment of elemental mercury in the subsurface is governed by two-phase flow phenomenon. The fundamental principles of two-phase flow and their implications for Hg distribution are reviewed in this section in order to understand the location and distribution of residual mercury. Air–water, NAPL–water, NAPL–air, oil–water, oil–air, mercury–water and mercury–air are few immiscible fluid pairs that are greatly influenced by the differences in their density, viscosity and interfacial tension. When two immiscible fluids are in contact with each other, they experience unbalanced forces at the interface giving rise to interfacial tension. As a result of this, the boundary between them exists as a curved surface which is the interface. Forces acting across the interface, called the interfacial forces, are responsible for the entrapment of the non-wetting fluid within the saturated zone. Water wets most aquifer materials relative to DNAPLs or air. Across the interface, there exists a pressure difference between both fluids, called capillary pressure (P_c). Mathematically, it is defined as

$$P_c = P_{nw} - P_w \qquad (1)$$

where P_c is the capillary pressure, P_{nw} is the pressure of the non-wetting fluid and P_w represents the pressure of the wetting fluid. The wetting fluid adheres to the soil surface forming a thin film around it and has the tendency to imbibe spontaneously into the pore space whereas the non-wetting fluid has to reach a breakthrough pressure to penetrate into the pore space.

Figure 1 shows the cross section of porous media with the non-wetting fluid trapped within the pore space under specific flow conditions. In mercury–water systems, Hg^0 is the non-wetting fluid and water is the wetting fluid. Groundwater displaces the non-wetting fluid leaving behind entrapped residuals of non-wetting fluid in the pore space. The amount of non-wetting fluid entrapped within the pore space is referred as residual non-wetting fluid saturation and is the ratio of the non-wetting fluid volume to the pore space volume.

Residual DNAPL saturations depend on many factors such as pore size distribution, wettability, fluid viscosity, density, interfacial tension and Darcy velocity (Wardlaw 1982). They are typically in the range of 10–50% and are found to be higher in the saturated zone. They are also higher in fine-grained porous media because of their dependence on inverse grain size (Mercer and Cohen 1990). Oostrom et al. (2002) observed residual saturation of carbon tetrachloride, to be higher in fine-grained sand compared to coarser materials. Moreover, residual saturation for a wetting fluid is different from that of a non-wetting fluid. Even among fluid systems where water is the wetting fluid, residual saturation may vary depending upon the density and interfacial tension values (Mercer and Cohen 1990).

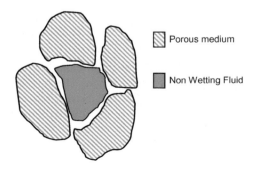

Fig. 1 Entrapped non-wetting fluid in porous media

3 Mechanisms of Entrapment

Capillary trapping is a function of wettability, contact angle as well as pore geometry. Capillary trapping may occur due to snap-off or bypassing mechanism. Snap-off occurs when a non-wetting fluid is forced by the wetting fluid from a large diameter pore body to a small diameter pore throat. This mechanism strongly depends on wettability and aspect ratio, i.e. the ratio of pore-body diameter to pore-throat diameter. Figure 2 illustrates the migration of non-wetting fluid through the porous media and its entrapment in the pore body by snap-off mechanism. Blobs trapped by snap-off typically occupy one or two pore spaces and are called singlets and doublets, respectively (Wilson et al. 1990).

The number of pore throats connected to each pore body also decides whether a non-wetting fluid will become entrapped by snap-off mechanism. In a soil with a high aspect ratio, a singlet is trapped by snap-off in each individual pore. For low aspect ratio soils, the non-wetting fluid is completely displaced without any entrapment. The combined effect of contact angle and pore aspect ratio determines the potential for snap-off mechanism. When wetting fluid tries to displace non-wetting fluid filled in pore space with different pore diameters, non-wetting fluid present in the smaller pore gets displaced first, disconnecting the non-wetting fluid present in the wider pore. This mechanism, known as bypassing, is shown in Fig. 3. Branched and more complex blobs can be expected to be trapped by means of bypassing mechanism. As shown in Fig. 3, water enters the narrower pore first and continues reaching the downstream end. The non-wetting liquid in the wider pore becomes disconnected from the main body of non-wetting phase liquid and has been bypassed by the wetting phase.

Snap-off would be the predominant mechanism of trapping in homogenous porous media. Morrow and Chatzis (1982) and Chatzis et al. (1983) observed singlets in their capillary trapping experiments. Powers (1992) and Mayer and Miller (1993) observed complex blobs occupying more pore spaces in addition to singlets. Powers (1992) accounted for nearly 30% of complex blobs in well-graded sand. According to Li and Wardlaw (1986), entrapment by snap-off does not occur for systems with contact angles greater than 70°. Even the intermediate contact angle of 90° causes the curvature of the interface to remain relatively small, and

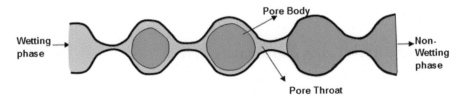

Fig. 2 Entrapment by snap-off mechanism (Adapted in part from Thomson 2007)

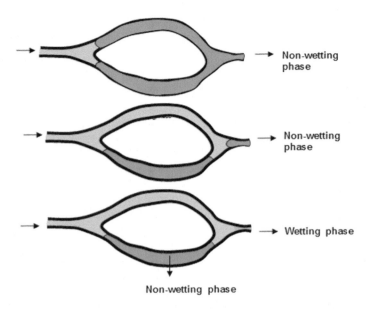

Fig. 3 Entrapment by bypass mechanism (Adapted in part from Thomson 2007)

therefore, there is no possibility of trapping. Hg^0 has a contact angle of 180° (Good and Mikhail 1981). It has a contact angle of 140° when measured on flat surfaces (Anderson 1987), yet a large percentage of singlets were found by the authors in their micromodel investigations using glass beads as porous media.

Figure 4 shows a singlet, doublet, triplet and a complex blob of Hg^0 entrapped in the micromodel experimental investigations. Having known that singlets are usually trapped by snap-off mechanism, it appears that interfacial tension is an influential parameter in entrapment mechanisms in addition to contact angle. Soil or heterogeneous condition encourages trapping through bypassing. This signifies that pore size distribution and heterogeneity also determine the method of trapping. As long as the aspect ratio does not exceed a value of six, the mercury ganglion remains continuous and mercury may leave the system of concern without getting entrapped. However, when this ratio is exceeded, mercury becomes entrapped by snap-off within the pore body (Rigby et al. 2003).

4 Capillary, Viscous and Gravity Forces

The three predominant forces governing immiscible fluid flow and its entrapment behaviour in the subsurface are capillary, viscous, and gravity or buoyancy forces. Capillary force is proportional to the interfacial tension at the fluid–fluid interface

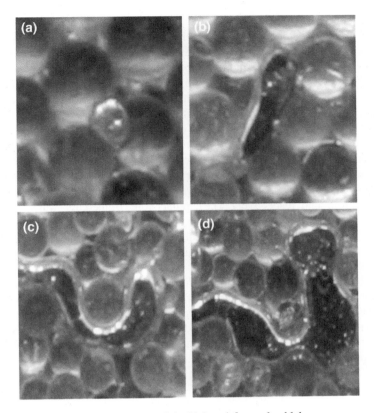

Fig. 4 a Singlet blob, **b** doublet blob, **c** triplet blob and **d** complex blob

and the contact angle between porous media and fluid–fluid interface. It is inversely proportional to the pore size. Mathematically it is represented by the Laplace equation as

$$P_c = \frac{2\sigma \cos \theta}{r} \qquad (2)$$

where σ is the interfacial tension between the non-wetting fluid and wetting fluid, θ is the contact angle the two fluids create at the pore wall and r is the mean radius of curvature (Bear 1972). The contact angle is the angle formed by the solid surface and the fluid–fluid interface, at the point where the three phases contact. It is an indicator of the wettability of the system and falls in the range between 0° and 180°. Wettability refers to the preferential spreading of one fluid over the porous media in the presence of another fluid. Wetting fluids have θ value between 0° and 90° whereas for non-wetting fluids have θ values range between 90° and 180°. Most materials are not wetted by mercury and θ value for mercury is generally in the range of 140° when measured on flat surfaces and 180° for porous media (Anderson 1987).

Viscous forces are directly proportional to the permeability and pressure gradient and determine the velocity of the displacing fluids and gravity or buoyancy forces are proportional to the density differences between the fluids. All the three forces make the fluid displacement more complex. At typical flow rates of the aquifer, capillary forces generally dominate the hydrodynamic forces like gravity and viscous forces and lead to entrapment of immiscible fluids, gravity and viscous forces. The domination of capillary forces over the hydrodynamic forces like gravity and viscous forces lead to entrapment of immiscible fluids. Depending upon the hydrogeology and nature of the fluids, these forces determine the residual saturation of an immiscible fluid. Compared to other NAPLs, Hg^0 is highly viscous, possess high interfacial tension and is an order of magnitude denser than other fluids. Gravity force of mercury–water system was 3.07×10^{-2} and 0.49×10^{-2} kg/s² for coarse and fine sand, respectively. The capillary force of mercury–water system was 1.22×10^4 and 7.65×10^4 times higher than the gravity force in for coarse sand and fine sand, respectively. High capillary forces in mercury–water system counteracted the high gravity forces and caused mercury entrapment (Devasena and Nambi 2010).

Capillary forces which are responsible for NAPL entrapment are also responsible for the difficulties associated with their clean up. Typical groundwater velocities cannot overcome capillary forces. Traditional pump and treat methods have proven to be unsuccessful in remediating NAPL contaminated sites (Taylor et al. 2001; Pennell et al. 1996). The larger interfacial tension between water and NAPL prevents the displacement and recovery of trapped NAPLs at normal groundwater velocities. Reducing the interfacial tension generally helps to enhance recovery of trapped fluids. Capillary force, being directly proportional to the interfacial tension, can be lowered by reducing the interfacial tension. Surfactants and cosolvents flushing generally reduce the interfacial tension and thereby (a) increase the solubility of NAPLs and (b) induce mobilization of trapped NAPL (Taylor et al. 2001; Pennell et al. 1996; Gupta and Mohanty 2001).

5 Measurement of Capillary Pressure as a Function of Water Saturation

Entrapment of immiscible fluids is largely controlled by capillary forces. Entrapment and distribution of immiscible fluids in the subsurface are best understood by the Laplace equation (Eq. 2) where the relationship between water saturation and capillary pressure is described. This constitutive relationship is best estimated by constructing P_c–S_w curves. Consider a completely saturated porous medium where water is the wetting fluid with water saturation, S_w as one. When non-wetting fluid enters a water-wet porous medium (Initiation of Primary drainage curve), the local capillary pressure is increased, wetting fluid is forced out of larger pressure by the non-wetting fluid and is reduced to irreducible water saturation

(S_{irw}). Irreducible water saturation is the water retained within the pore space after the cessation of the non-wetting fluid. As capillary pressure decreases during the main imbibition curve, water saturation again increases from S_{irw} to S_{nr}. At this critical residual saturation S_{nr}, the non-wetting phase is no longer interconnected and becomes entrapped and termed as residual non-wetting fluid saturation. The term drainage and imbibition are used with reference to water, the wetting phase. The mechanisms of water drainage (intrusion of non-wetting fluid) and water imbibition (extrusion of non-wetting fluid) are shown in Figs. 5 and 6, respectively.

Figure 7 illustrates the relevant features of P_c–S_w relationships for capillary force dominated systems. Such a P_c–S_w curve produced by incrementally increasing and decreasing the pressure on the non-wetting fluid is hysteric in nature. The primary drainage curve and main imbibition curve form an envelope for infinite number of hysteric curves formed during either increasing or decreasing saturation. $P_c$rence to water, the wetting phaseS_w relationship is the main tool for characterizing the immiscible fluid distribution in the subsurface (Ishakoglu and Baytas 2005). Devasena and Nambi (2010) estimated residual Hg^0 saturations as 0.04 for coarse sand and 0.07 for fine sand from two-phase capillary pressure–water

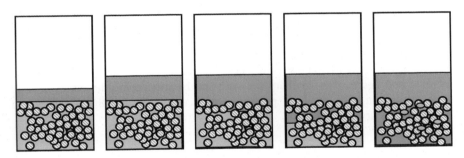

Fig. 5 Two-phase flow process—drainage (Adapted in part from Thomson 2007)

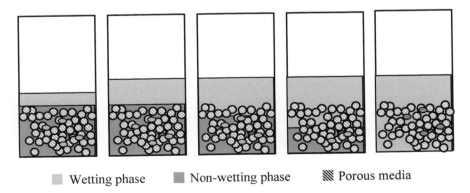

Fig. 6 Two-phase flow process—imbibition (Adapted in part from Thomson 2007)

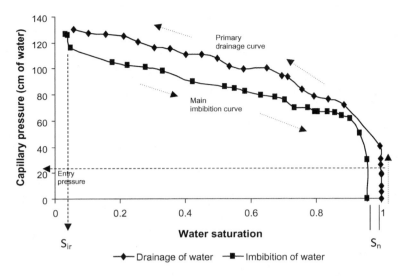

Fig. 7 Typical capillary pressure–water saturation curve (Devasena and Nambi 2010)

saturation experimental investigations on Hg^0–water systems. The experimental results of Hg^0–water system and PCE–water system brings out the influence of fluid properties on residual saturation values.

Drainage and imbibition curves were established by subjecting the 1D column to increase and decrease in capillary pressure and by monitoring the corresponding water saturation. The capillary pressure was determined with the densities and heads of mercury and water.

$$P_c = \rho_m g h_m - \rho_w g h_w \tag{3}$$

ρ_m and ρ_w are the densities of Hg^0 and water, respectively. h_m and h_w are the heights of Hg^0 and water measured from the bottom of the sand column, respectively, and g is the gravitational force. P_c can be converted to capillary pressure head h_c through Eq. 4.

$$h_c = P_c / \rho_w g \tag{4}$$

In most of the previous studies, NAPL displacement experiments were conducted in horizontal direction. Very few studies (Morrow and Songkron 1981; Dawson and Roberts 1997) have considered vertical displacement of NAPLs to find the influence of gravity forces on residual saturations. Irrespective of flow direction, the interface would be unconditionally stable when a denser and more viscous fluid displaces a less viscous fluid. The reverse of the above-mentioned condition creates an unconditionally unstable interface. The stability of the interface is affected by the buoyancy forces; the interface is stable when an LNAPL is injected at the top of a denser fluid and the interface is considered unstable when a DNAPL is injected at the top of a less dense fluid. Interface stability has also been defined by the velocity of the interface during vertical two-phase flow. The interface is stable when $U < U_c$ where U is the velocity of interface motion and U_c is the critical velocity of interface motion during dominant gravity force. On the other hand, the interface is stable for $U > U_c$ when viscosity dominates (Chen et al. 1995). Any fluid with high density and viscosity forms an unconditionally stable interface and ends up with minimum residual saturation (Powers 1992).

Mercury, being a dense and more viscous fluid, forms an unconditionally stable interface during both drainage and imbibition. The stable interface prevents the water pressure from altering the interface, and hence, water flows from the pore throat to the pore body and results in least residual mercury saturation. In the case of other DNAPLs such as PCE, an unstable interface is formed since their viscosity is less than water. The interface becomes highly curved as water enters the pore throat during imbibition and severs the interface formed in the pore throat resulting in high PCE residual saturation. During the main water drainage, it was observed that at least 18.5 cm of capillary pressure head was required to initiate water drainage (Fig. 7). At the end of the drainage cycle, water saturation reaches a minimum value of 0.07 referred as irreducible water saturation. In the main imbibition pathway, water saturation increases and reaches a value of 0.96 although the mercury head was lowered well below zero capillary pressure. This shows that 4% of mercury was entrapped in the pore space.

One-dimensional column experiments do not help in close observation of the entrapped blobs and their distribution. Two-dimensional micromodels are promising in understanding the distribution of entrapped Hg. An initially water-saturated micromodel was flooded with Hg^0 at a particular rate to simulate the migration of Hg^0 into the water-saturated zone. Then, Hg^0 was displaced by water flooding and residual Hg^0 was established over a range of capillary numbers. Images taken during the experiment were processed with image analysis software SigmaScan Pro 5.0. Hg^0 was found to be trapped as discrete blobs of varying sizes and shapes ranging from spherical shaped singlets/doublets occupying one/two pore bodies to complex multipore blobs.

Figure 8a shows entrapment of Hg in glass bead micromodel studies, and the threshold image of Fig. 8a is shown in Fig. 8b. In post-image analysis (Fig. 8b), Hg^0 appears dark red, and water-wetted glass beads appear black. The number of individual blobs entrapped was 103, and mean length for the entrapped blobs was 2.52 mm. Entrapped blobs were found as singlets, some blobs were larger than a singlet or doublet, and some blobs were more complex and extended in length. Similarly, singlets comprised approximately 20% of the total number of blobs for experiments conducted under low flow rate (low capillary number) and around 30% under high flow rate experiments (high capillary number) showing that as capillary number increases, singlets dominate the entrapment process.

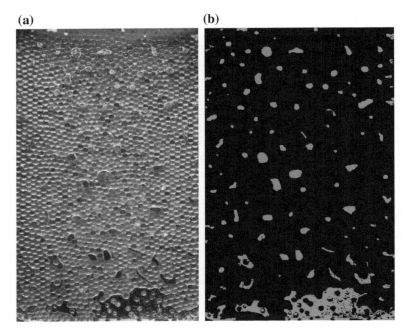

Fig. 8 Distribution of entrapped Hg^0 **a** before thresholding and **b** after thresholding

6 Empirical Models of Capillary Pressure–Saturation Curves

Mathematical models developed to predict NAPL entrapment and its migration need fundamental input parameters from P_c–S_w experiments (Ishakoglu and Baytas 2005). P_c–S_w data generated from the experiments are generally parameterized using empirical relations such as Brooks and Corey (1964) and van Genuchten (1980) equation. The entrapment parameters serve as useful inputs to two-phase and multiphase flow models. In addition, entry pressure or displacement pressure, irreducible wetting phase saturation and residual non-wetting phase saturation are the other crucial experimental data obtained from P_c–S_w experiments. Brooks and Corey (1964) developed an empirical relationship which takes into account the displacement pressure (P_d), pore size distribution index (λ) and effective water saturation (S_e).

For drainage,

$$S_e = \frac{S_w - S_{irw}}{1 - S_{irw}} \quad (5)$$

For imbibition,

$$S_e = \frac{S_w - S_{irw}}{1 - S_{irw} - S_{nr}} \tag{6}$$

$$S_e = (P_d/P_c)^\lambda \tag{7}$$

Small values of λ indicate well-graded media. However, the Brooks–Corey exponential curve is invalid at $P_c < P_d$ and does not include the slightest decrease in saturation at pressures higher than displacement pressure.

The empirical equation proposed by van Genuchten is robust and has gained favourable acceptance. It is given by

$$S_e = (1 + (\alpha h_c)^n)^{-m} \tag{8}$$

$$m = 1 - \frac{1}{n} \tag{9}$$

where α, n and m are constants. The parameter n is a fitting parameter associated with both drainage and imbibition as n_d and n_i, respectively. It is related to the pore size distribution index and it controls the shape of the curve ($n = \lambda + 1$). α_d is the measure of the pore-throat size, and α_i is the measure of the pore-body size. h_c is the capillary pressure head corresponding to the degree of saturation. Water saturation is expressed as S_e (effective saturation) for the convenience of normalizing the data between aqueous phase saturation values of 0 and 1 (Powers and Tamblin 1995). When compared to Brooks and Corey model, van Genuchten model does not utilize P_d directly, and hence, initial desaturation corresponds to any $P_c > 0$. The retention curve model (RETC) can be used to find the capillary entrapment parameters such as 'α_d', 'n_d', 'α_i' and 'n_i'. Figure 9 shows the experimentally observed and mathematically fitted P_c–S_w curve for mercury–water system where an average capillary pressure head of 18.5 cm of water was required by mercury–water system.

7 Dimensionless Numbers

The NAPL–water interface moves only when either the viscous force or the gravitational force surmounts the capillary force that is responsible for retaining the NAPL. When capillary forces are minimized, NAPL starts mobilizing (Morrow 1979; Mayer and Miller 1993; Gioia and Urciuolo 2006). The relative strength of the forces acting on a single residual NAPL blob is given by capillary number and Bond number. Capillary number (N_C) and Bond number (N_B) are dimensionless numbers that show the influence of capillary, gravity and viscous forces on non-wetting fluid entrapment. N_C measures the relative strength of the viscous to the capillary forces. N_C may be varied either by changing the aqueous flow rate or

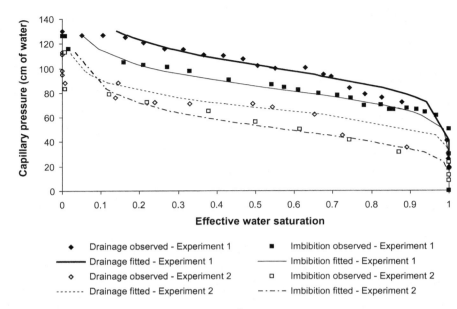

Fig. 9 Observed and fitted P_c–S_w curves for Hg^0–water system (Devasena and Nambi 2010)

by using surfactants that reduce the interfacial tension. N_B measures the relative strength of the gravity to the capillary forces. N_C can be varied by varying the aqueous flow velocity. N_B variation can be achieved by using different NAPLs, thereby changing NAPL density and interfacial tension from experiment to experiment (Mayer and Miller 1993) or by changing the porous media size. Manipulation of the parameters involved in N_C and N_B can be done either by increasing the aqueous flow rate or by varying the NAPL or flushing with alcohol/ surfactants so that the NAPL properties such as density, viscosity and interfacial tension are altered. Capillary number N_C is given by

$$N_C = \frac{\mu_w U_w}{\sigma} \qquad (10)$$

$$U_w = \frac{k\rho_w g i}{\mu_w} \qquad (11)$$

U_w is the superficial velocity of the wetting phase (m/s), k is the media permeability (m²), ρ_w is the density of the wetting phase (kg/m³), μ_w is its viscosity (kg/ms), σ is the interfacial tension between the two fluids (dynes/cm) g is the acceleration due to gravity (m/s²) and i is the hydraulic gradient. The viscous force is given by μ_w times U_w (kg/s²). Bond number N_B is given by

$$N_B = \frac{(\rho_{nw} - \rho_w)gr^2}{\sigma} \tag{12}$$

ρ_{nw} is the density of the non-wetting fluid (kg/m^3), ρ_w is the density of the wetting fluid (kg/m^3), and r is the radius of the grain size (m). The gravity force is given by $\Delta\rho$ times g times square of the radius of the grain size (kg/s^2). Pennell et al. (1996) found another dimensionless number referred as trapping number N_T (vectoral sum of N_C and N_B) under low N_C and high N_B conditions. A critical trapping number of 1×10^{-3} is required for complete NAPL mobilization whereas N_T in the range of 2×10^{-5} to 5×10^{-5} was required to initiate NAPL mobilization. For vertical flow, N_T is given by

$$N_T = \sqrt{N_C^2 + N_B^2} \tag{13}$$

For the case of horizontal flow,

$$N_T = |N_C + N_B| \tag{14}$$

The results of Pennell et al. (1996) are similar to critical capillary numbers attained by Morrow and Chatzis (1982) in sandstone packs. But Morrow and Chatzis (1982) did not consider the effect of N_B. Therefore, N_T can be used for systems with or without significant buoyancy effects.

Residual non-wetting fluid saturation is a function of both N_C and N_B (Morrow and Songkran 1981). The extent of influence of N_C and N_B on capillary trapping of non-wetting fluids was investigated by Conrad et al. (1992), Morrow et al. (1988), Meakin et al. (2000) and Ovdat and Berkowitz (2007). An increase in either of these dimensionless numbers decreases the residual non-wetting fluid saturation. Dombrowski and Brownell (1954) proved that increase in N_C resulted in significant decrease in residual saturation only when N_C was greater than 10^{-2}. Ng et al. (1978) showed that at low fluid velocities with $N_C \leq 2 \times 10^{-5}$, capillary forces dominated viscous forces and residual saturation either increased or became invariant. When N_C exceeded 2×10^{-5}, viscous forces became significant and residual non-wetting fluid saturation decreased. Morrow and Songkran (1981) found that residual saturation varied from 14% at low N_C and to almost zero at higher N_C for experiments conducted with bead packs. Moreover, intentionally varying the N_C either by reducing the interfacial tension or by increasing the aqueous flow rate is one of the NAPL remediation techniques. N_C greater than 10^{-5} is typically achieved by introducing surfactants/cosolvents which lower the interfacial tension between the immiscible fluids. Chatzis and Morrow (1984) found a critical N_C above which complete NAPL mobilization took place.

An inverse relationship between NAPL saturation and N_B is commonly reported (Morrow et al. 1988; Gupta and Mohanty 2001). The relationship between DNAPL saturation and N_B depends on the sign of N_B. DNAPL–water systems are usually operated in downward displacement mode with a positive N_B in order to minimize

Table 2 Force analysis of Hg^0–water experiments

Parameters	Porous media						
	Coarse sand Hg^0–water Exp 1	Coarse sand Hg^0–water Exp 2	Average	Fine sand Hg^0–water Exp 1	Fine sand Hg^0–water Exp 2	Average
Viscous force (kg/s²)	2.16×10^{-6}	1.47×10^{-6}	1.8×10^{-6}	0.47×10^{-6}	0.4×10^{-6}	0.45×10^{-6}
Gravity force (kg/s²)	3.07×10^{-2}	3.07×10^{-2}	3.1×10^{-2}	0.49×10^{-2}	0.49×10^{-2}	0.49×10^{-2}
N_C	5.7×10^{-9}	3.92×10^{-9}	4.8×10^{-9}	1.26×10^{-9}	1.12×10^{-9}	1.2×10^{-9}
N_B	8.2×10^{-5}	8.2×10^{-5}	8.2×10^{-5}	1.3×10^{-5}	1.3×10^{-5}	1.3×10^{-5}

fingering and achieve 100% NAPL saturation during its initial invasion. Negative Bond numbers are generally associated with LNAPLs or DNAPLs when operated in upward displacement mode. Dawson and Roberts (1997) showed a direct relationship between NAPL saturation and N_B where DNAPL saturation was measured through an upward displacement mode. The upward displacement mode may result in fingering and random variation in residual NAPL saturations, however, it simulates natural migration of DNAPL during a spill. DNAPL saturation is, therefore, a function of a linear combination of N_C and N_B, provided the relative permeability of water is also considered. In concise, for fluids with high density and high N_B, gravity force dominates and for highly viscous fluids, viscous forces govern. NAPL entrapment and mobilization very strongly depend upon the predominant forces.

Understanding the independent role of N_C and N_B becomes essential to decide whether entrapped mercury can be hydraulically mobilized. Table 2 provides gravity and viscous forces along with Capillary and Bond number measured for mercury–water systems. Capillary number was found in the order of 10^{-9} much lesser than the Bond number. N_C is usually increased to the range of 10^{-4} either by increasing the aqueous flow rate or by adding surfactants that reduce the interfacial tension. Chatzis and Morrow (1984) conducted numerous experiments under low N_B and N_C conditions to explore the relationship between N_C and residual saturation. A critical N_C of 2×10^{-5} was established in order to initiate mobilization of trapped NAPL blob and a critical N_C of 1.3×10^{-3} to completely mobilize all the trapped blobs. Pennell et al. (1996) varied N_C from 9.21×10^{-7} to 6.10×10^{-4} by using different surfactants and by increasing the flow rate. They also increased the N_B from 2.62×10^{-7} to 1.31×10^{-4} by reducing the interfacial tension of the displacing fluid and concluded that N_C alone cannot be used to predict NAPL mobilization especially when the N_B is higher. N_C in mercury–water systems is in the order of 10^{-9} which is an order of magnitude less than PCE–water system (10^{-8}). Such low N_C confirms that entrapment of mercury is more dependent on the combined effect of gravity and capillary forces than the viscous forces. Low N_C values of mercury–water system strongly specify that it may not be possible to raise the N_C by several orders of magnitude to mobilize entrapped mercury blobs. Figure 10 shows the variation of residual saturation of non-wetting fluids with respect to N_C. N_C for Hg^0–water system is between 7.5×10^{-11} to 4.08×10^{-10}. Corresponding N_C for PCE–water and air–water systems are between 5.91×10^{-10} to 3.2×10^{-9} and 3.93×10^{-9} to 2.15×10^{-9}, respectively. Hg^0–water systems showed clearly an order of magnitude lower N_C.

Theoretically, residual non-wetting saturation and N_B are inversely proportional during downward displacement of NAPL. Mercury–water systems displayed a large density difference and a high interfacial tension Compared to PCE–water system, gravity force was 20 times higher in mercury–water system. Gravity force, nevertheless a predominant control factor in the migration of highly dense mercury, was counteracted by not less trivial capillary force. The much higher capillary forces surmounted the gravity forces and lead to mercury entrapment. The capillary force was 1.22×10^4 times higher than the gravity force in mercury–water system for coarse sand. For the same coarse sand, the capillary force was 3.09×10^4 times higher in PCE–water system. Similarly, the capillary force was much higher in fine

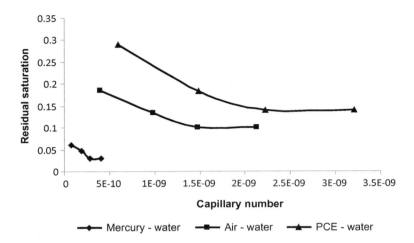

Fig. 10 Variation of residual saturation with N_C

sand for both mercury–water (7.65×10^4) and PCE–water (1.93×10^5). N_B for mercury–water was found to be 8.2×10^{-5} (coarse sand) and 1.3×10^{-5} (fine sand), approximately 2.5 times higher than PCE–water systems. N_B for PCE–water was 3.23×10^{-5} and 5.17×10^{-6} for coarse and fine sand, respectively. A 2.5-fold increase in N_B for mercury–water system compared to the PCE–water system resulted in low residual mercury saturation of 0.04 (coarse sand). Hence, at low N_C, gravity and capillary forces are the significant causes of mercury entrapment.

In the micromodel investigations, at low capillary numbers, stable, continuous, tortuous paths were observed. As the capillary number was further increased, complex ganglia became shorter, reaching the size of a singlet. Singlets composed approximately 20% of the total number of blobs for low NC (2.47×10^{-12}) experiments and around 30% for high N_C (2.47×10^{-10}) experiments showing that as capillary number increases, singlets dominate the entrapment process. Complex blobs composed 35% and 17% for low and high N_C, respectively. The mean shape factor of the distributed blobs was found to be 0.39 at low N_C (2.47×10^{-12}) and 0.6 at high N_C (2.47×10^{-10}). Shape factor is the measure of the divergence of the blob shape from that of a reference object such as sphere. The results indicate that the average blob is non-spherical at low N_C and sphericity increases as N_C increases.

Figure 11 shows the cumulative frequency distribution of shape factors for each experiment with varying N_C. About 17.8% of Hg^0 blobs at low N_C and 7.5% at high N_C had a shape factor of 0.1, while 9.3% and 26% had a shape factor of 0.9 at both low and high N_C, respectively. Similar results were found by Pan et al. (2006) where the shape factor shifted gradually towards 1 at low residual NAPL saturations (high N_C). They found that around 13% of blobs at high residual saturation and 9% of blobs at low residual saturation had a shape factor of 0.1. As the residual saturation decreased, the shape factor in their study reached around 0.8.

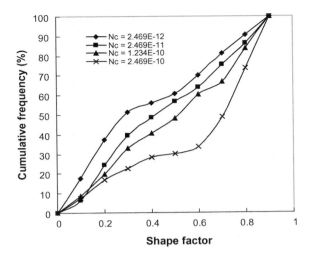

Fig. 11 Cumulative frequency distribution of shape factor of entrapped blobs

8 Conclusions

Elemental mercury, a widespread contaminant, stands unique with distinct properties such as very high density, high surface tension and low aqueous solubility. These properties not only allow migration of Hg^0 deeper into the subsurface but also entrap it in the pore spaces. Current remediation technologies are more pertinent to various inorganic and organic mercury compounds. Subsurface contamination due to spills and disposals of Hg^0, quantification of Hg^0 entrapment phenomena at pore scale and the effect of residual Hg^0 on remediation have not been studied extensively. With a significant increase in the number of Hg^0 spill sites and lack of studies on entrapment and entrapment models, the need for understanding Hg^0 migration and its entrapment and remediating entrapped Hg^0 becomes inevitable. The results of controlled capillary pressure–water saturation experiments and pore-scale glass bead micromodel experiments were discussed in this chapter. The influence of each of the forces involved during migration and entrapment and the effect of capillary number and Bond number on residual Hg^0 saturation were also discussed. The influence of porous media and fluid properties on residual Hg^0 saturation was also elicited. Pore-scale glass bead micromodels and associated image analysis studies brought forth the effect of N_C on the size and shape distribution of entrapped Hg^0. Theoretical investigations showed that fluid properties were found to have an equal effect on the process of non-wetting fluid entrapment and residual non-wetting fluid saturation especially for Hg^0, which has a high interfacial tension and density. Theoretical investigations also proved that migration of Hg^0 was greatest under the influence of gravity and capillary forces and to some extent was influenced by viscous forces. Hg^0–water systems exhibited high N_B and low N_C indicating that the critical N_C and N_T established for a common DNAPL such as PCE may not be directly applicable to Hg^0. The N_C cut off values established for PCE or TCE for its entrapment and mobilization cannot be applied for Hg^0 as its N_C is much lower.

Heterogeneities are expected to radically increase residual Hg^0 saturation. While the effect of heterogeneities on mercury migration, entrapment and distribution can be easily hypothesized, further capillary pressure–water saturation experiments are needed to quantify Hg^0 entrapment and generate entrapment parameters. Additional micromodel studies to quantify residual saturation and assess the distribution of Hg^0 on a wide variety of heterogeneities are also needed. The relationship between entrapped Hg^0 blob distributions and heterogeneities are recommended to be established before predicting the efficacy of any remediation effort. Capillary-trapped Hg^0 blob is expected to have a strong influence on mass transfer between phases such as liquid mercury to gaseous mercury. Follow-up work should involve experimental investigations in porous media to study mass transfer processes from entrapped Hg^0 and quantify steady-state volatilization rates from entrapped Hg^0 blobs. The probability and rate of Hg^0 speciation into organic or inorganic mercury should also be assessed from residual Hg^0 blobs. A thorough experimental investigation may be required to explore subsurface conditions and predict the effect of the aforementioned parameters on mercury mass transfer rates. An appropriate, constitutive and robust mathematical model to simulate Hg^0 release and its entrapment in the subsurface would be a valuable tool to develop. Entrapment parameters obtained from P_c–S_w experiments can be used to validate the model. The model could also be expanded to incorporate the fate and transport of mercury and to mimic the various remediation scenarios.

References

Anderson WG (1987) Wettability literature survey—part 4: Effects of wettability on Capillary pressure. J Petrol Technol 39(10):1283–1300
Anderson WG (1988) Wettability literature survey—part 6: the effects of wettability on water flooding. J Petrol Technol 39(12):1605–1622
Bear J (1972) Dynamics of fluids in porous media. Dover Publications, New York
Brooks RH, Corey AT (1964) Hydraulic properties of porous media. Colorado State University, Fort Collins
Chatzis I, Morrow NR, Lim HT (1983) Magnitude and detailed structure of residual oil saturation. Soc Petrol Eng J 23(2):311–325
Chatzis I, Morrow NR (1984) Correlation of capillary number relationships for sandstone. Soc Petrol Eng J 24(5):555–562
Chen G, Taniguchi M, Neuman SP (1995) An overview of instability and fingering during immiscible fluid flow in porous and fractured media. U.S. Nuclear Regulatory Commission, NRC Job Code L128.3
Chevalier LR, Fonte JM (2000) Correlation model to predict residual immiscible organic contaminants in sandy soils. J Hazard Mater B72:39–52
Conrad SH, Wilson JL, Mason WR, Peplinski WJ (1992) Visualization of residual organic liquid trapped in aquifers. Water Resour Res 28:467–478
Dawson HE, Roberts PV (1997) Influence of viscous, gravitational and capillary forces on DNAPL saturation. Ground Water 35(2):261–269

Demond AH, Lindner AS (1993) Estimation of interfacial tension between organic liquids and water. Environ Sci Technol 27(12):2318–2331

Devasena M, Nambi IM (2010) Migration and entrapment of elemental mercury in porous media. J Contam Hydrol 117:60–70

Dombrowski HS, Brownell LE (1954) Residual equilibrium saturation of porous media. Ind Eng Chem Res 46:1207–1219

Gillham RW, Rao PSC (1990) In significance and treatment of volatile organic compounds in water supplies. In: Ram NM, Russel FC, Contor KP (eds) Lewis Publishers, Chelsea, MI, pp 141–181

Gioia F, Urciuolo M (2006) Combined effect of Bond and capillary numbers on hydrocarbon mobility in water saturated porous media. J Hazard Mater B 133:218–225

Good RJ, Mikhail RS (1981) The contact angle in mercury intrusion porosimetry. Powder Technol 29:53–62

Gupta DK, Mohanty KK (2001) A laboratory study of surfactant flushing of DNAPL in the presence of macroemulsion. Environ Sci Technol 35:2836–2843

Ioannidis MA, Chatzis I, Payatakes AC (1991) A mercury porosimeter for investigating capillary phenomena and micro displacement mechanisms in capillary networks. J Colloid Interface Sci 143(1):22–36

Ishakoglu A, Baytas F (2005) The influence of contact angle on capillary pressure-saturation relations in a porous medium including various liquids. Int J Eng Sci 43:744–755

Kueper BH, Frind EO (1991a) Two phase flow in heterogeneous Porous Media 1. Model development. Water Resour Res 27(6):1049–1057

Kueper BH, Frind EO (1991b) Two phase flow in heterogeneous Porous Media 2. Model application. Water Resour Res 27(6):1059–1070

Li Y, Wardlaw NC (1986) The influence of wettability and critical pore throat aspect ratio on snap-off. J Colloid Interface Sci 109:461–472

Li Y, Abriola LM, Phelan TJ, Ramsburg CA, Pennell KD (2007) Experimental and numerical validation of the total trapping number for prediction of DNAPL mobilization. Environ Sci Technol 41:8135–8141

Mayer AS, Miller CS (1993) An experimental investigation of pore-scale distributions of non-aqueous phase liquids at residual saturation. Transp Porous Media 10:57–80

Meakin P, Wagner G, Vedvik A, Amundsen H, Feder J, Jossang T (2000) Invasion percolation and secondary migration: experiments and simulations. Mar Pet Geol 17:777–795

Mercer JW, Cohen RM (1990) A review of immiscible fluids in the subsurface: Properties, models, characterization and remediation. J Contam Hydrol 6:107–163

Morrow NR (1979) Interplay of capillary, viscous and buoyancy forces in the mobilization of residual oil. J Can Pet Geol 18:35–46

Morrow NR, Chatzis I (1982) Measurement and correlation of conditions for entrapment and mobilization of residual oil. Annual report, DOE/BC/10310-20, U.S.DOE

Morrow NR, Chatzis I, Taber JJ (1988) Entrapment and mobilization of residual oil in bead packs. SPE Reservoir Eng 3(3):927–934

Morrow NR, Songkran B (1981) Effect of viscous and buoyancy forces on non-wetting phase trapping in porous media. In: Surface phenomena in enhanced oil recovery. Plenum Press, New York, pp 387–411

Ng KM, Davis HT, Scriven LE (1978) Visualization of blob mechanisms in flow through porous media. Chem Eng Sci 33:1009–1017

Oostrom M, Hofstee C, Lenhard RJ, Wietsma TW (2002) Flow behavior and residual saturation formation of liquid carbon tetrachloride in unsaturated heterogeneous porous media. J Contam Hydrol 64:93–112

Ovdat H, Berkowitz B (2007) Pore-scale imbibition experiments in dry and prewetted porous media. Adv Water Resour 30:2373–2386

Pan C, Luo LS, Miller CT (2006) An evaluation of lattice Boltzmann schemes for porous medium flow simulation. Comput Fluids 35:898–909

Pennell KD, Jin M, Abriola LM, Pope GA (1994) Surfactant-enhanced remediation of soil columns contaminated by residual tetrachloroethylene. J Contam Hydrol 16:35–53

Pennell KD, Pope GA, Abriola LM (1996) Influence of viscous and buoyancy forces on the mobilization of residual tetrachloroethylene during surfactant flushing. Environ Sci Technol 30:1328–1335

Powers SE (1992) Dissolution of non-aqueous phase liquids in saturated subsurface systems. Ph. D. dissertation, University of Michigan

Powers SE, Tamblin ME (1995) Wettability of porous media after exposure to synthetic gasolines. J Contam Hydrol 19:105–125

Rigby SP, Fletcher RS, Riley SN (2003) Determination of the cause of mercury entrapment during porosimetry experiments on sol–gel silica catalyst supports. Appl Catal A 247(1):27–39

Taylor TP, Pennell KD, Abriola LM, Dane JH (2001) Surfactant enhanced recovery of tetrachloroethylene from a porous medium containing low permeability lenses 1. Experimental Studies. J Contam Hydrol 48:325–350

Thomson NR (2007) NAPLs course. University of Waterloo, Canada

U.S.DOE (2001) Mercury contaminated material decontamination methods: investigation and assessment. DE-FG21-95EW55094

Van Genuchten MT (1980) A closed form equation for predicting the hydraulic conductivity of unsaturated soils. Soil Sci Am J 44:892–898

Wardlaw NC (1982) The effects of geometry, wettability, viscosity and interfacial tension on trapping in single pore throat pairs. J Can Pet Technol 21(3):21–27

Wardlaw NC, McKellar M (1981) Mercury porosimetry and the interpretation of pore geometry in sedimentary rocks and artificial models. Powder Technol 29:127–143

Watts RJ (1996) Hazardous wastes: sources, pathways, receptors. Wiley, 764p

Wilson JL, Conrad SH, Mason WR, Peplinski W, Hagan E (1990) Laboratory investigation of residual liquid organics from spills, leaks and the disposal of hazardous wastes in groundwater. Final report, EPA/600/6-90/004, US Environmental Protection Agency, Washington, DC

New Insight into Immiscible Foam for Enhancing Oil Recovery

Mohammad Simjoo and Pacelli L. J. Zitha

1 Introduction

One of the most accepted and widely used methods for enhanced oil recovery (EOR) is gas flooding (Orr 2007). The main underlying mechanism for this EOR method is that the residual oil saturation for gas flooding is lower than that for water flooding (Lake 1989; Green and Willhite 1998). However, the efficiency of gas EOR processes in the reservoir conditions is often not promising due to very low viscosity and density of gas compared to water and crude oil. Thus, gas injection suffers from gravity segregation and viscous instabilities, leading to uneven oil displacement, early gas breakthrough, and low oil recovery factor.

Foaming of the injected gas is a potential solution to improve gas flooding performance by a considerable increase of gas viscosity and also by trapping a part of the gas inside the porous medium (Bernard et al. 1965; Hirasaki and Lawson 1985; Kovscek and Radke 1994; Rossen 1996; Li et al. 2010; Simjoo and Zitha 2015). Hereby, gas mobility substantially decreases and instead volumetric sweep efficiency increases during gas flooding into oil reservoirs. Thus, this has obvious potential benefits for enhancing oil recovery.

Bond and Holbrook (1958) were the first to propose that foam could be generated in oil reservoirs to obtain a favorable mobility ratio. This idea has been supported by several laboratory studies (Fried 1961; Chiang et al. 1980; Wellington and Vinegar 1988), field trials, and commercial projects (Patzek and Koinis 1990; Friedmann et al. 1994; Turta and Singhal 1998; Skauge et al. 2002). Although foam has been widely applied during various stages of oil production to control fluid

M. Simjoo (✉)
Sahand University of Technology, Tabriz, Iran
e-mail: simjoo@sut.ac.ir

P. L. J. Zitha
Delft University of Technology, Delft, The Netherlands
e-mail: p.l.j.zitha@tudelft.nl

mobility, there are still some important questions about the effects of oil on foam stability and foam propagation in reservoirs containing residual oil. Available experimental evidence resulted from bulk and porous media studies present varied results for foam–oil interaction. While several studies argued that the presence of oil could be detrimental on foam stability (Minssieux 1974; Jensen and Friedmann 1987; Schramm and Novosad 1992; Svorstøl et al. 1996; Arnaudov et al. 2001; Hadjiiski et al. 2001; Denkov 2004; Farajzadeh et al. 2012), others supported that stable foam could be effectively generated in the presence of oil by selecting appropriate foaming agents (Nikolov et al. 1986; Lau and O'Brien 1988; Suffridge et al. 1989; Schramm et al. 1993; Mannhardt et al. 1998; Aarra et al. 2002; Vikingstad and Aarra 2009; Emadi et al. 2011; Andrianov et al. 2012; Li et al. 2012; Simjoo et al. 2013a; Singh and Mohanty 2015). The latter idea has been discussed in several studies where the development of sufficiently stable foam leads to incremental oil recovery on top of water and gas flooding (Ali et al. 1985; Mannhardt et al. 1998; Yin et al. 2009; Andrianov et al. 2012). However, it was argued that before foam could propagate in porous media, oil saturation must be below a critical value (Jensen and Friedmann 1987; Svorstøl et al. 1996; Mannhardt and Svorstøl 1999). This was not supported by other studies in which foam was generated at relatively high oil saturation (Ali et al. 1985; Farajzadeh et al. 2010; Andrianov et al. 2012; Simjoo et al. 2013b). From the existing studies, one can infer that the topic of foam–oil indication is still a subject of debate and more systematic studies are needed to elucidate the corresponding mechanisms.

The goal of this study is to gain better insight into the effects of oil on foam stability and foam propagation in natural sandstone cores containing water-flood residual oil. More specifically, this study is concerned with the question of whether immiscible foam can be a tertiary oil recovery method. We also investigate the effect of surfactant concentration because it is one of the main physical parameters that directly affect the stability of foam films in the presence of oil. The use of a CT scanner enabled us to provide new insight into foam–oil interaction in porous media. We provide a detailed analysis of the incremental oil recovery by foam that has not been elaborated in the previous studies. This chapter proceeds with a description of the experimental part. Next, the results are presented and discussed. Then the main conclusions of this study are drawn.

2 Experimental Description

2.1 *Materials*

Brine was prepared by dissolving sodium chloride (NaCl, Merck) at a fixed concentration of 0.5 M in de-ionized water (pH = 6.8 ± 0.1). Density and viscosity of the brine at 21 °C were 1.02 ± 0.01 g/cm^3 and 1.18 ± 0.01 cP, respectively. The surfactant used to perform experiments was C_{14-16} alpha olefin sulfonate

Table 1 Physical properties of the core samples used in this study

Core sample	Bentheimer sandstone
Diameter (cm)	3.8 ± 0.1
Length (cm)	17.0 ± 0.1
Porosity (%)	21.0 ± 0.1
Pore volume (cm^3)	40.5 ± 0.5
Absolute permeability to brine (Darcy)	2.5 ± 0.1

(AOS, Stepan) with the molecular weight of 315 g/mole. It was provided as an aqueous solution containing 40.0 wt% active content and used as received without further treatment. The critical micelle concentration of the AOS solution in the presence of 0.5 M NaCl was 4.0×10^{-3} wt%. Normal hexadecane (n-C_{16}, Sigma-Aldrich) with a purity of 99.99% was used as model oil. Density and viscosity of the oil at 21 °C were 0.77 ± 0.01 g/cm^3 and 3.28 ± 0.01 cP, respectively. Nitrogen gas with a purity of 99.98% was used to conduct the experiments.

Bentheimer sandstone cores were used to perform the experiments. This sandstone contains up to 97% quartz and is consolidated and nearly homogeneous (Peksa et al. 2015). The core samples were drilled from a cubical block and then sawn to the desired dimensions using a water-cooled diamond saw. Next, the cores were dried in an oven at 60 °C for 48 h. Then they were encapsulated in a thin layer of low X-ray attenuation Araldite self-hardening glue to avoid possible bypassing along the side of the core. From the CT scans of the dry core, it was estimated that the glue penetrates about 1.0 mm of the core sample. The effective diameter was used for the calculation of the total pore volumes of the cores. After hardening, the glued core was machined to ensure that the core fits precisely into the core holder. The physical properties of the core samples are presented in Table 1.

The core holder was made of polyether ether ketone (PEEK), a synthetic material that combines good mechanical properties with a low X-ray attenuation. It transmits X-ray within a narrow energy window, which significantly minimizes the beam hardening artifact due to the polychromaticity of the X-ray beam. Several holes were drilled through the glue layer into the core surface along the core length for pressure measurements. The pressure connectors were also made of PEEK to prevent interference of the pressure lines with the CT scanning.

2.2 Experimental Setup

The setup used to conduct the core-flooding experiments is shown in Fig. 1. It consists of a core-holder in line with a double effect piston displacement pump (Pharmacia Biotech P-500) in parallel with a gas mass flow controller (Bronkhorst) and on the other end, a back pressure regulator and a collector for the produced fluids. The pump was used to inject brine, oil, and surfactant solution. Nitrogen gas

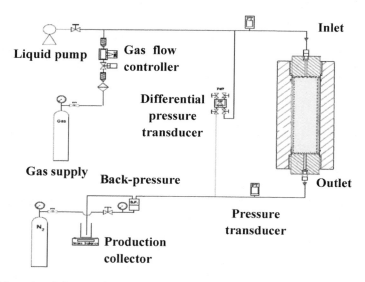

Fig. 1 Schematic of the experimental setup used to perform core-flooding experiments. The core-holder was placed vertically on the couch of CT scanner. There were additional lines not shown in the schematic that allowed the injection of fluids from the bottom of the core

was supplied by a 200 bar cylinder equipped with a pressure regulator (KHP Series, Swagelok) and connected to the core inlet through a mass flow controller. A differential pressure transducer was used to record the overall pressure drop along the core length. A data acquisition system (National Instruments) was used to record pressure, liquid production, and gas and liquid injection rates.

2.3 CT Scanner

The CT scans were obtained using a third-generation SAMATOM Volume Zoom Quad slice scanner. The core-flooding setup was placed on the couch of the CT scanner and core-holder was fixed vertically to the edge of the couch using polymethyl methacrylate stand, which is equally transparent to X-rays. The X-ray tube of the CT scanner operated at a voltage of 140 kV and a current of 250 mA. The sequential scan mode was used for image acquisition. The thickness of each CT slice was 1 mm and one series of scans included eight slices. The B40 medium filter was used for the reconstruction of the images. A typical slice image consists of 512×512 pixels with the pixel size of 0.3 mm $\times$ 0.3 mm. Since the noise for CT images typically ranges from 3 to 20 Hounsfield units (HU), the accuracy in the measured fluid saturation is within $\pm 2\%$. The following equation was used to obtain oil saturation from the measured attenuation coefficients in Hounsfield units, eliminating the contribution of rock by the subtraction:

$$S_o = \frac{1}{\varphi}\left(\frac{HU - HU_{wet}}{HU_o - HU_w}\right) \quad (1)$$

where φ is porosity of the core, HU is the attenuation coefficient obtained during oil injection into the brine-saturated core, HU_{wet} is the attenuation coefficient of the brine-saturated core, HU_o is the attenuation coefficient of the oil phase, and HU_w is the attenuation coefficient of the water (brine, 0.5 M NaCl) phase. To describe the distribution of fluids during foam (or gas) flooding, the CT images were converted into the total liquid saturation ($S_{liq} = S_o + S_w$) profiles. The reason to obtain total liquid saturation rather than oil and water saturations is that for the latter case one needs to CT scan the core using a true dual-energy method. This option was available in the CT scanner, but the contrast of the images resulting from the pair of energies was not good enough to discriminate fluids at three-phase flow conditions. Therefore, total liquid saturation was reported in terms of the summation of oil and water saturations using the following equation:

$$S_o = \frac{HU_{foam} - HU_{dry}}{HU_{pre-flush} - HU_{dry}} \quad (2)$$

where subscripts foam, pre-flush, and dry stand for core with foam, core at the end of surfactant pre-flush, and dry core, respectively.

2.4 Experimental Procedure

The basic sequence used to conduct core-flooding experiments is shown in Table 2. First, air was removed by flushing the core with CO_2 at 5 bar injection pressure. Then the dry core was saturated by injecting at least 10 pore volumes (PV) of brine while maintaining a back pressure of 25 bar. This was done to dissolve any CO_2 present in the core and thus to ensure a complete saturation of the core with brine. Next, primary drainage was performed by injecting model oil at 0.5 cm^3/min under gravity stable conditions (from top of the core). Subsequently, imbibition was done by injecting brine from the bottom of the core and continued until no more oil was produced and the pressure drop over the core became constant. Then 2.0 PV of the surfactant solution was injected into the core to satisfy its adsorption capacity. Next, nitrogen and surfactant solution were co-injected downward to generate foam at different surfactant concentrations (0.1, 0.5, 1.0 wt%). Foam flooding was performed at a fixed superficial velocity of 4.58 ft/day and with foam quality of 91% at a back pressure of 20 bar and ambient temperature (21 ± 1 °C). One baseline gas flooding experiment was performed at a superficial velocity of 4.2 ft/day at the same experimental conditions.

Table 2 Basic sequence used to conduct core-flooding experiments

Step	Description	Back pressure (bar)	Flow rate (cm^3/min)	Injection direction
1	CO_2 flushing	–	5.0	Down
2	Core saturation	25	1.0	Up
3	Oil injection	–	0.5	Down
4	Water flooding	–	0.5–1.0	Up
5	Surfactant pre-flush	–	1.0	Up
6	Foam or gas flooding	20	1.1	Down

At each stage of the experiment, the core was CT scanned to determine the distribution of fluid saturations in the porous medium and to reveal the propagation of the foam front. The results of core-flooding experiments were examined in terms of CT scan images, total liquid saturation, mobility reduction factor (as a ratio of foam to single-phase water flow pressure drops at the same superficial velocity), incremental oil recovery (oil recovered during gas or foam flooding divided by oil initially in place, OIIP, after primary drainage), and dynamic capillary desaturation curve.

3 Results and Discussion

3.1 Primary Drainage and Imbibition

Since the results of primary drainage and imbibition were similar for all core-flooding experiments, only one prototypical experiment will be discussed in detail. Figure 2 shows a series of CT images taken at different injected pore volumes during the primary drainage and imbibition. The image at zero PV represents a core fully saturated with brine, which is characterized by a dark red color. As oil is injected from top of the core, the color of images changes from red to yellow, indicating that oil drains water from the porous medium. Oil breakthrough occurred at 0.74 ± 0.02 PV, and thereafter the color in the image changes only slightly. After the breakthrough, a relatively high intensity of the red color is evident near the outlet face. This is due to the capillary end effect: the wetting phase saturation must satisfy the zero capillary pressure conditions prevailing at outlet boundary. Further oil injection partially eliminates the capillary end effect. Subsequently, imbibition was done by injecting brine upward. Change in the color of the images from yellow to orange indicates the displacement of oil by the brine. Water breakthrough occurred slightly before 0.36 ± 0.02 PV, and thereafter the color of the images remained practically unchanged. The last image in Fig. 2 represents the end of the surfactant pre-flush, where 2.0 PV of the AOS solution was injected at the same flow rate as brine during water flooding. Only a small amount of oil of about

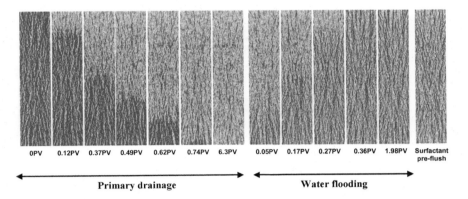

Fig. 2 CT images during primary drainage (left), water flooding (middle), and surfactant pre-flush (right). The image at zero PV corresponds to the core fully saturated with brine. During primary drainage, brine (dark red) is displaced by oil (yellow). Oil production by water flooding is evident by a change in the color from yellow to orange

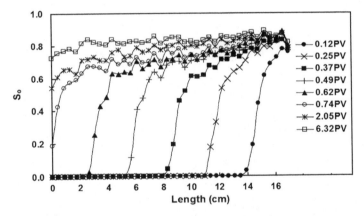

Fig. 3 Oil saturation profile during primary drainage, obtained from the corresponding CT images shown in Fig. 2. Oil was injected from the top of the core, which is located on the right side of the figure. The average oil saturation at the end of drainage was 0.85 ± 0.02

$2.0 \pm 0.5\%$ of the OIIP was produced during surfactant pre-flush. We recall that the main purpose of surfactant pre-flush was to quench the adsorption capacities of the rock sample to reduce surfactant loss as much as possible during foam flooding.

For further quantitative analysis, we plotted oil saturation during primary drainage and imbibition for different injected pore volumes as shown in Figs. 3 and 4. The general behavior of the oil saturation profiles, Fig. 3, during primary drainage is reminiscent of a Buckley–Leverett displacement: a sharp shock region followed by a rarefaction wave. As mentioned before, after breakthrough time oil saturation remained low at the outlet face due to the capillary end effect, but it increased

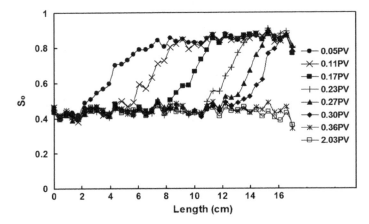

Fig. 4 Oil saturation profile during water flooding, obtained from the corresponding CT images shown in Fig. 2. Water was injected from the bottom of the core, which is located on the left side of the figure. The average oil saturation at the end of water flooding was 0.46 ± 0.02

gradually by injecting more oil. The average oil saturation along the core at the end of the primary drainage was 0.85 ± 0.02, equal to a connate water saturation of $S_{wc} = 0.15 \pm 0.02$. This is consistent with the value obtained from the measured special core analysis (SCAL) data for Bentheimer sandstone, where S_{wc} varied from 0.14 to 0.18 (Andrianov et al. 2012).

Figure 4 shows oil saturation profile during water flooding. The general behavior of the advancing front is similar to that of primary drainage, although the broadening of the frontal region due to capillary pressure is more pronounced. Most of the oil was produced before water breakthrough and thereafter only a tiny amount of extra oil was recovered. The remaining oil was distributed uniformly through the core length and exhibited an average value of 0.46 ± 0.02. This value is higher than those obtained from the SCAL measurements, which is about 0.28 ± 0.05 (Andrianov et al. 2012). However, several pore volumes of brine were injected after water breakthrough, but very little oil was produced. Instead, water cut (i.e., fraction of water in total produced liquid) increased considerably and reached nearly unity. Therefore, we concluded that under the experimental condition the measured oil saturation at the end of the imbibition is water-flood residual oil saturation.

3.2 Gas Flooding

Before performing foam flooding experiments, a baseline gas flooding experiment was performed by injecting nitrogen directly after water flooding from the top of the core. The gas injection was done under gravity stable conditions according to the

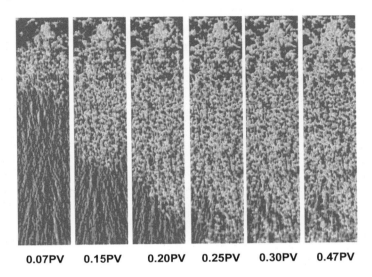

Fig. 5 CT images obtained during gravity stable gas flooding in a water-flooded core. Gas breakthrough occurred at the early time only after 0.25 ± 0.02 PV of gas injection

Dietz stability analysis (Dietz 1953) by which the critical injection velocity was obtained 4.2 ft/day, which was the same as the injected velocity. Figure 5 shows the CT scan images taken during gas flooding. As gas is injected downward, there is a change in the image color from orange to yellowish green, indicating that a part of the liquid is displaced by gas. During this period only water moving ahead of the gas front is produced. The gas breakthrough was observed early after the beginning of gas injection at 0.25 ± 0.02 PV. This time coincided also with the start of oil production at the core outlet. Visual inspection of the effluents revealed that after gas breakthrough a small amount of oil was recovered in a highly dispersed form (tiny droplets).

For further analysis of gas flooding, total liquid saturation (S_{liq}) profiles were calculated using Eq. (2) at different injected pore volumes. As shown in Fig. 6, only a very small fraction of total liquid (and thereby a very small fraction of water-flood residual oil) was produced by gas. After gas breakthrough, the average liquid saturation along the core was 0.86 ± 0.02. This shows that the performance of gas flooding, even under gravity stable conditions, is not satisfactory due to the high mobility of the injected gas and thus most of the initial liquid including water-flood residual oil is left in the porous medium. This high amount of remaining oil is a potential target for foam EOR as will be discussed in the next section.

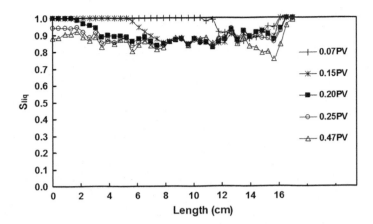

Fig. 6 Total liquid saturation profiles during gas flooding, obtained from the CT images shown in Fig. 5. Gas was injected from the top of the core, which is located on the right side of the figure

3.3 Foam Flooding

3.3.1 CT Scan Images

To investigate the potential performance of foam to recover incremental oil, nitrogen and the AOS surfactant solution were co-injected into the water-flooded core. Figures 7, 8 and 9 present a series of CT scan images obtained during foam flooding for different AOS concentrations (0.1, 0.5 and 1.0 wt%). In the CT images, the region with orange color corresponds to the liquid phase consisting of residual oil plus surfactant solution, representing a two-phase flow region. As gas and surfactant solution are co-injected from the top of the core, the intensity of orange color diminishes progressively in favor of more blue/green, representing a three-phase flow region. This gives a first qualitative impression about the change in fluid saturations in the core due to the foam flow. Four main regions can be distinguished from the CT images before foam breakthrough time. The first one is located over the first 2.0 ± 0.2 cm from the core inlet. In this region, orange color persists for large numbers of injected pore volumes. This indicates the presence of high liquid saturation in the inlet region. This can be explained by the fact that the injected gas needs to travel a certain distance into the core before the foam is fully developed. As a result gas mobility remains high, resulting in a weak liquid displacement. Moreover, if the oil phase is present in the pore spaces, it might slow down the net rate of foam development and thus leading to even higher gas mobility. The second region is the upstream of the advancing front, characterized by a clear change in the image color from orange to blue/green. This region grows over the core length as the foam front moves toward the core outlet. The third region is a transition zone between swept and un-swept parts of the core. This transition zone is not as sharp as the one for foam flow in the absence of oil

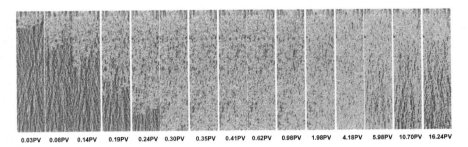

Fig. 7 CT images obtained during 0.1 wt% AOS foam flooding. Co-injection of gas and surfactant solution was done under gravity stable conditions. Orange color stands for residual oil plus surfactant solution. Green/blue color indicates three-phase region. Foam breakthrough occurred at 0.28 ± 0.02 PV

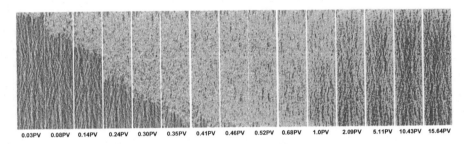

Fig. 8 CT images obtained during 0.5 wt% AOS foam flooding. Foam breakthrough occurred at 0.41 ± 0.02 PV

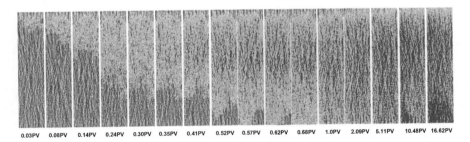

Fig. 9 CT images obtained during 1.0 wt% AOS foam flooding. Foam breakthrough occurred at 0.57 ± 0.02 PV

(Simjoo et al. 2013c). This is most likely due to the fact that oil mobilized by foam forms a diffuse oil bank with high oil saturation ahead of the advancing foam front, which could partially destabilize foam. The fourth region is downstream of the advancing front, shown by orange color indicating liquid saturation equal to unity.

A detailed examination of the CT images, however, reveals that the intensity of orange color in the region displaced by foam decreases with increasing surfactant concentration. Let us consider the CT image at 0.30 PV for different surfactant concentrations. For 0.1 wt% AOS foam orange color is more visible in the flow domain. However, for 0.5 and 1.0 wt% AOS foams the number of orange spots diminishes and instead a progressive appearance and spreading of blue color are evident. Recall that the higher intensity of blue color in the CT image indicates the presence of a stronger foam and thus more liquid desaturation from the liquid-filled pores. Therefore, one can conclude that foam development improves considerably when surfactant concentration increases from 0.1 to 1.0 wt%.

Foam breakthrough time is another indicator of foam performance: breakthrough time for 0.1 wt% foam was 0.28 ± 0.02 PV, which is slightly longer than that for baseline gas flooding (about 0.25 ± 0.02 PV). Foam breakthrough time increases with surfactant concentration: 0.41 ± 0.02 PV for 0.5 wt% foam and 0.57 ± 0.02 PV for 1.0 wt% foam. The above results reveal that the foam propagation rate decreases with increasing surfactant concentration, which indicates a better macroscopic sweep of the core. For a longer time of foam injection, the CT images of 1.0 wt% AOS foam show that a new secondary foam front emerges at the downstream of the core and propagates upward against the main flow direction (see the image at 16.62 PV in Fig. 9). The appearance of this new front was visualized by a higher intensity of the blue-colored zone, indicating that strong foam was generated in the downstream of the core. The general feature of this new secondary front is qualitatively similar to that observed for 1.0 wt% AOS foam in the absence of oil (Simjoo et al. 2013c). However, the secondary foam front in absence of oil was developed earlier, after 1.5 PV compared to 16 PV for foam in the presence of oil.

3.3.2 Saturation Profiles

For further quantitative analysis of foam propagation in the presence of oil, we examine total liquid saturation (S_{liq}) at different injected pore volumes. Figures 10, 11, and 12 show S_{liq} profiles obtained for three surfactant concentrations investigated. Let us first consider 0.1 wt% foam flow (Fig. 10). The four regions observed before foam breakthrough can be characterized further as follows: in the inlet region, S_{liq} decreases from 0.90 ± 0.02 to 0.80 ± 0.02 at a distance of 2.0 ± 0.2 cm from the core inlet. In the upstream region displaced by foam, average S_{liq} is about 0.80 ± 0.02. This region is followed by a transition zone at which S_{liq} increases from 0.80 ± 0.02 to unity. After foam breakthrough at 0.28 ± 0.02 PV, the remaining liquid saturation is distributed uniformly through the core and exhibits an average value of 0.78 ± 0.02, except the inlet region (see the S_{liq} profile at 0.98 PV in Fig. 10). As more foam is injected, liquid saturation decreases continuously through the core: average S_{liq} decreases further from 0.78 ± 0.02 to 0.55 ± 0.02 after injecting 5.98 PV of foam. Beyond this pore volume, liquid saturation continues to decrease but at a slower rate: S_{liq} obtains an

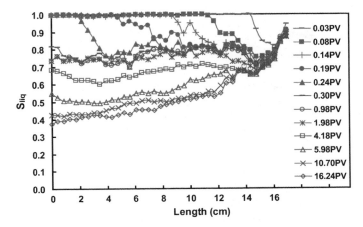

Fig. 10 Total liquid saturation profiles for 0.1 wt% foam obtained from the CT scan images shown in Fig. 7. Gas and surfactant solution were co-injected from the top of the core, which is located on the right side of the figure. Closed and open symbols indicate liquid saturation profiles before and after foam breakthrough

average value of 0.43 ± 0.02 after 16 PV injected. Note that even after injecting 16 PV, S_{liq} in the inlet region remains high. This reflects the fact that the injected gas needs some distance, which is 2.0 ± 0.2 cm for the conditions investigated, before it is fully developed into the foam. Now let us consider the S_{liq} profiles for foam at higher surfactant concentrations (Figs. 11 and 12). The overall trend of the S_{liq} profiles for 0.5 and 1.0 wt% foams is qualitatively similar to that for 0.1 wt% foam: a high amount of S_{liq} in the inlet region, followed by a reduction in S_{liq} in the upstream region, and then a transition zone through which S_{liq} increases to unity. However, when the surfactant concentration increases from 0.1 to 1.0 wt%, more liquid desaturation occurs by foam and thus a smaller S_{liq} is obtained in the upstream region. Let us compare the average S_{liq} before foam breakthrough for different surfactant concentrations. For 0.1 wt% foam average S_{liq} is 0.80 ± 0.02 while for 0.5 and 1.0 wt% foams it decreases further to 0.70 ± 0.02 and 0.50 ± 0.02, respectively. Note that the corresponding average S_{liq} for the baseline gas flooding was 0.88 ± 0.02.

The effect of the surfactant concentration on the S_{liq} profiles is also evident after foam breakthrough. Let us compare S_{liq} after 1.0 PV of foam injection. While the average S_{liq} for 0.1 wt% foam is 0.78 ± 0.02, it decreases further to 0.55 ± 0.02 for 0.5 wt% foam and 0.47 ± 0.02 for 1.0 wt% foam. The results above prove that foaming of the injected gas, even at surfactant concentration as low as 0.1 wt%, provides a better mobility control compared to gas flooding and, correspondingly, diminishes liquid saturation more effectively. Also, as the surfactant concentration increases from 0.1 to 1.0 wt%, foam flow exhibits better macroscopic sweep efficiency due to the development of a stronger foam.

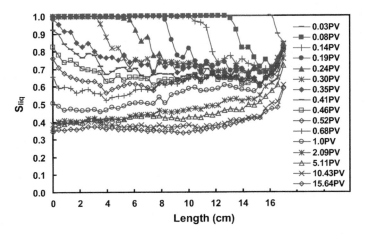

Fig. 11 Total liquid saturation profiles for 0.5 wt% foam obtained from the CT scan images shown in Fig. 8

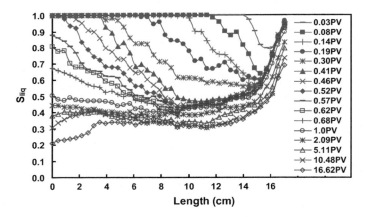

Fig. 12 Total liquid saturation profiles for 1.0 wt% foam obtained from the CT scan images shown in Fig. 9

Table 3 gives a summary of the effect of the surfactant concentration on foam propagation rate and foam breakthrough time in the absence and presence of oil. The data shows that for all concentrations investigated the foam propagation rate is higher in the presence of oil due to the partial destabilization of foam. This leads to an earlier foam breakthrough time compared to foam flow in the absence of oil. For instance for 1.0 wt% concentration when oil is present in the porous medium, foam propagation rate increases from 0.22 ± 0.01 to 0.30 ± 0.01 m/PV, corresponding with a reduction in foam breakthrough time from 0.76 ± 0.2 to 0.57 ± 0.2 PV. Remarkably, although the generated foam becomes partially destabilized by oil,

Table 3 Effect of surfactant concentration on foam propagation rate and foam breakthrough time in the absence and presence of oil

Concentration (wt%)	In the absence of oil		In the presence of oil	
	Propagation rate (m/PV, ±0.01)	Breakthrough time (PV, ±0.02)	Propagation rate (m/PV, ±0.01)	Breakthrough time (PV, ±0.02)
0.1	0.29	0.58	0.61	0.28
0.5	0.24	0.71	0.42	0.41
1.0	0.22	0.76	0.30	0.57

Foam was generated at a fixed quality of 91% at 20 bar back pressure and temperature of 21 ± 1 °C

the foam front is still strong enough to induce a reduction in liquid saturation as noted in the CT images and saturation profiles (Figs. 9 and 12). Comparing the magnitude of the foam propagation rate in the absence and presence oil shows that the effect of the surfactant concentration is more pronounced when oil is present in the porous medium: foam propagation rate in the presence of oil decreases twice as the surfactant concentration increases from 0.1 to 1.0 wt%. These results once again show that surfactant concentration is a key physical parameter to adjust the properties of the advancing foam front for a given EOR application.

3.3.3 Foam Mobility Reduction Factor

Figure 13 shows mobility reduction factor (MRF) obtained for three surfactant concentrations investigated. Recall that MRF was defined as a ratio of foam to single-phase water pressure drops over the core length. MRF for 0.1 wt% foam remains low during the first 4.0 PV injected: it increases very slowly from 18 ± 5 to 40 ± 5, corresponding to a pressure gradient of 1.9 bar/m. This slow increase is consistent with liquid saturation remaining high (about 0.68 ± 0.02) in the porous medium. Beyond 4.0 PV, MRF rises progressively and then levels off to 335 ± 5. As the surfactant concentration increases, MRF grows rapidly in the first injected pore volume, reaching 300 ± 5 for 0.5 wt% foam and 470 ± 5 for 1.0 wt% foam. The sharp increase of MRF is consistent with the substantial reduction in liquid saturation within the first injected pore volume (Figs. 11 and 12). The rate of increase of MRF is faster (about two times) for 1.0 wt% foam compared to 0.5 wt% foam. It seems that MRF for 0.5 wt% foam continues to increase very slowly such that, after the initial rise, it reaches 630 ± 5. MRF for 1.0 wt% foam, however, tends to decrease slowly after peaking at 700 ± 5. The above results combined with the liquid saturation profiles show that increasing the surfactant concentration enhances foam stability in the presence of oil, giving a large MRF. This improves the sweep efficiency and displaces more liquid (including part of residual oil) from the core.

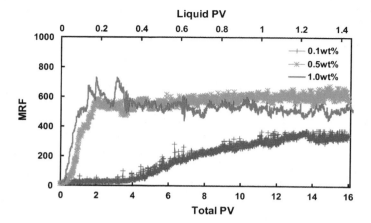

Fig. 13 Mobility reduction factor during foam flooding for different surfactant concentrations. A higher surfactant concentration leads to a larger MRF in the presence of oil

3.3.4 Incremental Oil Recovery by Foam

Figures 14 and 15 show oil cut profile and incremental oil recovery due to foam flooding for the three surfactant concentrations investigated. The amount of produced oil was obtained from the material balance on the core and the analysis of the effluents. Oil cut was defined as the fraction of oil in the produced liquid. The incremental oil recovery was defined as a ratio of the produced oil to the oil initially in place. Let us first consider 0.1 wt% foam. In the first 0.3 PV, oil is produced at a high rate such that a jump in the oil cut profile is evident, which is consistent with the formation of an oil bank. Then the oil cut profile decreases significantly to as low as 3.0% up to 2.0 PV. Thereafter, no more oil is produced until 4.0 PV. Beyond that oil production resumes with a modest increase in the oil cut profile to 8.0% until 10 PV and then it continues with an almost constant oil cut for the rest of the experiment. We note that increase in oil production after 4.0 PV coincides with a progressive rise of MRF from 40 ± 5 to 300 ± 5 and also with a reduction in average liquid saturation from 0.68 ± 0.02 to 0.50 ± 0.02 (see Figs. 10 and 13). Beyond 10.0 PV, when the oil production rate becomes constant, the MRF profile also levels off to a plateau value of 335 ± 5. When the surfactant concentration increases to 0.5 and 1.0 wt%, a qualitatively different oil recovery is observed. First of all, for both concentrations oil breakthrough is delayed due to a good mobility control provided by foam. At short times (within the first 1.5 PV) the oil production rate tends to increase as can be judged from the oil cut profiles, while during long times it decreases progressively. Therefore, oil is produced first by the formation of an oil bank followed by a long tail production. The size of the oil bank increases with surfactant concentration: after 3.0 PV the oil cut for 0.5 wt% foam is $15 \pm 1\%$ compared to $25 \pm 1\%$ for 1.0 wt% foam. The tailing oil production is, however, less sensitive to the surfactant concentration: the oil cut for 0.5 and 1.0 wt% foams

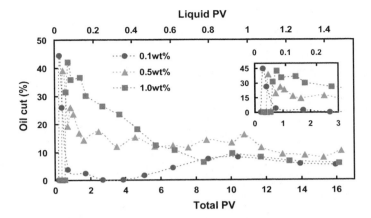

Fig. 14 Oil cut profile during foam flooding for different surfactant concentrations. The first peak corresponds to the production of the generated oil bank. The size of oil bank is extended for a longer PV as surfactant concentration increases

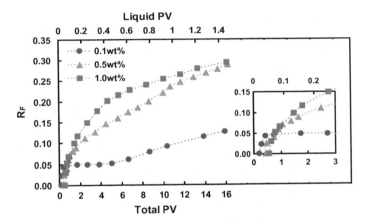

Fig. 15 Incremental oil recovery during foam flooding for different surfactant concentrations

is $7 \pm 1\%$ and $9 \pm 1\%$ respectively. For both 0.5 and 1.0 wt% foams the oil bank production coincides with a sharp increase in MRF, while the tailing production occurs without drastic changes in MRF and also in liquid saturation profiles.

Table 4 gives a summary of the incremental oil recovery by foam for the three surfactant concentrations investigated at two different pore volumes, namely 3.0 and 16.0 PV (respectively equal to the injection of 0.27 and 1.46 PV of surfactant solution) representing the short and long time of foam injection. The results show that an increase in the surfactant concentration leads to a substantially higher oil recovery consisting of a larger MRF (see Fig. 13). After 3.0 PV the incremental oil

Table 4 Incremental oil recovery by foam at different surfactant concentrations

Concentration (wt%)	Incremental oil recovery (% of OIIP)	
	After 3.0 PV of foam injection	After 16 PV of foam injection
0.1	5.0 ± 0.5	13.0 ± 0.5
0.5	12 ± 2	28 ± 2
1.0	16 ± 2	29 ± 2

recovery by 0.1, 0.5 and 1.0 wt% foams was, respectively, 5.0 ± 0.5%, 12 ± 2%, and 16 ± 2% of the OIIP. Note that the corresponding oil recovery by gas flooding was 4.0 ± 0.5% of the OIIP. As foam is injected for a longer time, oil recovery continues to increase, but at a slower rate. For example for 1.0 wt% foam beyond 16 ± 2% oil recovery obtained after 3.0 PV, an additional oil production of 13 ± 2% was recovered for the next 13 PV. Table 4 also shows that oil recovery by 0.1 wt% foam after 16 PV is about half of the 0.5 and 1.0 wt% foams. The lower oil recovery (13.0 ± 0.5% of the OIIP) by 0.1 wt% AOS foam is consistent with its lower stabilized MRF.

3.3.5 Foam EOR Mechanism

In the previous sections, we provided tangible evidence that stable foam can be obtained by co-injection of surfactant (AOS in brine) and nitrogen in sandstone cores pre-flushed by surfactant solution. This was established by a careful analysis of the CT scan images and the liquid saturation profiles in Figs. 7, 8, 9, 10, 11, and 12 and the mobility reduction factor in Fig. 13. We demonstrated that foam is generated at a surfactant concentration as low as 0.1 wt%, but its mobility reduction factor remained low and correspondingly the incremental oil recovery was slightly higher than that of gas flooding. We also found that the foam mobility reduction factor increases considerably upon increasing the surfactant concentration to 0.5 and 1.0 wt%. The increase of surfactant concentration also resulted in higher oil recovery by foam as shown in Fig. 15. We identified that the incremental oil by foam is obtained first by a formation of an oil bank followed by a long tail production.

In order to explore further the foam-flooding mechanism, we have plotted oil saturation as a function of the capillary number N_c, as shown in Fig. 16. This curve can be viewed as a dynamic picture of the evolution of oil saturation with respect to N_c during foam flow. The capillary number N_c was defined as follows:

$$N_c = \frac{k k_{rf} \Delta P}{\sigma_{ow} L} \quad (3)$$

where k is absolute permeability of the core sample, k_{rf} is foam relative permeability, ΔP is pressure drop over the core, σ_{ow} is the interfacial tension between oil and water, and L is the core length. For simplicity, it was assumed that foam is a

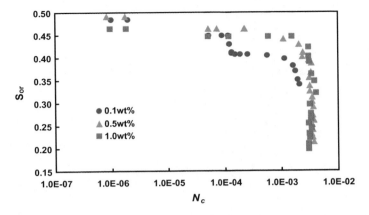

Fig. 16 Dynamic capillary desaturation curve for core-flooding experiments with foam at different surfactant concentrations. The first two points stand for water flooding with N_c in the order of 10^{-7} to 10^{-6}. The third point stands for surfactant pre-flush. The rest data points correspond to foam flooding where N_c increases with surfactant concentration

single-phase fluid such that k_{rf} is equal to one. This implies that an increase in the capillary number amounts essentially to an increase of viscous forces. Each N_c profile in Fig. 16 starts with water flooding at two injection velocities of 2.1 and 4.2 ft/day. The corresponding N_c changes in the order of 10^{-7} to 10^{-6} at which capillary forces are dominant to trap the oil phase. The oil recovery factor due to water flooding was about $44 \pm 2\%$ of the OIIP. The third point in Fig. 16 stands for the surfactant pre-flush by which IFT between oil and water decreases to 1.9 ± 0.1 mN/m under the experimental conditions. Although N_c increases by one order of magnitude when we switch from water to surfactant, capillary forces are still dominant and keep most of the water-flood residual oil trapped. During the surfactant pre-flush, only a small amount of oil of about $2.0 \pm 0.5\%$ of the OIIP was recovered. The next point at each N_c profile corresponds to the beginning of foam flooding, which shows that N_c increases during foam flooding. This can be explained by a larger magnitude of viscous forces as a result of pressure build-up due to the foam development. An increase in viscous forces along with a modest reduction in IFT, due to the presence of surfactant, ensures partial mobilization of the trapped oil. Our results also show that N_c increases with the surfactant concentration. For 0.1 wt% foam, N_c remains low at the early time hardly exceeding 1.0×10^{-4}. Thereafter, N_c increases progressively reaching 2.0×10^{-3} after 16.0 PV of foam injection. For 0.5 and 1.0 wt% foams N_c instead increases from 5.0×10^{-5} at the end of the surfactant pre-flush to 3.3×10^{-3} after 3.0 PV of foam injection and then remains fairly constant at this value for the rest of the injection time. The last point of the dynamic capillary desaturation curve for 1.0 wt% foam shows that $43 \pm 2\%$ of the water-flood residual oil was produced by the foam flood. However, if one would continue foam flooding for longer time, the oil saturation would decrease further as can be inferred from the oil recovery profiles in

Fig. 15. According to Maldal et al. (1997) for Bentheimer sandstone at the N_c at the order of 10^{-3} about 40% of the water-flood residual oil is produced.

The sharp increase of N_c in the presence of oil (Fig. 16) is perhaps the best indication of foam development and its ability to recover incremental oil. This confirms that we can adjust the magnitude of foam mobility by selecting an adequate surfactant at an appropriate concentration. However, we should note that the use of higher surfactant concentrations could be limited due to the fact that it might lead to more complex phase behavior of the surfactant with the appearance of liquid crystal phases (Davis 1996).

The following important conclusion can be drawn from the dynamic capillary desaturation curve in Fig. 16 and also from the MRF profiles in Fig. 13: for 0.5 and 1.0 wt% surfactant concentrations, the foam is created (MRF goes to a larger value) long before all the oil is produced. This supports the idea that foam is formed at the water-flood residual oil which leads to increase in the capillary number N_c and oil displacement out of the core. Therefore, it seems not necessary that oil is first reduced from the water-flood residual oil in order for foam to be formed, as has been suggested earlier (Ali et al. 1985; Nikolov et al. 1986; Mannhardt et al. 1998; Yin et al. 2009; Andrianov et al. 2012). This confirms the possibility of generating strong foam in the presence of oil and then incremental oil recovery by the foam. However, for 0.1 wt% concentration, it appears that foam strength increases as oil saturation decreases.

Figure 17 summarizes the incremental oil recovery for the three surfactant concentrations investigated. As can be seen during foam flooding, incremental oil is first produced at high rates and then the oil production rate decreases slowly. Visual inspection of the effluents together with the CT scan images and foam mobility data have already indicated that early incremental oil is obtained by a generated oil bank

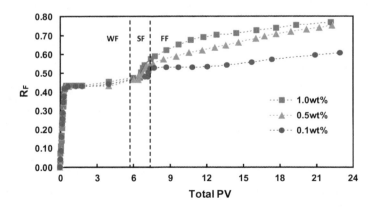

Fig. 17 Oil recovery factor profile for core-flooding experiments performed in this study. Each profile starts with water flooding (WF) followed by surfactant pre-flush (SF) and then foam flooding (FF) at different surfactant concentrations. Oil recover factor by water flooding was about 44 ± 2% of the OIIP. A small amount of oil of about 2.0 ± 0.5% of the OIIP was produced during surfactant pre-flush. Additional oil was produced during foam flooding

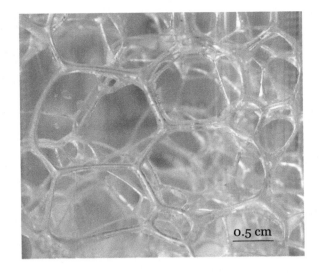

Fig. 18 Distribution of oil droplets inside 0.5 wt% AOS foam structure. The model oil was n-hexadecane, which was colored red for visualization purposes (Simjoo et al. 2013a)

while the later oil is due to the transport of tiny oil droplets within the flowing foam. This is consistent with the fact that at a long time of foam injection, oil saturation decreases further under a fairly constant N_c. Previous studies (Nikolov et al. 1986; Jensen and Friedmann 1987; Raterman 1989; Mannhardt et al. 1998; Schramm and Novosad 1990; Yin et al. 2009) demonstrated only final oil recoveries without mentioning in detail how such produced oil was obtained over the time of foam injection. Mannhardt et al. (1998) reported additional oil production by foam over several tens of injected pore volumes with a mechanism by which emulsified oil droplets are carried inside the foam structure. This proposed mechanism agrees qualitatively with other micro-models and bulk studies of foam–oil interaction (Nikolov et al. 1986; Manlowe and Radke 1990; Koczo et al. 1992). The physical picture emerging from our core-flood analyses at a long time of foam injection and also bulk foam stability in the presence of oil (see Fig. 18) are in agreement with the work of Mannhardt et al. (1998). As shown in Fig. 18 for AOS bulk foam in the presence of normal hexadecane, in which the model oil was colored red, the oil phase is mostly accumulated in the Plateau borders (Simjoo et al. 2013a). Such oil transporting property might work for carrying oil droplets by foam in the porous medium. However, we emphasize that the oil recovery mechanism by foam transport becomes important only at a later time of foam injection, namely after producing the oil bank.

Based on the above discussion, the question arises whether the standard definition of the capillary number as a ratio of viscous to capillary forces is an appropriate representation of the active mechanism(s) in the decrease of oil saturation during foam flooding at a constant N_c. In fact more precisely, the question arises whether foam flow is just water or surfactant flooding at higher gas saturation. To answer these questions, more detailed studies focusing on the microscopic mechanisms of foam flooding are required. The first step could be to perform foam flooding experiment with the aid of an appropriate dual-source CT scanner to provide three-phase saturation maps. In this way, we would be able to track the local distribution of the oil phase to validate the proposed oil mechanisms (formation of oil bank and long tail production) and also to determine precisely at which period during foam injection these mechanisms play a main role. In addition, micro-model studies could be performed while ensuring that pore structures are as close as possible to those of granular porous media. These results could provide more insight into the microscopic aspects of the transport of oil droplets inside the foam structure.

This study demonstrated that stable foam is obtained in natural sandstone cores containing water-flood residual oil. Foam flooding provided an incremental oil recovery ranging from 13 ± 0.5% to 29 ± 2% of the OIIP as the surfactant concentration increases from 0.1 to 1.0 wt% (Fig. 17). This provides the first-hand experimental results showing that immiscible foam can increase oil recovery, thus supports the concept that foam could be potentially employed for EOR application. However, to implement foam EOR process in the field further studies are required to address more realistic situations such as the effects of crude oil and temperature on foam stability and also the way to generate foam to obtain an appropriate foam strength. It seems that existing foam models need only slight modifications to capture early oil recovery, but for the later oil recovery, i.e., transport of dispersed oil, one requires a novel phenomenological framework.

4 Conclusions

Flow of nitrogen foam stabilized by C_{14-16} alpha olefin sulfonate (AOS) in natural sandstone cores containing water-flood residual oil was studied. Foam flooding was performed under gravity stable conditions at a fixed superficial velocity of 4.58 ft/day and foam quality 91% at a back pressure of 20 bar and ambient temperature (21 ± 1 °C). Effect of surfactant concentration (0.1, 0.5 and 1.0 wt%) on the foam strength and foam propagation was examined. The main conclusions from this study are as follows:

- CT scan saturation maps, foam mobility profile and effluent measurements confirm that stable foam can be obtained in the presence of water-flood residual oil.
- By increasing the surfactant concentration from 0.1 to 1.0 wt%, foam exhibited a better front-like displacement, characterized by a longer breakthrough time and a further reduction in liquid saturation as well as in oil saturation.
- Incremental oil production by foam was obtained at two distinct mechanisms in line with CT scan images and visual inspection of the effluents. The first one was due to the generation of an oil bank and occurred in the first few pore volumes injected. The second one was a tailing mechanism where the incremental oil production was obtained due to the transport of dispersed oil within

the flowing foam, which could be consistent with a picture of tiny oil droplets carried in the Plateau borders. The size of the oil bank increased with the surfactant concentration, but the second mechanism was found to be less sensitive to the surfactant concentration.
- The two distinct oil recovery mechanisms observed during foam flooding were described by a dynamic capillary desaturation curve. In the oil bank mechanism, reduction of oil saturation was accompanied by an increase in the capillary number indicating that viscous forces are high enough to mobilize a part of the trapped oil. In the tailing mechanism, the capillary number remains fairly constant despite the continuous decrease of oil saturation. This could be due to the displacement of oil in a dispersed form.
- Under the experimental conditions of this study, AOS foam flooding provided an incremental oil recovery ranging from $5 \pm 0.5\%$ of the OIIP for 0.1 wt% foam to $12 \pm 2\%$ of the OIIP for 1.0 wt% foam after injection of 3.0 PV (equal to the injection of 0.27 PV of surfactant solution). At the same injection time, gas flooding recovered only $4.0 \pm 0.5\%$ of the OIIP. For a long time of foam injection, after 16.0 PV, oil recovery factors ranged from $13 \pm 0.5\%$ for 0.1 wt % foam to $29 \pm 2\%$ for 1.0 wt% foam.

Acknowledgements The authors thank Shell Global Solutions International for sponsoring this study through the Shell GameChanger program on immiscible foam. M. Simjoo acknowledges the financial support of Iran Ministry of Science, Research, and Technology for the research study on foam.

References

Aarra MG, Skauge A, Martinsen HA (2002) FAWAG: a breakthrough for EOR in the North Sea. SPE 77695, SPE annual technical conference and exhibition, San Antonio, Texas

Ali J, Burley RW, Nutt CW (1985) Foam enhanced oil recovery from sand packs. Chem Eng Res and Des 63(2):101–111

Andrianov A, Farajzadeh R, Nick MM, Talanana M, Zitha PLJ (2012) Immiscible foam for enhancing oil recovery: bulk and porous media experiments. Ind Eng Chem Res 51(5):2214–2226

Arnaudov L, Denkov ND, Surcheva I, Durbut P (2001) Effect of oily additives on foamability and foam stability. 1. Role of interfacial properties. Langmuir 17(22):6999–7010

Bernard GG, Holm LW, Jacob LW (1965) Effect of foam on trapped gas saturation and on permeability of porous media to water. SPE J 5(4):295–300

Bond DC, Holbrook OC (1958) Gas drive oil recovery process. US Patent, No. 2,866,507

Chiang JC, Sawyal SK, Castanier LM, Brigham WE, Sufi A (1980) Foam as a mobility control agent in steam injection processes. SPE 8912, SPE California regional meeting, 9–11 Apr, Los Angeles, California

Davis HT (1996) Statistical mechanics of phases, interfaces and thin films. VCH, New York

Denkov ND (2004) Mechanisms of foam destruction by oil-based antifoams. Langmuir 20(22):9463–9505

Dietz DN (1953) A theoretical approach to the problem of encroaching and by-passing edge water. In: Proceedings of Akad. van Wetenschappen, Amsterdam, vol 56-B, p 83

Emadi A, Sohrabi M, Jamiolahmady M, Ireland S, Robertson G (2011) Mechanistic study of improved heavy oil recovery by CO_2-foam injection. SPE 143013, SPE enhanced oil recovery conference, Kuala Lumpur, Malaysia

Farajzadeh R, Andrianov A, Zitha PLJ (2010) Investigation of immiscible and miscible foam for enhancing oil recovery. Ind Eng Chem Res 49(4):1910–1919

Farajzadeh R, Andrianov A, Hirasaki GJ, Krastev R, Rossen WR (2012) Foam-oil interaction in porous media: implications for foam assisted enhanced oil recovery. SPE 154197, SPE EOR conference at oil and gas, West Asia, Muscat, Oman

Fried AN (1961) The foam drive process for increasing the recovery of oil. Report of investigations 5866, U.S. Department of the Interior, Bureau of Mines, Washington, D.C.

Friedmann F, Smith ME, Guice WR, Gump JM, Nelson DG (1994) Steam foam mechanistic field trial in the midway-sunset field. SPE Res Eng 9(4):297–304

Green DW, Willhite GP (1998) Enhanced oil recovery, SPE textbook series, vol 6, Richardson, TX, USA

Hadjiiski A, Tcholakova S, Denkov ND, Durbut P (2001) Effect of oily additives on foamability and foam stability. 2. Entry barriers. Langmuir 17(22):7011–7021

Hirasaki GJ, Lawson J (1985) Mechanisms of foam flow in porous media: apparent viscosity in smooth capillaries. SPE J 25(2):176–190

Jensen JA, Friedmann F (1987) Physical and chemical effects of an oil phase on the propagation of foam in porous media. SPE 16375, California regional meeting, Ventura, California

Koczo K, Lobo L, Wasan DT (1992) Effect of oil on foam stability: aqueous film stabilized by emulsions. J Colloid Interface Sci 150(2):492–506

Kovscek AR, Radke CJ (1994) Fundamentals of foam transport in porous media. In: Schramm LL (ed) Foams: fundamentals and applications in the petroleum industry. ACS, Washington D.C.

Lake LW (1989) Enhanced oil recovery. Prentice-Hall, Englewood Cliffs

Lau HC, O'Brien SM (1988) Effects of spreading and nonspreading oils on foam propagation through porous media. SPE Res Eng 3(3):893–896

Li RF, Yan W, Liu S, Hirasaki GJ, Miller CA (2010) Foam mobility control for surfactant enhanced oil recovery. SPE J 15(04):928–942

Li RF, Hirasaki GJ, Miller CA, Masalmeh SK (2012) Wettability alteration and foam mobility control in a layered, 2D heterogeneous sandpack. SPE J 17(4):1207–1220

Maldal T, Gulbrandsen AH, Gilje E (1997) Correlation of capillary number curves and remaining oil saturations for reservoir and model sandstones. In Situ Rep 21(3):239–269

Manlowe DJ, Radke CJ (1990) A pore-level investigation of foam/oil interactions in porous media. SPE Res Eng 5(4):495–502

Mannhardt K, Svorstøl I (1999) Effect of oil saturation on foam propagation in Snorre reservoir core. J Pet Sci Eng 23(3–4):189–200

Mannhardt K, Novosad JJ, Schramm LL (1998) Foam-oil interactions at reservoir conditions. SPE 39681, SPE/DOE improved oil recovery symposium, Tulsa, OK, USA

Minssieux L (1974) Oil displacement by foams in relation to their physical properties in porous media. J Pet Tech 26(1):100–108

Nikolov AD, Wasan DT, Huang DW, Edwards DA (1986) The effect of oil on foam stability: mechanisms and implications for oil displacement by foam in porous media. SPE 15443, SPE annual technical conference and exhibition, New Orleans, Louisiana, USA

Orr FM (2007) Theory of gas injection processes. Tie-Line Publications, Copenhagen

Patzek TW, Koinis MT (1990) Kern River steam-foam pilots. J Pet Tech 42(4):496–503

Peksa AE, Wolf KHA, Zitha PLJ (2015) Bentheimer sandstone revisited for experimental purposes. Mar Pet Geol 67:701–719

Raterman KT (1989) An investigation of oil destabilization of nitrogen foams in porous media. SPE 19692, SPE annual technical conference and exhibition, San Antonio, Texas

Rossen WR (1996) Foams in enhanced oil recovery. In: Prud'homme RK, Khan S (eds) Foams: theory, measurements and applications. Marcel Dekker, New York

Schramm LL, Novosad JJ (1990) Micro-visualization of foam interactions with a crude oil. J Colloids Surf 46:21–43

Schramm LL, Novosad JJ (1992) The destabilization of foams for improved oil recovery by crude oils: effect of the nature of the oil. J Pet Sci Eng 7(1):77–90

Schramm LL, Turta AT, Novosad JJ (1993) Microvisual and coreflood studies of foam interactions with a light crude oil. SPE Res Eng 8(3):201–206

Simjoo M, Zitha PLJ (2015) Modeling of foam flow using stochastic bubble population model and experimental validation. Transp Porous Med 107(3):799–820

Simjoo M, Rezaei T, Andrianov A, Zitha PLJ (2013a) Foam stability in the presence of oil: effect of surfactant concentration and oil type. Colloids Surf A 438:148–158

Simjoo M, Dong Y, Andrianov A, Talanana M, Zitha PLJ (2013b) CT scan study of immiscible foam flow in porous media for enhancing oil recovery. Inds Eng Chem Res 52(18):6221–6233

Simjoo M, Dong Y, Andrianov A, Talanana M, Zitha PLJ (2013c) Novel insight into foam mobility control. SPE J 18(3):416–427

Singh R, Mohanty KK (2015) Synergy between nanoparticles and surfactants in stabilizing foams for oil recovery. Energy Fuels 29(2):467–479

Skauge A, Aarra MG, Surguchev LM, Martinsen HA, Rasmussen L (2002) Foam-assisted WAG: experience from the Snorre field. SPE 75157, SPE/DOE improved oil recovery symposium, Tulsa, Oklahoma

Suffridge FE, Raterman KT, Russell GC (1989) Foam performance under reservoir conditions. SPE 19691, SPE annual technical conference and exhibition, San Antonio, Texas

Svorstøl I, Vassenden F, Mannhardt K (1996) Laboratory studies for design of a foam pilot in the Snorre field. SPE 35400, SPE/DOE improved oil recovery symposium, Tulsa, Oklahoma

Turta AT, Singhal AK (1998) Field foam applications in enhanced oil recovery projects: screening and design aspects. SPE 48895, international oil and gas conference and exhibition in China, Beijing, China

Vikingstad AK, Aarra MG (2009) Comparing the static and dynamic foam properties of a fluorinated and an alpha olefin sulfonate surfactant. J Pet Sci Eng 65(1–2):105–111

Wellington SL, Vinegar HJ (1988) Surfactant-induced mobility control for carbon dioxide studied with computerized tomography. In: Smith DH (ed) Surfactant based mobility control: progress in miscible flood enhanced oil recovery. ACS, Washington D.C.

Yin G, Grigg RB, Yi S (2009) Oil recovery and surfactant adsorption during CO_2 foam flooding. SPE 19787, offshore technology conference, Houston, Texas

Part II
Numerical Modeling of Fluid Flow Under Heterogeneous Conditions

Numerical Simulation of Flows in a Channel with Impermeable and Permeable Walls Using Finite Volume Methods

Z. F. Tian, C. Xu and P. A. Dowd

1 Introduction

Thermal heat from enhanced geothermal systems (EGS) has the ability to generate and dispatch baseload electricity without storage and with low carbon emissions. EGS reservoirs in hot dry rocks (HDR) are, in general, located at significant depths, commonly more than 3 km below the surface, and close to radiogenic heat sources (Mohais et al. 2011a). In their natural state, these rocks typically have temperatures at around 300 °C and usually have very low permeability. Extracting heat from an EGS requires a connected network of fractures in the rock mass through which fluid can be circulated and brought to the surface as very hot water. The fracture network is usually created by hydraulic fracturing, which creates new fractures and causes existing fractures to propagate (Mohais et al. 2011a, b). Cold working fluid, usually water, is injected into the fracture network through an injection well, flows through the fracture network, where it is heated by the surrounding rock and is then extracted through a production well. The heat in the extracted water can be used to generate electricity or can be used as a heat source for other applications.

As a result of the hydro-fracturing process and the granular composition of the rocks, the walls of the fracture channels have permeable properties due to cracks and fissures of varying sizes in the channel wall (Christopher and Armstead 1978; Mohais et al. 2011b; Phillips 1991). The efficiency of the geothermal reservoir is highly dependent on the permeability of the rock fractures within the reservoir (Natarajan and Kumar 2012) and on the flows within the fracture channels, as

Z. F. Tian (✉)
School of Mechanical Engineering, The University of Adelaide,
Adelaide, SA 5005, Australia
e-mail: zhao.tian@adelaide.edu.au

C. Xu · P. A. Dowd
School of Civil, Environmental and Mining Engineering,
The University of Adelaide, Adelaide, SA 5005, Australia

© Springer Nature Singapore Pte Ltd. 2018
N. Narayanan et al. (eds.), *Flow and Transport in Subsurface Environment*,
Springer Transactions in Civil and Environmental Engineering,
https://doi.org/10.1007/978-981-10-8773-8_4

demonstrated in Xu et al. (2015). Figure 1 shows idealised three-dimensional (3D) and two-dimensional (2D) schematic views of a flow channel with porous upper and lower walls. The walls are porous due to the cracks and fissures generated by the fracturing process. As the span-wise dimension of the fracture channels (z-direction in Fig. 1a) is always much larger than the height of the channel (y-direction in Fig. 1a, b) and variation of flow field in z-direction is assumed to be negligible, the channel flow is generally considered as a 2D flow as shown in Fig. 1b.

Several studies of fluid flows in channels with porous walls have been reported in the literature. In early studies of the flow at the interface between the channel and the porous medium, the velocity u in the x-direction in Fig. 1b, at the channel-porous medium interface was usually assumed to be zero (Mikelic and Jäger 2000). In fluid dynamics, this is known as the no-slip boundary condition.

Beavers and Joseph (1967) pioneered the investigation of the slip velocity at channel-porous medium interfaces. In an experiment, they compared fluid flow through a porous block with flow through a channel formed by an upper wall without flow through the wall and a lower permeable wall formed by the upper surface of the porous block. Flows were compared for various samples of two types of permeable material. This type of channel differs slightly from those in EGS in which the upper and lower walls channels (fractures) are porous (Fig. 1).

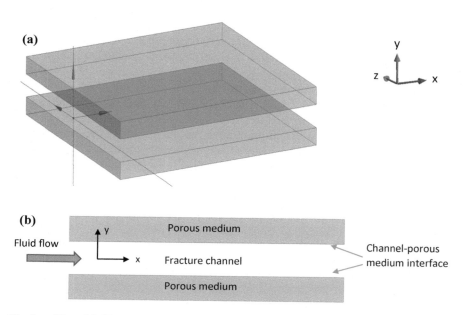

Fig. 1 **a** 3D and **b** 2D schematic view of channel flow with two porous medium walls

Beavers and Joseph (1967) derived the boundary condition of the interface wall between the channel and the porous medium as

$$\left.\frac{du}{dx}\right|_{y^+=0} = \frac{\alpha}{\sqrt{k}}(u_{\text{interface}} - u_m), \quad (1)$$

where α is a dimensionless coefficient termed the slip coefficient or Beavers–Joseph coefficient. In Eq. (1), 0+ is a boundary limit point and $y = 0+$ means that du/dy values are calculated using the velocity u (velocity in x-direction) on the channel side but not the porous medium side. In Eq. (1), k is the absolute permeability of the porous medium (m^2); $u_{\text{interface}}$ is the fluid velocity u at the interface between the channel and the porous medium (m s^{-1}); u_m is the fluid velocity in the x-direction in the porous medium (m s^{-1}), given by Darcy's law as below

$$-\frac{dp_m}{dx} = \frac{\mu}{k} u_m, \quad (2)$$

where μ is the dynamic viscosity of the fluid (Pa s) and p_m is the pressure of the fluid in the porous medium (Pa).

Saffman (1971) further developed the generic velocity boundary condition for the fluid-porous medium interface based on the Beavers–Joseph boundary condition (Eq. 1). Saffman's boundary condition (Nield 2009; Saffman 1971) is

$$u_{\text{interface}} = \frac{\sqrt{k}}{\alpha} \frac{\partial u_{\text{interface}}}{\partial n} + O(k), \quad (3)$$

where $O(k)$ is the average velocity in the porous medium (m s^{-1}), which can be neglected (Nield 2009). In Eq. (3), n denotes the direction normal to the fluid-porous medium interface. Again $\partial u_{\text{interface}}/\partial n$ is calculated on the channel side but not the porous medium side. The Beavers–Joseph boundary condition is a special case of Saffman's boundary condition (Nield 2009; Saffman 1971) for channel flows. Saffman's modification of the Beavers–Joseph condition has been further confirmed by theoretical studies such as Mikelic and Jäger (2000).

In a later study, Jones (1973) pointed out that the Beavers–Joseph boundary condition for generic cases should be

$$\left.\left(\frac{\partial u}{\partial y} + \frac{\partial v}{\partial x}\right)\right|_{y^+=0} = \frac{\alpha}{\sqrt{k}}(u_{\text{interface}} - u_m) \quad (4)$$

meaning that the left-hand side of the equation should be the shear strain rate for generic cases. In Eq. (4), v is the velocity component in y-direction. It can be seen that the Beavers–Joseph condition is a special case of Jones' condition, for a one-dimensional flow, $\partial v/\partial x = 0$.

For the non-dimensional slip coefficient α in Eqs. (1), (3) and (4), Beavers and Joseph (1967) found that values of α depend on the structure of the porous material at the fluid-porous medium interface and materials with similar permeability may have, significantly, different slip coefficients but are independent of the fluid viscosity (Nield 2009). In the Beavers and Joseph (1967) study, for foametal with average pore sizes of 0.00406 m (0.016 in.), 0.00864 m (0.034 in.) and 0.0114 m (0.045 in.), the values of α are 0.78, 1.45 and 4.0, respectively; while, for aloxide with average pore sizes of 0.0033 m (0.013 in.) and 0.00686 m (0.027 in.), the value of α is 0.1, for both pore sizes. The influence of the fluid (water, oil and gas) on the value of α was found to be insignificant, whereas α is very sensitive to the nature of the porous interface (Beavers et al. 1974). A later numerical study by Larson and Higdon (1986) concluded that the value of α is sensitive to microscopic changes in the definition of the interface implying that it is not possible to define a consistent value of α for any media. The numerical studies of Sahraoui and Kaviany (1992) shows that α depends on porosity, Reynolds number (Re), channel height, choice of interface, bulk flow direction and interface structure. Another numerical study (Liu and Prosperetti 2011) shows that α values for pressure-driven and shear-driven flows are somewhat different and α values depend on the Reynolds number.

Several numerical studies of channel flows with porous media boundary conditions have been reported in the literature. One of the earliest, Berman (1953), used the perturbation method to solve the Navier–Stokes equations to describe the flow in a channel with a rectangular cross section and two equally porous walls. The velocity u at the channel-porous medium interface is taken as zero (no-slip condition) and the vertical velocity at the interface is assumed to be constant. Terrill and Shrestha (1965) used the perturbation method to solve the Navier–Stokes equations for laminar flows through parallel and uniformly porous walls. The boundary conditions in their study are similar to those in Berman (1953); the u velocities at the channel-porous medium interface are zero and the vertical velocities are constant but different for the two porous walls. Granger et al. (1989) obtained the analytical solutions of the Navier–Stokes equations for both a rectangular channel with one porous wall and a porous tubular channel. They found that the velocity u profile of the flow in porous channels may be considered parabolic and there is no pressure profile across the width of the channel. Recently, Herschlag et al. (2015) obtained the analytical solution for the flow in a channel with high wall permeability. The boundary conditions at the channel walls are no-slip for axial velocity and Darcy's law for the vertical velocity.

The analytical studies reviewed above are for cases in which there are vertical flows at the channel-porous medium interfaces and the axial velocities at the channel-porous medium interface are assumed to be zero. This no-slip velocity condition is not realistic for fracture channels in EGS or similar systems. Mohais et al. (2011a, b) used the perturbation method to solve the Navier–Stokes equations to provide an analytical solution for laminar flow in a channel with porous walls and non-zero axial velocity at the fluid-porous media interface. For a channel with walls that contain small fissures, cracks and granular material, the axial velocity profile in the channel can be affected by factors such as the slip boundary coefficient, permeability and the channel width (Mohais et al. 2011a, b; Tian et al. 2012).

This chapter reports computational fluid dynamics (CFD) simulations of fluid flows in a single horizontal fracture sandwiched between two equally porous media, in the context of an EGS. Little research has been reported in modelling this problem using the finite volume approach.

The first objective is to compare the velocity profiles predicted by CFD with analytical solutions (Mohais et al. 2011a, b). To the authors' knowledge, validations of these analytical solutions of flow velocity profiles in a channel (Mohais et al. 2011a, b) have not been investigated using numerical models. These validations will assist the development of other complex models of channel flows in future research.

The second objective is to investigate the influence of parameters such as slip coefficient α, Reynolds number, permeability and channel height on the velocity profiles and pressure drops in the channel flows. Of particular interest are the pressure drops in the channel flows under different conditions as they are not available in the analytical solutions reported in Mohais et al. (2011a, b).

2 Numerical Methods

2.1 CFD Domain and Boundary Conditions

In this study, laminar water flow is simulated in a 2D channel with a height of $2h$. The channel is contained between two equally porous media, as illustrated in Fig. 1b. ANSYS/Designmodeler 17.2 was used to generate the CFD domain. Figure 2 is a schematic diagram of the domain and boundary conditions of the single fracture channel model. To reduce the computational time for the simulations, the CFD domain is half of a single channel with a symmetric boundary at the bottom, as shown in Fig. 2, and thus the height of the half-channel modelled is h. Following the work of Mohais et al. (2011b), two values of h(0.001 m and 0.0001 m) were tested in the study; these values are commonly used in modelling channel flows in EGS reservoirs. The Reynolds numbers for the flows vary from 0.5 to 7.0 as the analytical solutions of Mohais et al. (2011a, b) hold for $Re < 7$. The Reynolds number of the flow, Re, is defined in this case as

$$Rew = \frac{2h\rho u_{ave}}{\mu}, \tag{5}$$

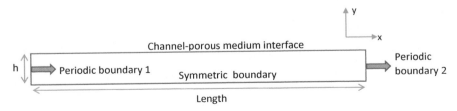

Fig. 2 The boundary conditions of the CFD model of a single fracture (not to scale)

where u_{ave} is the average velocity u of the flow in the channel (m s^{-1}), $2h$ is the full height of the channel (m) and ρ is the density of fluid (kg m^{-3}) (water in this study).

In the x-direction, the periodic boundary condition is imposed with a mass flow rate of u_{ave}. The values of u_{ave} are calculated using Eq. (5) with corresponding Reynolds number Re and channel height $2h$, for each case. The length of the channel L is 0.08 m for $h = 0.001$ m and 0.01 m for $h = 0.0001$ m. The fluid in the channel is water at 25 °C. The dynamic viscosity of water is 0.0008899 Pa s and the density of water is 997 kg m^{-3}.

At the channel-porous medium interface, the velocity u of the fluid is imposed using the analytical equation developed by Mohais et al. (2011b). The porous walls are assumed to be saturated, i.e. there is no fluid flow across the interface boundary. The boundary conditions at the channel-porous medium interface used in the study are

$$v_{\text{interface}} = 0 \tag{6}$$

$$u_{\text{interface}} = u_{ave} f'(y^*)|_{y^*=1} = u_{ave} \left(f_0'(y^*) + Rew f_{01}'(y^*) \right)|_{y^*=1}, \tag{7}$$

where $v_{\text{interface}}$ is the fluid velocity in the y-direction at the channel-porous medium interface and $u_{\text{interface}}$ is the fluid velocity in the x-direction at the channel-porous medium interface. In Eq. (7), y^* is the normalised distance in y-direction defined as $y^* = y/h$. Rew in Eq. (7) is the Reynolds number of fluid at the channel-porous medium interface and is calculated as in Mohais et al. (2011b)

$$Rew = \frac{2h \rho u_{\text{interface}}}{\mu}. \tag{8}$$

In Eq. (7), $f_0(y^*)$ and $f_1(y^*)$ are determined as in Mohais et al. (2011b)

$$f_0(y^*) = y^{*3} \left(\frac{-1}{2(1+3\phi)} + y^* \frac{3+6\phi}{2+6\phi} \right) \tag{9}$$

$$f_1(y^*) = -\frac{y^{*7}}{2520} \left(\frac{9}{(1+3\phi)^2} \right) + \frac{y^{*3}}{6} \left(\frac{9(7\phi+1)}{140(1+3\phi)^3} \right) + y^* \left(\frac{1}{280(1+3\phi)^2} - \frac{3(7\phi+1)}{280(1+3\phi)^3} \right), \tag{10}$$

where $\phi = \sqrt{k}/(\alpha h)$.

The CFD package, ANSYS/CFX 17.2, was used for all steady-state simulations based on the Navier–Stokes equations. The steady state Navier–Stokes equations for incompressible flows are

$$\mathbf{V} \cdot \mathbf{U} = \mathbf{0} \tag{11}$$

$$\mathbf{V} \cdot (\mathbf{UU}) = -\frac{\mathbf{V}p}{\rho} + \mathbf{V} \cdot \left\{ \frac{\mu}{\rho} \left[\mathbf{V}\mathbf{U} + (\mathbf{V}\mathbf{U})^T - \frac{2}{3} \mathbf{V} \cdot \mathbf{I}\mathbf{U} \right] \right\} + \mathbf{G}, \tag{12}$$

where **G** is the source term due to gravity and **I** is the identity matrix. **U** in Eqs. (11) and (12) is velocity vector. Following the work of Mohais et al. (2011b), gravity has negligible effects on the flows and therefore is neglected in this chapter. ANSYS/CFX is a finite volume solver. The fourth-order Rhie–Chow option was used for the velocity pressure coupling. The high-resolution scheme was used for advection terms. A double precision version of the solver was employed to ensure the accuracy of the CFD results. The maximum residual target was set at 1.5×10^{-6} for all simulations.

2.2 CFD Mesh and Mesh-Independent Test

ANSYS/meshing was used to generate the CFD mesh. A grid independence test was conducted for the case for $h = 0.001$ m, $Re = 0.5$ and $\alpha = 1$. An initial structured mesh of 700 (in the streamwise direction, x) × 50 (height, y) was generated and then refined to a mesh of 1000 × 80, and further refined to a mesh of 1500 × 120. Mesh independence was checked by comparing the fluid velocity profile along periodic boundary 1 (indicated as the red line in Fig. 3).

Figure 4 shows the comparison of normalised velocity u profiles for $h = 0.001$ m, $\alpha = 1$, $Re = 0.5$ and $k = 10^{-8}$ m^2, for three mesh systems. The velocity u is normalised by the average velocity u_{ave}.

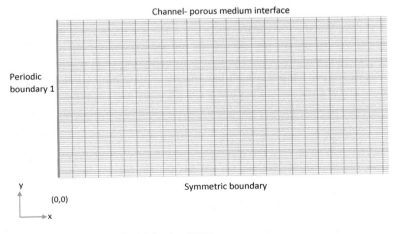

Fig. 3 Mesh at periodic boundary 1 for $h = 0.001$ m

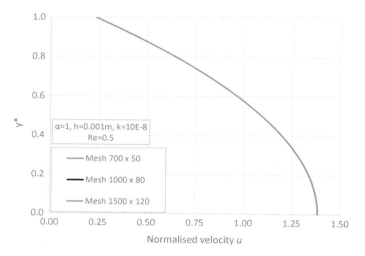

Fig. 4 Mesh independence test based on $h = 0.001$ m, $\alpha = 1$, $Re = 0.5$ and $k = 10^{-8}$ m^2

All three meshes give almost identical results. The maximum difference in the results obtained from the 1000 × 80 mesh and the 1500 × 120 mesh is less than 1%. For $h = 0.001$ m, the 1000 × 80 structured mesh was used for all other simulations. Figure 3 shows the mesh nodes at periodic boundary condition 1 for this case. For $h = 0.0001$ m, a similar mesh independence test was conducted and the 1200 × 80 structured mesh was used for the simulations.

3 Results and Discussion

The CFD flow profiles for the channel with impermeable walls are reported in Sect. 3.1. The no-slip boundary ($u_{\text{interface}} = 0$) condition was imposed on the channel-porous medium interface. The purpose of reporting results for this boundary condition is twofold. The first is to verify the CFD results by comparing the predicted velocity profiles with the analytical solutions of Potter et al. (2016), and the second is to examine the differences between the pressure drop values of the slip boundary cases with those of the no-slip boundary conditions.

In Sects. 3.2–3.6, a series of parametric studies are reported for the flows in the channels with permeable walls. The slip boundary conditions calculated using Eq. (7) in Mohais et al. (2011b) were used to calculate the slip velocity ($u_{\text{interface}}$) at the channel-porous medium interface. The key parameters influencing the velocity profiles and pressure drop in the channel, including Re number, slip coefficient α, permeability k and channel height h are investigated using the CFD model.

In Sect. 3.7, we compare the slip velocities predicted by Eq. (7) and by Saffman's boundary condition (Eq. 3) using the $\partial u/\partial n$ values predicted by CFD.

Table 1 CFD predicted $-\Delta p/L$ values for different cases

Case	$h = 0.0001$ m, $Re = 5$	$h = 0.0001$ m, $Re = 0.5$	$h = 0.001$ m, $Re = 5$	$h = 0.001$ m, $Re = 0.5$
CFD predicted $-\Delta p/L$ (Pa m^{-1})	5940	594	5.94	0.594

3.1 Flow and Pressure Drop in Channels with No-Slip Walls

Flows in channels with impermeable walls are simulated by imposing the no-slip boundary condition at the channel-porous medium interface. Table 1 lists the pressure drop per unit length, $-\Delta p/L$ (Pa m^{-1}), predicted by CFD for the no-slip boundary condition, which is defined as

$$\frac{-\Delta p}{L} = \frac{-(p_2 - p_1)}{L}, \qquad (13)$$

where p_2 and p_1 are the area-averaged static pressures at periodic boundary 2 and 1 (Fig. 2), respectively. In Eq. (13), L denotes the length of the channel (m). Note, in our model, p_1 is always higher than p_2 in this pressure-driven flow.

The velocity profiles predicted by CFD in the channel with the no-slip boundary condition are verified by comparing them with the analytical solutions for laminar channel flows (Potter et al. 2016)

$$u(y) = \frac{1}{2\mu}\frac{dy}{dx}(y^2 - 2hy). \qquad (14)$$

Figure 5 shows a comparison of the CFD-predicted velocity profile with the analytical solution of Eq. (14) for the case $Re = 0.5$ and $h = 0.0001$ m. The CFD model performs very well; the velocity profiles obtained from the two methods are almost identical with the maximum difference between them of about 0.7%. The same conclusion can be drawn for the other three cases: $Re = 5$ and $h = 0.0001$ m, $Re = 0.5$ and $h = 0.001$ m and $Re = 5$ and $h = 0.001$ m.

3.2 Effect of Slip Coefficient on the Velocity Profiles

Figure 6 shows the CFD-predicted velocity u profiles normalised by the average velocity u_{ave} with varying values of α and $Re = 0.5$, $h = 0.001$ m and $k = 10^{-8}$ m^2. In this case, the CFD results are compared with the solution of the analytical model given in Mohais et al. (2011b). The analytical model of Mohais et al. (2011b) is

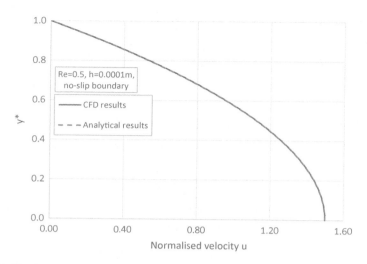

Fig. 5 Verification of CFD-predicted velocity profile against the analytical solution of Eq. (14) (Potter et al. 2016), no-slip boundary, $h = 0.0001$ m and $Re = 0.5$

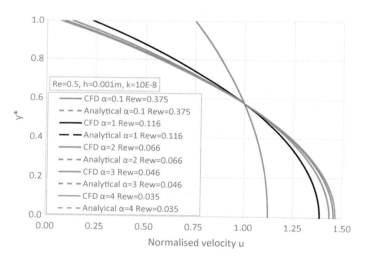

Fig. 6 Comparison of CFD and analytical results (Eq. 3) for flows with walls with varying α values and $Re = 0.5$, $h = 0.001$ m, $k = 10^{-8}$ m^2

$$f(y^*) = f_0(y^*) + Re w f_1(y^*), \qquad (15)$$

where $f_0(y^*)$ and $f_1(y^*)$ are given in Eqs. (9) and (10), respectively.

The CFD predictions and the analytical solutions are almost identical for all cases shown in Fig. 6. The maximum difference between the CFD results and the analytical solutions for the cases shown in Fig. 6 is less than 1%. This very good agreement confirms the analytical solutions of velocity profiles in the channel derived by Mohais et al. (2011b).

As shown in Fig. 6, all the velocity curves are parabolic as expected for the channel flow (Mohais et al. 2011b). As the values of α increases from 0.1 to 4, the normalised velocity u at the channel-porous medium interface decreases from 0.75 ($\alpha = 0.1$) to 0.0706 ($\alpha = 4$). The normalised velocity at the channel centre line decreases as the value of α increases.

Figure 7 compares the CFD prediction with the analytical solutions of the profiles of normalised fluid velocity u with varying values of α and $Re = 5$, $h = 0.001$ m and $k = 10^{-8}$ m^{-2}. The variations in $u_{interface}$ for different α values and $Re = 5$ are very similar to those for $Re = 0.5$, with the normalised velocity u at the wall decreasing from 0.756 when $\alpha = 0.1$ to 0.0786 when $\alpha = 4$. The velocity at the channel-porous medium interface for $\alpha = 0.1$ and $Re = 5$ in Fig. 7 is 0.757, compared with 0.751 for the $Re = 0.5$. The slightly higher interface velocity for $Re = 5$ is due to its higher corresponding Rew (Eq. 15), which is 3.78 compared with 0.375 for $Re = 0.5$ (shown in Fig. 6). The normalised velocity u at the channel centre line decreases as the value of α increases. This is not surprising as the mass flow rates of all the cases in Fig. 6 are the same. For these incompressible flows, the increase in velocity u at the region near the channel-porous medium interface needs to be balanced by the decrease of velocity u at the centre.

Figure 8 compares the CFD and analytical results for flows in a narrower channel with $h = 0.0001$ m at $Re = 0.5$ and varying α values. When $\alpha = 0.1$, the velocity profile is close to a vertical line, i.e. the fluid velocity at the channel-porous medium interface is close to the flow velocity at the channel centre line, similar to

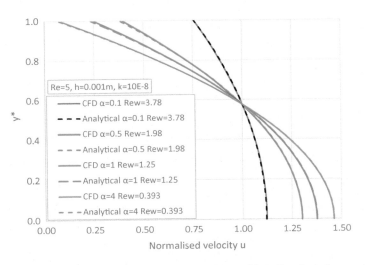

Fig. 7 Comparison of CFD and analytical results for flows with walls of varying α values and $Re = 5$, $h = 0.001$ m, $k = 10^{-8}$ m^2

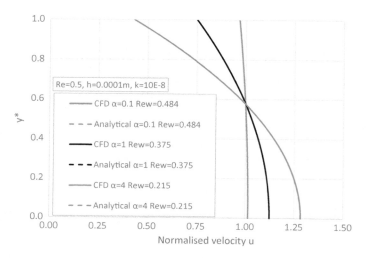

Fig. 8 Comparison of CFD and analytical results for flows with walls of varying α values and $Re = 0.5$, $h = 0.0001$ m, $k = 10^{-8}$ m^2

the profile of a plug flow. When α increases to 1 and then to 4, the fluid velocity at the interface decreases and the flow at the channel centre line increases. Compared with the results shown in Fig. 6, for $h = 0.001$ m and the same Re number, k value and α value, the fluid velocities at the channel-porous medium interface for $h = 0.0001$ m are higher. Again, this is caused by the higher Rew values for the cases shown in Fig. 8 than those shown in Fig. 6. The profiles of normalised velocity for $h = 0.0001$ m at $Re = 5$ with varying α values are given in Fig. 9. These profiles are very similar to those of the $Re = 0.5$ cases shown in Fig. 8, suggesting that the velocity profiles become insensitive to the Reynolds number.

3.3 Effect of α Values on the Pressure Drops

The effect of the slip coefficient, α, on the pressure drop per unit length, $-\Delta p/L$, in the channel is investigated for channel heights of $h = 0.001$ m and $h = 0.0001$ m. As shown in Fig. 10, for both $Re = 5$ and $Re = 0.5$ with $h = 0.001$ m, the $-\Delta p/L$ value is higher for a higher value of α. In other words, the higher the α value, the higher is the head loss in the channel flow. Please note that the pressure drops in the porous walls are not calculated in the current chapter. To calculate the total energy required to push the water through the channel, the porous walls should be included in the CFD domain.

For $Re = 0.5$ and $h = 0.001$ m, the $-\Delta p/L$ value predicted by CFD for the no-slip boundary case is 0.594 Pa m^{-1} (in Table 1). As shown in Fig. 10, the $-\Delta p/L$ value of 0.15 Pa m^{-1} for $\alpha = 0.1$ is 25% of that for the corresponding

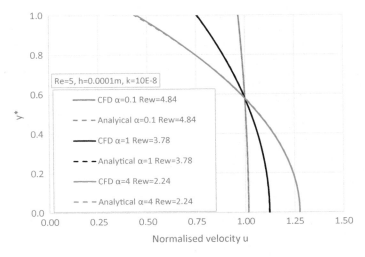

Fig. 9 Comparison of CFD and analytical results for flows with walls of varying α and $Re = 5$, $h = 0.0001$ m, $k = 10^{-8}$ m^2

Fig. 10 Effect of α values on the pressure drop per unit length $-\Delta p/L$ when $h = 0.001$ m, $k = 10^{-8}$ m^2

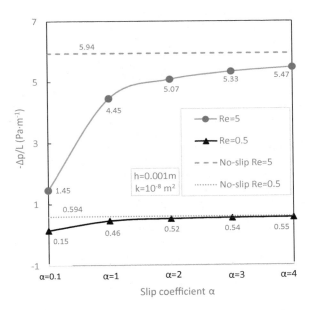

no-slip boundary case. The $-\Delta p/L$ value of 0.46 Pa m^{-1} for $\alpha = 1$ is 77.4% of that for the no-slip boundary case. The $-\Delta p/L$ value of 0.55 for $\alpha = 4$ is about 93% of that for the no-slip boundary case.

For the $Re = 5$ cases in Fig. 10, the $-\Delta p/L$ value predicted by CFD for the no-slip boundary, when $h = 0.001$ m, is 5.94 Pa m^{-1}. With the slip boundary condition, the $-\Delta p/L$ value is 1.45 Pa m^{-1} for $\alpha = 0.1$, which is 24.4% of that of the no-slip boundary case. The $-\Delta p/L$ value approaches 5.94 Pa m^{-1} as α increases and when $\alpha = 4$, the value is 5.47 Pa m^{-1}, which is 92% of that of the no-slip boundary case.

The effect of α values on the pressure drop per unit length in the channel is much more pronounced when $h = 0.0001$ m (Fig. 11). For $Re = 5$ and $h = 0.0001$ m, the $-\Delta p/L$ value of the no-slip boundary is 5940 Pa m^{-1} (Table 1). With the slip boundary condition, for $\alpha = 0.1$, the value is 191 Pa m^{-1}, which is 3.2% of that of no-slip boundary case, and for $\alpha = 4$, the $-\Delta p/L$ value is 3277 Pa m^{-1}, about 55% of that of no-slip boundary case. Similar trends can be observed for $Re = 0.5$ and $h = 0.0001$ m. The $-\Delta p/L$ value of the no-slip boundary is 594 Pa m^{-1}, while the value of $-\Delta p/L$ for $\alpha = 0.1$ is 19 Pa m^{-1} and for $\alpha = 4$ is 338 Pa m^{-1}, which are 3.2 and 57%, respectively, of the no-slip boundary case.

For both $Re = 5$ and $Re = 0.5$ cases, when $h = 0.001$ m (Fig. 10), the pressure drop per unit length under the slip boundary condition can result in up to a 75.6% ($\alpha = 0.1$) reduction compared with the pressure drop per unit length for the non-slip boundary cases. When $h = 0.0001$ m, however, the slip boundary condition can result in up to a 96.8% ($\alpha = 0.1$) reduction in pressure drop per unit length compared with the non-slip boundary condition results. The reduction in pressure drop per unit length for the slip boundary condition can be attributed to the reduction in shear stress at the channel-porous medium interface. For this channel flow, the shear stress at the channel-porous medium interface can be calculated as

$$\tau_{\text{interface}} = \mu \frac{\partial y}{\partial x}\bigg|_{\text{interface}}. \tag{16}$$

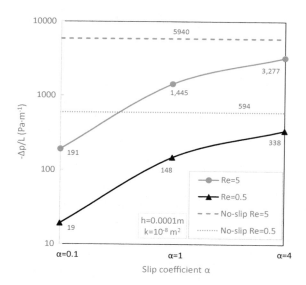

Fig. 11 Effect of α values on the pressure drop per unit length $-\Delta p/L$ when $h = 0.0001$ m, $k = 10^{-8}$ m^2

It is obvious from Figs. 6, 7, 8 and 9 that lower values of α lead to higher slip velocities at the channel-porous medium interface and lower velocity at the centre. This leads to 'flatter' velocity profiles and lower velocity gradients $\partial u/\partial y$ at the channel-porous medium interface and hence, based on Eq. (16), lower shear stress at the interface. This reduction in shear stress contributes to the reduction of skin friction and, in turn, reduction in the pressure drop per unit length along the channel.

3.4 Effect of Permeability k on the Velocity Profiles

Figure 12 shows the velocity profiles of flows in the channel for different permeability k values of the porous medium and for $Re = 0.5$, $h = 0.001$ m and $\alpha = 1$. For lower permeability values ($k = 10^{-10}$ m^2 and $k = 10^{-12}$ m^2), the normalised u velocities at the channel-porous medium interface are as small as 0.033 and 0.003, respectively, leading to very small Rew values of 0.0149 and 0.00163, respectively. As the permeability k increases from $k = 10^{-8}$ m^2 to $k = 10^{-6}$ m^2, the normalised velocity u at the channel-porous medium interface increases from 0.2327 to 0.7506, while the normalised velocity u at the channel centre line decreases from 1.38 to 1.12 due to the conservation of mass flow rate.

When the Re number increases from 0.5 to 5, the effect of permeability values on the velocity profiles is similar to the $Re = 0.5$ cases, i.e. higher permeability causes higher slip velocity at the channel-porous medium interface (Fig. 13). It is noticeable that for cases with the same permeability, the normalised slip velocity for $Re = 5$ (Fig. 13) is higher than that for $Re = 0.5$ (Fig. 12).

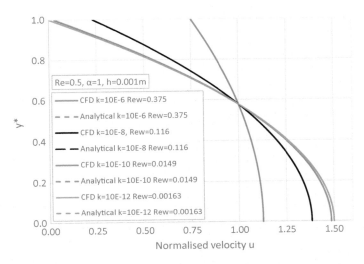

Fig. 12 Effect of permeability on the velocity profiles for $Re = 0.5$, $h = 0.001$ m, $\alpha = 1$

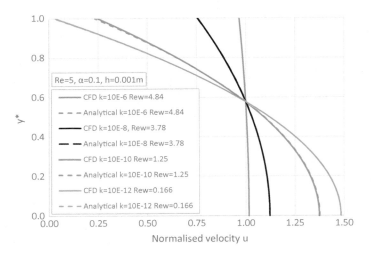

Fig. 13 Effect of permeability k on the velocity profiles for $Re = 5$, $h = 0.001$ m, $\alpha = 0.1$; units of k (m^2)

The effects of permeability k on the velocity profiles were also investigated for cases when $h = 0.0001$ m. Figure 14 shows the velocity profiles of low Re (i.e. $Re = 0.5$) and different permeability values. The profiles in Fig. 14 are comparable with those in Fig. 13 even though the latter are for a high Reynolds number of $Re = 5$ in a wider channel with $h = 0.001$ m. The reason for these similar profiles is

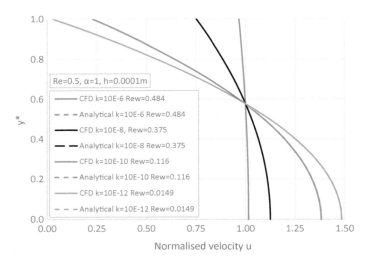

Fig. 14 Effects of permeability k on the velocity profiles for $Re = 0.5$, $h = 0.0001$ m, $\alpha = 1$; units of k (m^2)

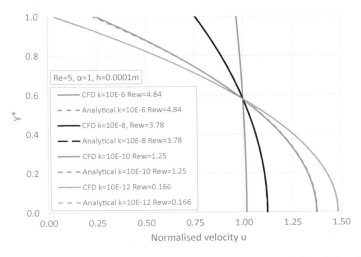

Fig. 15 Effects of permeability k on the velocity profiles for cases of $Re = 5$, $h = 0.0001$ m, $\alpha = 1$; units of k (m^2)

that the same ϕ value $\left(\phi = \sqrt{k}/(\alpha h)\right)$ is obtained for all cases shown in these two figures even they have different Re and h values, which theoretically would lead to the same profile (see Eq. 7). On the other hand, the effect on the velocity profiles of Re ranging from 0.5 to 5 is not significant. This can be confirmed by comparing the velocity profiles for a higher $Re(Re = 5)$, shown in Fig. 15, to those of a lower $Re(Re = 0.5)$, shown in Fig. 14, with all other parameters unchanged. Nevertheless, more discussion about the effect of Re on the velocity profiles can be found in Sect. 3.6.

3.5 Effect of Permeability k on the Pressure Drop Per Unit Length

Figures 16 and 17 show the variation of pressure drop per unit length $-\Delta p/L$ for the channel with permeable walls and varying permeability, k. As the permeability k decreases, there is an increase in pressure drop per unit length, $-\Delta p/L$, for all the cases shown. This again can be explained by the fact that lower permeability results in lower slip velocity at the interface and hence a higher wall shear stress and skin friction, and the increase in skin friction consequently leads to an increase in the pressure drop, as discussed above.

It is also noticeable in Figs. 16 and 17 that, when the permeability is less than a certain threshold (and any further decrease in k will not significantly increase the pressure drop per unit length along the channel), the pressure drop per unit length, $-\Delta p/L$, tends to approach that of the corresponding no-slip boundary case shown in

Fig. 16 Effect of permeability k on the pressure drop per unit, $-\Delta p/L$, in the channel $h = 0.001$ m, $\alpha = 1$

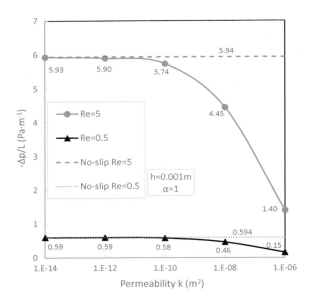

Fig. 17 Effect of permeability k on the pressure drop per unit, $-\Delta p/L$, in the channel $h = 0.0001$ m, $\alpha = 1$

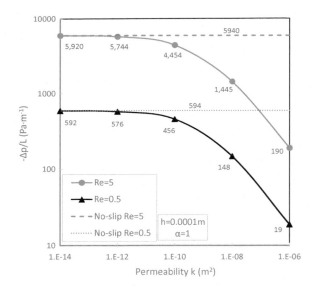

Table 1. For example, in Fig. 16, $-\Delta p/L$ reaches 0.59 Pa m^{-1} when k is 10^{-12} m^2 ($Re = 0.5$ and $h = 0.001$ m), which is approximately the same as the value of $-\Delta p/L$ for $k = 10^{-14}$ m^2 for the corresponding no-slip boundary case in Table 1. This observation is consistent with the predicted velocity profiles shown in Fig. 12, where the slip velocity is very small when $k = 10^{-12}$ m^2 and the shear stress at the interface is very close to that of the no-slip boundary case.

Figures 16 and 17 show the variation of pressure drop per unit length $-\Delta p/L$ for the channel with permeable walls and varying permeability k. As the permeability k decreases, there is an increase in pressure drop per unit length $-\Delta p/L$, for all the cases shown. This again can be explained by the fact that lower permeability results in lower slip velocity at the interface and hence, a higher wall shear stress and skin friction, and the increase in skin friction consequently leads to an increase in the pressure drop, as discussed above.

It is also noticeable in Figs. 16 and 17 that, when the permeability is less than a certain threshold (and any further decrease in k will not significantly increase the pressure drop per unit length along the channel), the pressure drop per unit length $-\Delta p/L$ tends to approach that of the corresponding no-slip boundary case shown in Table 1. For example, in Fig. 16, $-\Delta p/L$ reaches 0.59 Pa m^{-1} when k is 10^{-12} m^2 ($Re = 0.5$ and $h = 0.001$ m), which is approximately the same as the value of $-\Delta p/L$ for $k = 10^{-14}$ m^2 for the corresponding no-slip boundary case in Table 1. This observation is consistent with the predicted velocity profiles shown in Fig. 12, where the slip velocity is very small when $k = 10^{-12}$ m^2 and the shear stress at the interface is very close to that of the no-slip boundary case.

3.6 Effect of Re Number on the Velocity Profiles and Pressure Drop Per Unit Length

As discussed in Sect. 3.4, the effect of Re on the velocity profiles is insignificant. This is confirmed by a further parametric study shown in Fig. 18, which shows the normalised velocity profiles of the flows with different Re ranging from 0.5 to 7, with other parameters constant at $\alpha = 1$, $h = 0.001$ and $k = 10^{-8}$ m^2 as shown in Fig. 12. It can be seen that the velocity profiles of flows with different Re are very similar, although there are small differences at the interface surface and in the centre line zone. As shown in the figure, the Rew numbers are different in these cases; for example, the Rew number for the $Re = 0.5$ case is 0.116, whereas for the $Re = 5$ case it is 1.25. The difference in the Rew values can be attributed to the different $u_{interface}$ values as calculated by Eq. (7). Nevertheless, these differences in the normalised $u_{interface}$ values are very small for the Re range studied. The same conclusion can be drawn for the other cases, suggesting that the effect of Re on the normalised velocity profiles is negligible for cases that have the same slip coefficient, channel height and wall permeability.

The effect of Re on the pressure drops per unit length is much more pronounced as demonstrated in Figs. 10, 11, 16 and 17. The pressure drop per unit length for high Re cases ($Re = 5$) are an order of magnitude higher than those for the low $Re(Re = 0.5)$ cases.

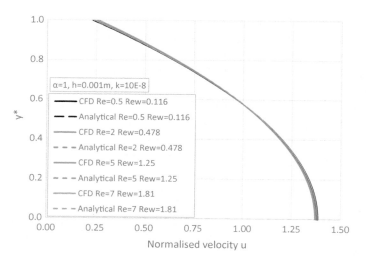

Fig. 18 The normalised velocity u at the interface surface with different Re numbers

3.7 Comparison with Saffman's Boundary Condition

At this point, it is interesting to compare the slip velocity predicted by the model (Eq. 7) and by Saffman's boundary condition (Eq. 3). The values of $\partial u/\partial n$ derived by CFD at the fluid-porous medium interface are used in Eq. (3) to calculate $u_{\text{interface}}$, which is then compared with the analytical values of $u_{\text{interface}}$ calculated by Eq. (7). Table 2 lists the normalised velocity on the interface surface calculated by both methods for different cases. Note that $O(k)$ is neglected in Eq. (3) following Nield (2009) and Mohais et al. (2011b). The absolute values of the differences in

Table 2 Comparison of normalised velocity u at interface surface by using different equations

Re	h (m)	α	k (m^2)	Normalised velocity u at interface by Eq. (7)	Normalised velocity u at interface by Eq. (3) ($O(K)$ neglected)	Difference (%)
0.5	0.001	1	10^{-8}	0.231	0.227	−1.69
			10^{-6}	0.751	0.739	−1.52
		0.1	10^{-8}	0.0698	0.0688	−1.42
		4	10^{-8}	0.751	0.739	−1.58
	0.0001	0.1	10^{-8}	0.0968	0.0956	−1.26
5	0.001	4	10^{-8}	0.0704	0.0683	−3.11
		0.1	10^{-8}	0.755	0.721	−4.58
7	0.001	0.1	10^{-8}	0.757	0.713	−5.83

the normalised u velocity values predicted by the two methods are less than 2% when $Re = 0.5$, but increase as Re increases. When $Re = 7$, the absolute value of the difference is as high as 5.83% for $h = 0.001$ m, $\alpha = 0.1$ and $k = 10^{-8}$ m^{-2}.

4 Conclusions

The flows in a 2D channel with permeable and impermeable walls were studied using CFD techniques.

The velocity profiles based on CFD were compared with analytical solutions in the literature (Potter et al. 2016) for impermeable wall conditions (no-slip boundary condition) and analytical solutions developed by Mohais et al. (2011a, b) for permeable wall conditions (slip boundary conditions). There is good agreement between the CFD results and the analytical solutions with an observed maximum difference of less than 1% for all cases investigated.

The effects of key parameters—Re, slip coefficient α, permeability k and channel height h—for the case of permeable walls were investigated using a CFD model. The results show that, in general, when the slip coefficient α decreases from 4 to 0.1, the slip velocity at the channel-porous medium interface increases, but the velocity at the channel centre decreases, leading to a 'flatter' velocity profile, similar to that of a plug flow. This 'flatter' velocity profiles are a result of lower shear stress at the interface and lower skin friction. The lower skin friction leads to lower pressure drops per unit length along the channel.

When the wall permeability k decreases, the slip velocity at the channel-porous medium interface also decreases which leads to an increase in shear stress, and hence the increase in pressure drop along the channel. As demonstrated in the results, when the permeability is less than a certain threshold, the slip velocity at the interface becomes very small and approaches to zero, and therefore the shear stress and pressure drop values approach those of the no-slip boundary cases.

The effect of Re on the velocity profiles is small for the range of Re values (0.5–7) investigated in this chapter. However, the effect of Re on the pressure drops per unit length in the channel is much more pronounced. The pressure drops per unit length for high Re cases ($Re = 5$) are an order of magnitude higher than those for low Re($Re = 0.5$) cases.

We are developing a CFD model that includes both the channel and the porous walls to further validate the velocity solutions of Mohais et al. (2011a, b) at the channel-porous medium interfaces. The CFD model will be used to predict the pressure drops per unit length in the porous walls that is not included in the current chapter.

Dedication The authors dedicate this chapter to their former colleague and friend, Dr. Rosemarie Mohais. A short but courageous battle with cancer ended Rosemarie's life at a tragically young age in March 2014. This chapter is an extension of the work that Rosemarie conducted during her membership with our research group at the University of Adelaide and it is with gratitude and sadness that we acknowledge her contribution.

References

Beavers GS, Joseph DD (1967) Boundary conditions at a naturally permeable wall. J Fluid Mech 30:197–207

Beavers G, Sparrow E, Masha B (1974) Boundary condition at a porous surface which bounds a fluid flow. AIChE J 20:596–597

Berman AS (1953) Laminar flow in channels with porous walls. J Appl Phys 24:1232–1235

Christopher H, Armstead H (1978) Geothermal energy: its past, present, and future contributions to the energy needs of man. E and FN Spon, London

Granger J, Dodds J, Midoux N (1989) Laminar flow in channels with porous walls. Chem Eng J 42:193–204

Herschlag G, Liu J-G, Layton AT (2015) An exact solution for stokes flow in a channel with arbitrarily large wall permeability. SIAM J Appl Math 75:2246–2267

Jones I (1973) Low Reynolds number flow past a porous spherical shell. In: Mathematical proceedings of the Cambridge philosophical society, vol 1. Cambridge Univ Press, pp 231–238

Larson R, Higdon J (1986) Microscopic flow near the surface of two-dimensional porous media. Part 1. Axial flow. J Fluid Mech 166:449–472

Liu Q, Prosperetti A (2011) Pressure-driven flow in a channel with porous walls. J Fluid Mech 679:77–100

Mikelic A, Jäger W (2000) On the interface boundary condition of Beavers, Joseph, and Saffman. SIAM J Appl Math 60:1111–1127

Mohais R, Xu CS, Dowd PA (2011a) An analytical model of coupled fluid flow and heat transfer through a fracture with permeable walls in an EGS. In: The proceedings of the 2011 Australian geothermal energy conference, Melbourne, Australia, 2011, pp 175–179

Mohais R, Xu CS, Dowd PA (2011b) Fluid flow and heat transfer within a single horizontal fracture in an enhanced geothermal system. J Heat Transf 133:112603

Natarajan N, Kumar GS (2012) Evolution of fracture permeability due to co-colloidal bacterial transport in a coupled fracture-skin-matrix system. Geosci Frontiers 3:503–514

Nield D (2009) The Beavers-Joseph boundary condition and related matters: a historical and critical note. Transp Porous Med 78:537–540

Phillips OM (1991) Flow and reactions in permeable rocks. Cambridge University Press

Potter MC, Wiggert DC, Ramadan BH (2016) Mechanics of fluids. Cengage Learning

Saffman PG (1971) On the boundary condition at the surface of a porous medium. Stud Appl Math 50:93–101

Sahraoui M, Kaviany M (1992) Slip and no-slip velocity boundary-conditions at interface of porous, plain media. Int J Heat Mass Transf 35:927–943

Terrill RM, Shrestha GM (1965) Laminar flow through parallel and uniformly porous walls of different permeability. Z Angew Math Phys ZAMP 16:470–482

Tian ZF, Mohais R, Xu CS, Zhu X (2012) CFD modelling of the velocity profile within a Single Horizontal Fracture in an Enhanced Geothermal System. In: Proceedings of the 18th Australasian fluid mechanics conference, Launceston, Australia, 2012, Paper no 225

Xu CS, Dowd PA, Tian ZF (2015) A simplified coupled hydro-thermal model for enhanced geothermal systems. Appl Energ 140:135–145

A Comparative Analysis of Mixed Finite Element and Conventional Finite Element Methods for One-Dimensional Steady Heterogeneous Darcy Flow

Debasmita Misra and John L. Nieber

1 Introduction

For Darcian flow in porous media, the specific discharge, q is governed by the Darcy law, which states that the volume of water flowing through a unit cross-sectional area normal to the direction of flow per unit time is mathematically represented as

$$q = -K\nabla\phi, \qquad (1)$$

where K is the hydraulic conductivity of the fluid in the porous media, $\phi = (h + z)$ is the piezometric head with h as the pressure head, and z is the elevation head of the fluid in the media and $\nabla = \left(\frac{\partial}{\partial x}, \frac{\partial}{\partial y}, \frac{\partial}{\partial z}\right)$. For steady-state and incompressible flow, the law of conservation of mass is expressed as a dot product of ∇ and q, as

$$\nabla \cdot q = f, \qquad (2)$$

where f represents sources or sinks in the flow system.

To successfully model contaminant transport in soil and groundwater, it is necessary to have an accurate representation of q. Although analytical solutions are

---

D. Misra (✉)
Department of Mining and Geological Engineering, College of Engineering and Mines, University of Alaska Fairbanks, P.O. Box 755800, Fairbanks, AK 99775-5800, USA
e-mail: dmisra@alaska.edu

J. L. Nieber
Department of Biosystems and Agricultural Engineering, 203 Biosystems and Agricultural Engineering Building, University of Minnesota, 1390 Eckles Avenue, St. Paul, MN 55108-6005, USA
e-mail: nieber@umn.edu

available for the set of Eqs. 1 and 2, these are limited to highly simplified flow conditions. Hence, for complex flow conditions, the solutions are obtained numerically using methods such as the finite difference or the finite element method.

Complex flow situations can occur in soils that are completely saturated with water when the porous media is highly heterogeneous or contains sources or sinks which leads to high flux gradients in the flow field. An example of this condition may occur in a situation such as a clay lens within an aquifer. It is especially important to obtain accurate flux profiles for these conditions.

Finite element methods are popular because they are capable of solving problems with complex flow geometry by deforming the mesh and updating the element matrix coefficients to accommodate the geometry of the solution region. It was applied initially to saturated groundwater flow problems by Javandel and Witherspoon (1968) and Zienkiewicz and Parekh (1970). At the same time, this technique had been introduced to subsurface flow problems in the petroleum engineering discipline by Price et al. (1968) who investigated steep front solutions to the linear convection–diffusion equation. A thorough discussion on the application of the finite element method in the fields of hydrology and petroleum engineering and the difference in the approaches to discretize the flow domain, types of basis functions, and methods used to integrate the residual differential equation has been provided by Huyakorn and Pinder (1983).

Allen et al. (1985) have commented that the task of simulating contaminant flows in porous media is computationally demanding because of the difficulty associated with computing accurate fluid fluxes. While using the Darcy law, one must differentiate heads or pressures to get fluxes, and this leads to at least two related mathematical problems. First, any pathologic behavior in pressure or head translates severely to the computed flux field. Second, standard numerical solutions of the flow equations commonly result in discrete approximations of the pressure or head. Differentiating these approximations to compute fluxes incurs a loss of accuracy that is typically one order in the spatial grid mesh. These difficulties can cause non-convergent approximations near fields of high flux gradients or can result in poor prediction of flux values, which is crucial in predicting contaminant transport.

The governing equation used in conventional analyses is given by

$$\nabla \cdot (K \nabla \phi) + f = 0 \qquad (3)$$

which is obtained by substituting Eq. 1 in 2. Numerical solutions of Eq. 3 are applied to a discretized flow domain to obtain ϕ. The Darcy flux is then computed by taking the derivative of the approximate pressure head across each element of the discretized flow domain. This flux distribution is continuous within the element but results in discontinuity at the element boundaries of the flow domain. This leads to a local and/or global violation of the conservation of mass. Many researchers who have modeled both linear and nonlinear groundwater flow problems have recognized such mass imbalance in the solution. Yeh (1981) found mass balance errors of 24–30% in a complex groundwater flow simulation where the Darcy flux was evaluated using the conventional technique. He proposed an alternate approach to compute the Darcy velocity (flux) at the node points. The essence of his approach is to use the same finite element method to integrate the weighted residuals of Darcy's law to distribute the elemental fluxes to their corresponding nodes after the computation of the pressure head. The application of his procedure ensured continuity of the velocity vector at the nodes and the boundaries. He found that this procedure reduced the mass balance errors from 24–30 to 2–9%.

The errors found by Yeh (1981) were caused by his treatment of the boundary conditions as shown by Lynch (1984). Some error might have been introduced by his improper treatment of the nonlinear form of the equation. Lynch (1984) instead included the natural boundary surface integrals in computing the pressure head. This resulted in a highly accurate global mass balance despite the discontinuity of the fluxes in the inner elements.

Other scientists have recognized the problem of mass imbalance in the finite element solution and hence inaccuracy in the computation of Darcy flux (Cordes and Kinzelback 1992; Kaluarachchi and Parker 1987; Srivastava and Yeh 1992; Van Genuchten 1983; and Yeh et al. 1993). These scientists have tried to circumvent the problem in many different ways. Van Genuchten (1983) used Hermite basis functions to overcome the discontinuity of the velocity field across interelement boundaries. His formulation led to continuous pressure gradients at the nodes, however, the solutions obtained showed oscillations in flux values. He had to use an extremely fine grid to overcome the problem of oscillations. Kaluarachchi and Parker (1987) have commented, *"Because the major mechanism of chemical transport through vadose and saturated zones is advection, accurate models for water flow under variably saturated conditions are a prerequisite for modeling solute movement"*. They have used numerical quadrature and influence coefficient methods in an attempt to obtain accuracy in their solution of pressure head and flux. The solutions obtained were accurate for homogeneous flow conditions but were less accurate for heterogeneous flow conditions. Cordes and Kinzelbach (1992) extended the method proposed by Pollock (1988) to compute the flux-conserving pathlines using the computed fluxes between the cells of a block-centered finite difference scheme to model saturated groundwater flow problems. Their method (conforming finite element method) yielded an exact balance of water mass and proved to be superior to traditional Lagrangian techniques used for flow simulations. Their method is an attractive choice for saturated groundwater flow problems in more than one dimensions, however, extension of the method to nonlinear flow problems is yet to be tested.

1.1 The Mixed Finite Elements

Independent and prior to the works of the above researchers, groundwater hydrologists and petroleum engineers had applied concepts developed in structural mechanics using the dual principles of minimum potential energy and minimum complementary energy by use of Hellinger–Reissner's variational principle to the groundwater flow problems. This class of approximation techniques is known by the generic name of *Mixed Finite Element Methods* which were originally introduced by Reissner (1950). The method has found increased interest in linear steady-state and transient groundwater flow problems due to its accuracy in simulating the Darcy flux (Meissner 1973; Segol et al. 1975). Oden and Carey (1983) comment that the mixed methods result naturally from finite element approximations of any variational boundary value problem with constraints. The method has circumvented the previously discussed problems, and there are many successful applications reported in recent literature. Allen et al. (1985) have perhaps provided the best summary of the concept of mixed finite elements (MFE) and its advantage in simulating groundwater flow. It reads: "... *The essential idea of the mixed method is that, by solving the second-order equation governing groundwater flow as a set of coupled first order equations in velocity and head, one can compute both fields explicitly without sacrificing accuracy in the velocity through differentiation...*".

The name **mixed finite elements** originates from the fact that two quantities are computed simultaneously, in this case the hydraulic head and the specific discharge (Darcy flux) in Eqs. 1 and 2. Different orders of approximation are used to define the two quantities in the numerical scheme. These orders of approximation are determined from the desired level of accuracy of the variable that is important for the problem under consideration. Ackerer et al. (1996) have commented: *The mixed finite element method is actually a family of methods from which one can pick a scheme of arbitrarily high accuracy.* To obtain higher accuracy in the velocity profile, a higher order approximation is used to describe the specific discharge in the finite element formulation. The method also ensures continuity of the specific discharge across element boundaries. Additionally, the method avoids the post-processing used to compute Darcy velocity either by conventional methods or by the method proposed by Yeh (1981). Carey and Oden (1986) comment that for flow problems that need higher order of accuracy in the hydraulic potential, the choice of using MFE methods is of little practical value. The conventional Galerkin finite element method is preferable in such cases. A detailed description of the MFE method as applied to potential flow problems in porous media has been provided by Russell and Wheeler (1983), Chavent and Roberts (1991) and Misra (1994) among others. For a detailed theoretical treatment of the MFE method, the readers are referred to Poceski (1992).

Applications of the method to linear subsurface flow problems (Allen et al. 1985; Sovich 1988; Chiang et al. 1989; Dougherty 1990; Kaasschieter 1990, 1995; Beckie et al. 1993; Durlofsky 1994; Mosé et al. 1994) and multiphase flow problems (Russell and Wheeler 1983; Chavent et al. 1987; Meyling et al. 1990; Mulder and Meyling 1991) have shown some increased interest. While the advantages of using the MFE method to compute accurate Darcy flux have been advocated by many, there is a contrary response by others (Cordes and Kinzelback 1992, 1996; Srivastava and Brusseau 1995). Cordes and Kinzelbach (1996) have explicitly commented: "... *a large amount of research has been devoted to the application of mixed finite elements (MFE) to groundwater flow problems. However, there is still a lack of understanding of the merits of the method, of its relation to alternative approaches, and of the effects of the geometry on the quality of the results*". We are in complete agreement with their observation. Additional studies are needed to compare MFE with the conventional finite element (CFE) schemes. A noteworthy study of this type is that by Durlofsky (1994) where an extensive comparison of the MFE and the control volume finite element method is given and the merits of the MFE method are discussed. Another comparison of MFE and CFE schemes is provided by Mosé et al. (1990) wherein they have compared the computed flux distribution in a three-dimensional steady-state saturated flow field using both methods. They have shown that the distribution of flux obtained from the MFE scheme provides better accuracy than those obtained from the CFE scheme.

Chavent and Roberts (1991) wonder why the idea of MFE methods has not been more frequently exploited. In their explanation they provide three reasons. The first two reasons relate to the computational efficiency of the method, which is beyond the scope of this discussion. The third reason is stated as: "...*the mathematical theory of mixed finite elements is quite involved, and much of the literature on mixed and mixed-hybrid finite elements gives a mathematically rigorous presentation of these elements or a treatment highly specialized for engineering applications which in either case is difficult to read by people who have not been specially trained*".

The objectives of this paper are (i) to provide a clear and practical conceptual description of the MFE methods, (ii) to compare solutions obtained using the MFE methods to those obtained using CFE methods with linear and higher order basis functions as applied to one-dimensional steady linear flow problems, and (iii) to study the properties associated with the results obtained using both the methods. For this paper, our interest lies in the computational accuracy of MFE methods and not in the computational efficiency issues.

2 Development of the MFE Method

The MFE method, as discussed above, provides an alternative scheme to improve the accuracy of predicted velocities. The underlying principle is to solve the conservation of mass equation and Darcy's law simultaneously using numerical techniques. Meissner (1973) used Hellinger–Reissner's variational principle to formulate a mixed finite element model for potential flow problems, which provided enhanced convergence of the solution.

Carey and Oden (1986) have presented the mixed finite element formulation for flows represented by a velocity-potential system. They have used the concept developed by them earlier to apply Lagrange multipliers for seeking a minimum of a functional which is constrained by another equation (see Carey and Oden 1983). In groundwater flow problems, either the conservation of mass equation or Darcy's law can be treated as the constraint equation. The choice is dependent upon the level of accuracy desired for either the hydraulic head or the Darcy flux.

2.1 The Governing Differential Equations

In the formulation of the flow problem for this paper, the interest is to attain higher order accuracy in flow velocity. Thus, the conservation of mass equation will be used as the constraint equation.

The residual equations for the MFE method for one-dimensional flow are as follows:

Darcy's Law:

$$\Re_1(\phi, q) = \frac{\partial \phi}{\partial x} + \frac{q}{K} = 0 \tag{4}$$

Conservation of Mass:

$$\Re_2(q) = -\frac{\partial q}{\partial x} + f(x) = 0 \tag{5}$$

2.2 Choice of Approximating Functions

The choice of the approximating functions for the variables ϕ and q is critical in achieving success in stability of the solution and thus accuracy. Carey and Oden (1986) refer to the discrete Babuska–Brezzi conditions for linear problems to achieve stability in the MFE solutions. Hartman (1986) and Kolar (1992) provide a simple description of the Babuska–Brezzi condition as, *"The Babuska-Brezzi condition is closely connected with Lagrange multiplier methods and mixed methods. A mixed method results when a differential equation is split into a system of two differential equations of lower order. Lagrange multipliers are introduced when side conditions must be eliminated. Common to both methods is that the number of unknown functions raises from 1 to 2 (ϕ and q) and that the variational problem becomes a saddle point problem."*

The choice of the interpolant for q in Eqs. 4 and 5, with the conservation of mass being the constraint equation, has an extra degree of smoothness more than the interpolant for ϕ. Raviart and Thomas (1977) have proposed piecewise polynomial trial spaces for two-dimensional second-order elliptic problems in MFE applications. These polynomial trial functions obtained from the Raviart–Thomas (R-T) space can be used to approximate q and ϕ in Eqs. 4 and 5 in a way that Babuska–Brezzi conditions are satisfied.

From a similar perspective of the discussion in the previous paragraph, the key to the MFE method as described in simple terms by Dougherty (1990) is the choice of the approximations for the hydraulic head (ϕ) to be in the same space as $\nabla \cdot q$. Thus, if ϕ is piecewise constant then $i \cdot q$ should be piecewise linear in x and piecewise constant in y and $j \cdot q$ must be piecewise linear in y and piecewise constant in x. This is the zero-order R-T space ($r = 0$ method). The first-order R-T space has a choice of piecewise bilinear approximation for ϕ and $i \cdot q$ is piecewise quadratic in x and piecewise linear in y with $j \cdot q$ being piecewise quadratic in y and piecewise linear in x ($r = 1$ method). With the above approximations, the method offers increased accuracy for velocity with the expense of some loss in accuracy in the hydraulic head.

In one-dimensional flow, the zero-order ($r = 0$ method) R-T MFE approach approximates q as piecewise linear functions and ϕ as piecewise constant over the flow domain. Thus, q has a C^0 continuity at element interfaces and ϕ has a C^{-1} continuity. These approximations satisfy the Babuska–Brezzi conditions. The first-order ($r = 1$ method) approach approximates q as a piecewise quadratic function and ϕ as piecewise linear thus exhibiting C^1 continuity for q and C^0 continuity for ϕ at element interfaces. For each case of approximating functions, the lower order of accuracy in approximation of ϕ usually results in a poorer simulation of the hydraulic head distribution in comparison to the conventional finite element method where a higher order interpolant is used. However, the velocity is predicted quite accurately with this approximation.

At this point one might wonder, "why not use the same degree of accuracy in approximation of the two variables"? In fact, this is the choice in the method proposed by Yeh (1981; henceforth to be known as Yeh's method) for interpolation of both ϕ and q, when linear functions are used for both the variables. The difference lies in the simultaneous solution for ϕ and q in the mixed finite element method as opposed to the Yeh's method, where q is computed using the solution of ϕ obtained from the combined flow equation. We state here without proof that a choice of the same degree in accuracy of approximation for q and ϕ for elliptic boundary value problems in MFE formulation leads to redundancy in the solution matrix and results in trivial solutions. The approximation has been used in non-elliptic partial differential equations where non-convergence of the solution was observed even in homogeneous porous media conditions (e.g., Kolar 1992). We reserve any further comments on the outcome of using same order of approximation for q and ϕ in MFE formulation because this approximation does not satisfy the Babuska–Brezzi conditions. This approximation is therefore not used in this study for the MFE method. In the following sections, we will separately develop the MFE formulation for the $r = 0$ and the $r = 1$ methods.

2.3 Zero-Order Raviart–Thomas Approach

Using the zero-order approximation for the variables, the residual equations can be written as

$$\Re_1(\tilde{\phi}, \tilde{q}) = \frac{d\tilde{\phi}}{dx} + \frac{\tilde{q}}{K} = \varepsilon_1 \neq 0 \tag{6}$$

$$\Re_2(\tilde{q}) = -\frac{d\tilde{q}}{dx} + f(x) = \varepsilon_2 \neq 0 \tag{7}$$

Application of the Galerkin method of weighted residuals to Eqs. 6 and 7 results in the following pair of integral equations 8 and 9:

$$G_{1k} = \int_0^X N_{1k} \Re_1(\tilde{\phi}, \tilde{q}) dx = 0 \tag{8}$$

$$G_{2k} = \int_0^X N_{2k} \Re_2(\tilde{q}) dx = 0, \tag{9}$$

where N_{1k} and N_{2k} are the weighting functions. These functions are required to be different for the success of the method. The constraint equation uses a lower order weighting function than the equation to be minimized (Allen et al. 1992; Ewing and Wheeler 1983). Since we use the conservation of mass equation as the constraint equation, Eqs. 8 and 9 are integrated with N_{1k} as a piecewise linear weighting function and N_{2k} a piecewise constant weighting function.

Upon formulation of the MFE-Galerkin method on an element basis with the application of the divergence theorem to the Darcy's weighted residual equation, the following relationships are obtained:

$$G_{1k}^{(e)} = -\int_{X_i}^{X_j} \tilde{\phi} \frac{dN_{1k}}{dx} dx + N_{1k} \tilde{\phi} \big|_{X_i}^{X_j} + \int_{X_i}^{X_j} N_{1k} \frac{\tilde{q}}{K} dx \tag{10}$$

$$G_{2k}^{(e)} = -\int_{X_i}^{X_j} N_{2k} \frac{d\tilde{q}}{dx} dx + \int_{X_i}^{X_j} N_{2k} f(x) dx, \tag{11}$$

where the approximations used for the Darcy flux and the hydraulic head are as follows:

$$\tilde{q} = N_i q_i + N_j q_j \tag{12}$$

$$\tilde{\phi} = \phi_e = \text{Const.} \tag{13}$$

The constant weighting function is 1. The divergence theorem can also be applied to the weighted residual integral of the conservation of mass equation. But there is no necessity for this since q is continuous at the element interfaces. However, the use of divergence theorem is essential for the weighted residual integral of Darcy's law because the derivative of ϕ does not exist at element interfaces, ϕ is represented over the element and not at the node.

There are three terms in the $G_{1k}^{(e)}$ expression. The first two terms resulted from using the divergence theorem. The last term, $\int_{X_i}^{X_j} N_{1k} \frac{\tilde{q}}{K} dx$ is of special significance in the MFE formulation. Normally, the integration is performed by substituting Eq. 12 for $\tilde{q}$ (distributed q approach). Russell and Wheeler (1983) showed that a one-point Gauss quadrature to numerically integrate terms in Eq. 10 results in solution matrices that are exactly the same as the ones obtained by using a block-centered finite difference scheme. The same solution matrices are obtained using a lumped approach to integrate the third integral. This lumping formulation leads to

$$q_k^{(e)} = \frac{\int_e N_{1k} \tilde{q} dx}{\int_e N_{1k} dx}, \tag{14}$$

where $q_k^{(e)}$ is the average q over the element. It will be shown that this concept is easily extended to the MFE formulation using the first-order R-T approach.

Integration of Eqs. 10 and 11 results in a system of algebraic equations. We can develop two separate systems of algebraic equations using either the distributed q or the lumped q approach in integrating the term $\int_{X_i}^{X_j} N_{1k} \frac{\tilde{q}}{K} dx$ (Henceforth these treatments of q will be referred in the paper without any reference to the above equations as: q is lumped or distributed).

The second term in $G_{1k}^{(e)}$ contributes to the natural boundary condition. It is important to note here that this boundary condition is not a flux boundary condition as in the CFE methods. The interelement requirement discussed by Segerlind (1984, p. 32) is used to ensure that the residual becomes zero. The interelement requirement ensures that the contribution of the second term from two neighboring elements at their interface is forced to zero. A justification of using the interelement requirement for the MFE formulation is provided by Misra (1994, Appendix A). This treatment of the second term can contribute to significant error in prediction of the hydraulic head because the natural boundary conditions are of Dirichlet type, which are forced to zero in the inner nodes by the imposition of the interelement requirement.

The final system of algebraic equations obtained after integration of Eqs. 10 and 11 reads

$$G_{1k}^{(e)} = [\underline{\underline{B}}\,\underline{\phi} + \underline{\underline{C}}\,\underline{q} + \underline{\Phi}]^{(e)} \tag{15}$$

$$G_{2k}^{(e)} = [\underline{\underline{D}}\,\underline{q} + \underline{F}]^{(e)}, \tag{16}$$

where the coefficient matrices and vectors are described as follows:

$$\begin{aligned}
\underline{\underline{B}} &= \begin{bmatrix} +1 \\ -1 \end{bmatrix} \\
\underline{\underline{D}} &= \underline{\underline{B}}^{\mathrm{T}} \\
\underline{\Phi} &= \begin{Bmatrix} -\phi_i \\ \phi_j \end{Bmatrix} \\
\underline{F} &= f^{(e)} L^{(e)} \{1\} \\
\underline{\underline{C}} &= \frac{L^{(e)}}{6K^{(e)}} \begin{bmatrix} 2 & 1 \\ 1 & 2 \end{bmatrix} : \text{Distributed } q \\
\underline{\underline{C}} &= \frac{L^{(e)}}{2K^{(e)}} \begin{bmatrix} 1 & 0 \\ 0 & 1 \end{bmatrix} : \text{Lumped } q
\end{aligned} \tag{17}$$

where $L^{(e)}$ is the element length. The global matrices are formed by adding the contribution from each element matrix within the domain. The global form of Eqs. 15 and 16 is solved simultaneously for q and ϕ. The matrices obtained from Eqs. 15 and 16 are merged to form a single global matrix which is shown below:

$$\begin{bmatrix} \underline{\underline{C}} & \underline{\underline{B}} \\ \underline{\underline{D}} & \underline{\underline{0}} \end{bmatrix} \begin{Bmatrix} q \\ \phi \end{Bmatrix} = \begin{Bmatrix} \Phi \\ F \end{Bmatrix}, \tag{18}$$

where $\underline{\underline{0}}$ is a square matrix containing zeros. Ewing and Wheeler (1983) and Dougherty (1990) suggested using block Gauss elimination to convert Eq. 18 to the following form:

$$\begin{bmatrix} \underline{\underline{C}} & \underline{\underline{B}} \\ \underline{\underline{0}} & \underline{\underline{A}} \end{bmatrix} \begin{Bmatrix} q \\ \phi \end{Bmatrix} = \begin{Bmatrix} \Phi \\ R \end{Bmatrix}, \tag{19}$$

where matrix A and vector R are described by

$$\underline{\underline{A}} = -\underline{\underline{D}} \cdot \underline{\underline{C}}^{-1} \cdot \underline{\underline{B}}$$
$$\underline{R} = \underline{F} - \underline{\underline{D}} \cdot \underline{\underline{C}}^{-1} \cdot \underline{\Phi} \qquad (20)$$

The lower block of Eqs. 19 is solved for ϕ after which q is determined by backsubstitution. Here A is a symmetric, positive definite matrix. It is dense when q is distributed but is sparse (tridiagonal) when q is lumped. The major advantage discussed by Dougherty (1990) in performing the above conversion is in using preconditioned conjugate gradient (PCG) in the solution of ϕ. This approach reduces the number of equations to be solved, and is effective in terms of the amount of computational effort required for a given level of accuracy. Additional discussion on the use of PCG method is beyond the scope of this paper.

2.4 First-Order Raviart–Thomas Approach

The use of first-order R-T approach is limited (Allen et al. 1985; Ewing and Wheeler 1983). While Ewing and Wheeler (1983) discuss the computational aspects of this approach, Allen et al. (1985) have applied this approach to compute subsurface flow velocities. From both papers, it is obvious that this approach does not provide any greater accuracy in the solutions as compared to the $r = 0$ method, while at the same time, there is an increase in the computation costs.

The difference between $r = 0$ and $r = 1$ methods is the higher order of interpolants used as basis functions in the latter case. In this method, the basis function used for q is piecewise quadratic, while for ϕ the interpolant is piecewise linear. These interpolants are expressed as

$$\tilde{q} = N_i q_i + N_j q_j + N_k q_k \qquad (21)$$

$$\tilde{\phi} = N_i \phi_i + N_k \phi_k, \qquad (22)$$

where the element is bounded by the nodes i and k. The subscript j refers to a node at the center of the element. The MFE-Galerkin formulation with this approach is exactly the same as shown in Eqs. 10 and 11. The weighting function N_{1k} is quadratic and N_{2k} is linear. The quadratic basis functions are of the following form:

$$\begin{aligned} N_i &= \left(1 - \frac{2s}{L}\right)\left(1 - \frac{s}{L}\right) \\ N_j &= \frac{4s}{L}\left(1 - \frac{s}{L}\right) \\ N_k &= \frac{s}{L}\left(\frac{2s}{L} - 1\right) \\ s &= x - X_i \end{aligned} \qquad (23)$$

The description for the $r = 0$ approach presented in the previous section also applies in its entirety to $r = 1$ method. The final form of the matrix for simultaneous solution of q and ϕ is the same as Eq. 18 except the elements of the matrices and the vectors are different as shown in the following:

$$\underline{\underline{B}} = \begin{bmatrix} \frac{5}{6} & \frac{1}{6} \\ -\frac{2}{3} & \frac{2}{3} \\ -\frac{1}{6} & -\frac{5}{6} \end{bmatrix}$$

$$\underline{\underline{D}} = \underline{\underline{B}}^T$$

$$\underline{\Phi} = \begin{Bmatrix} -\phi_i \\ 0 \\ \phi_k \end{Bmatrix}$$

$$\underline{F} = \frac{f^{(e)} L^{(e)}}{2} \begin{Bmatrix} 1 \\ 1 \end{Bmatrix} \qquad (24)$$

$$\underline{\underline{C}} = \frac{L^{(e)}}{K^{(e)}} \begin{bmatrix} \frac{2}{15} & \frac{1}{15} & -\frac{1}{30} \\ \frac{1}{15} & \frac{8}{15} & \frac{1}{15} \\ -\frac{1}{30} & \frac{1}{15} & \frac{2}{15} \end{bmatrix} : \text{Distributed } q$$

$$\underline{\underline{C}} = \frac{L^{(e)}}{K^{(e)}} \begin{bmatrix} \frac{1}{6} & 0 & 0 \\ 0 & \frac{2}{3} & 0 \\ 0 & 0 & \frac{1}{6} \end{bmatrix} : \text{Lumped } q$$

Similar to the $r = 0$ scheme, the block Gauss elimination method can also be used in this case to solve the final system of equations.

3 Brief Overview of CFE Method

The residual form of Eq. 3 for one-dimensional flow problems, as shown below, is the starting point for the mathematical formulation of finite elements by application of the method of weighted residual.

$$\Re(\phi) = \frac{\partial}{\partial x}\left[K \frac{\partial \phi}{\partial x}\right] - f(x) \qquad (25)$$

With an exact mathematical description of ϕ on the right-hand side, $\Re(\phi)$ in Eq. 25 should be equal to zero. The weighted residual method requires the substitution of a trial or basis function for the hydraulic head ($\tilde{\phi}$) which may not exactly satisfy the governing differential equation and results in an error, as shown by

$$\Re(\tilde{\phi}) = \frac{\partial}{\partial x}\left[K\frac{\partial \tilde{\phi}}{\partial x}\right] - f(x) = \varepsilon(x) \neq 0 \qquad (26)$$

The basis functions that will be used in this paper are the linear $\left(\tilde{\phi} = N_i\phi_i + N_j\phi_j\right)$ [N_i and N_j are the shape functions which have been defined by Segerlind (1984)] and the Hermite cubic functions $\left(\tilde{\phi} = H_{0i}\phi_i + H_{0j}\phi_j + H_{1i}\frac{d\phi_i}{dx} + H_{1j}\frac{d\phi_j}{dx}\right)$ [These shape functions are defined appropriately by Lapidus and Pinder (1982) and van Genuchten (1982, 1983)]. The linear functions provide high computational efficiency, exhibit C^0 continuity at element interfaces, and are simple to use. The nodal values of ϕ that are obtained from the solution using the linear basis functions will be postprocessed in order to obtain the nodal value of the specific discharge (Yeh's method). The Hermite polynomial basis function is considered because of its unique property that these functions interpolate the coefficient function ϕ as well as its derivative between node points. Thus, the gradient of ϕ obtained can be inserted into Eq. 1 directly to obtain the specific discharge at the nodes. The Hermite cubic functions exhibit C^1 continuity.

The method of weighted residuals requires that the weighted average of the spatially dependent residuals generated due to the approximation of ϕ should be equal to zero. Mathematically, this is

$$\int_0^X W_i \Re(\tilde{\phi}) dx = 0, \qquad (27)$$

where $W_i = W_i(x)$ are the weighting functions associated with each unknown coefficient in the approximate solution. We shall use the Galerkin method of weighted residuals to formulate the finite element solution matrix from Eq. 27 in order to maintain uniformity in comparing the results of MFE and CFE schemes.

Integration of Eq. 27 over a single element and application of the divergence theorem (e.g., see Becker et al. 1981) to any second-order differential terms in the residual expression results in the following:

$$G_k^{(e)} = -\int_{X_i}^{X_j} K\frac{dN_k}{dx}\frac{d\tilde{\phi}}{dx} dx + N_k K \frac{d\tilde{\phi}}{dx}\bigg|_{X_i}^{X_j} - \int_{X_i}^{X_j} N_k f(x) dx \qquad (28)$$

Substitution of the linear basis function in place of the variable $\tilde{\phi}$ and as a result of integration of the three expressions in Eq. 28, the following matrix form of the equation for a single element can be obtained.

$$\left\{\begin{array}{c} G_i \\ G_j \end{array}\right\}^{(e)} = \frac{K^{(e)}}{L^{(e)}} \begin{bmatrix} -1 & +1 \\ +1 & -1 \end{bmatrix} \left\{\begin{array}{c} \phi_i \\ \phi_j \end{array}\right\} - f^{(e)} L^{(e)} \left\{\begin{array}{c} 1 \\ 1 \end{array}\right\} + \left\{\begin{array}{c} -K\frac{d\tilde{\phi}}{dx}\big|_{X_i} \\ K\frac{d\tilde{\phi}}{dx}\big|_{X_j} \end{array}\right\} \qquad (29)$$

Equation 29 can be rewritten in a different form:

$$G^{(e)} = [\underline{\underline{B}}\underline{\phi} - \underline{F} + \underline{f_q}]^{(e)}, \qquad (30)$$

where $\underline{\underline{B}}$ is called the stiffness matrix, $\underline{F}$ is the force vector, and $\underline{f_q}$ is the element contribution to the natural boundary condition.

Equation 30 can be solved for nodal values of ϕ. The interest of this research is not in solution of the values of hydraulic head. Rather the interest is focused on predicting accurate velocities in the porous media. Conventionally, the specific discharge is computed numerically by taking the derivative of the approximate hydraulic head in Eq. 1. This leads to an accuracy in the prediction of q which is one order less than the accuracy used for predicting the values of the hydraulic head at the nodes. This can lead to significant local mass balance errors. Yeh's method provides an alternative to obtain nodal values of the specific discharge that are of the same order of interpolation as the hydraulic head. Application of the Galerkin finite element method to the residual form of Eq. 1 results in the following equation for an element:

$$G^{(e)} = [\underline{\underline{C}}\underline{q} + \underline{H}]^{(e)}, \qquad (31)$$

where the coefficient matrix C and the vector H are given by

$$\begin{aligned} \underline{\underline{C}}^{(e)} &= L^{(e)} \begin{bmatrix} \frac{1}{3} & \frac{1}{6} \\ \frac{1}{6} & \frac{1}{3} \end{bmatrix} \\ \underline{H}^{(e)} &= \frac{q^{(e)} L^{(e)}}{2} \left\{\begin{array}{c} 1 \\ 1 \end{array}\right\} \end{aligned} \qquad (32)$$

Hermite polynomials are not as popular as the linear bases because of the increased computational effort required to generate the coefficient matrices. In linear polynomial approximation, the continuity of the derivative did not exist at the interface, which resulted in the flux being discontinuous at the interfaces. Thus, the need to use Yeh's method arose. However, with Hermite polynomials, we can get continuous pressure gradients at element boundaries and hence continuity in the Darcy flux if the conductivity is continuous at the boundary.

With Hermite cubic approximation, there are two unknowns at each node of the element, viz., ϕ and $\frac{\partial \phi}{\partial x}$. The Galerkin finite element method is used in the same manner as in the case of the linear interpolation function in the preceding section. However, the weighted residual integration shown in Eq. 27 is performed twice

A Comparative Analysis of Mixed Finite Element ...

with different sets of weighting functions because we have two unknowns at each node now. Hence,

$$G_{Ik} = \int_{X_i}^{X_j} H_{0k} \Re(\tilde{\phi}) dx$$

$$G_{IIk} = \int_{X_i}^{X_j} H_{1k} \Re(\tilde{\phi}) dx$$
(33)

Integration on an element basis and using the divergence theorem the following form of Eq. 33 is obtained:

$$G_{Ik}^{(e)} = -\int_{X_i}^{X_j} K \frac{dH_{0k}}{dx} \frac{d\tilde{\phi}}{dx} dx + H_{0k} K \frac{d\tilde{\phi}}{dx} \bigg|_{X_i}^{X_j} - \int_{X_i}^{X_j} H_{0k} f(x) dx$$

$$G_{IIk}^{(e)} = -\int_{X_i}^{X_j} K \frac{dH_{1k}}{dx} \frac{d\tilde{\phi}}{dx} dx + H_{1k} K \frac{d\tilde{\phi}}{dx} \bigg|_{X_i}^{X_j} - \int_{X_i}^{X_j} H_{1k} f(x) dx$$
(34)

It is not mandatory to apply the divergence theorem in Hermite cubic interpolation since the derivatives of ϕ have C^1 continuity between elements. However, it is advantageous because the Neumann boundary condition can be incorporated directly into the set of integral equations (see Lapidus and Pinder 1982). The integration of Eq. 34 is described in detail by Lapidus and Pinder (1982) and Huyakorn and Pinder (1983). Also, for a good description on the computation and incorporation of the Hermite polynomials, the readers are referred to Prenter (1975). Without going into the intermediate steps involved in deriving the final system of algebraic equations from Eq. 34, we present the outcome of the integration below:

$$\underline{G}^{(e)} = -\underline{\underline{B}}^{(e)} \left\{ \begin{array}{c} \phi \\ \frac{d\phi}{dx} \end{array} \right\}^{(e)} - \underline{F}^{(e)} + \underline{f_q}^{(e)},$$
(35)

where

$$\underline{\underline{B}}^{(e)} = -K^{(e)} \begin{bmatrix} \int_e H'_{0i} H'_{0i} dx & \int_e H'_{0i} H'_{0j} dx & \int_e H'_{0i} H'_{1i} dx & \int_e H'_{0i} H'_{1j} dx \\ \int_e H'_{0j} H'_{0i} dx & \int_e H'_{0j} H'_{0j} dx & \int_e H'_{0j} H'_{1i} dx & \int_e H'_{0j} H'_{1j} dx \\ \int_e H'_{1i} H'_{0i} dx & \int_e H'_{1i} H'_{0j} dx & \int_e H'_{1i} H'_{1i} dx & \int_e H'_{1i} H'_{1j} dx \\ \int_e H'_{1j} H'_{0i} dx & \int_e H'_{1j} H'_{0j} dx & \int_e H'_{1j} H'_{1i} dx & \int_e H'_{1j} H'_{1j} dx \end{bmatrix}$$
(36)

$$\underline{F}^{(e)} = \begin{Bmatrix} \int_e H_{0i} f(x) dx \\ \int_e H_{0j} f(x) dx \\ \int_e H_{1i} f(x) dx \\ \int_e H_{1j} f(x) dx \end{Bmatrix} \tag{37}$$

$$\underline{f_q}^{(e)} = \begin{Bmatrix} -K\frac{d\bar{\phi}}{dx}\Big|_{X_i} \\ K\frac{d\bar{\phi}}{dx}\Big|_{X_j} \\ -K\frac{d\bar{\phi}}{dx}\Big|_{X_i} \\ K\frac{d\bar{\phi}}{dx}\Big|_{X_j} \end{Bmatrix} \tag{38}$$

The terms H'_{0k} and H'_{1k} ($k = i, j$) are the derivatives of H_{0k} and H_{1k} with respect to x.

4 Example Application

The numerical simulations using MFE and CFE methods described above were carried out for the case of a flow in horizontal, confined aquifer of 10^5 m length (L). The boundary conditions used for the simulation are $\phi(x = 0) = 10$ m and $\phi(x = L) = 2$ m. Both homogeneous and heterogeneous aquifer properties have been tested during the simulation. For homogeneous condition, the value of the saturated hydraulic conductivity used was 1.0 m/day. Two types of heterogeneous conditions have been treated to assess the predictability of the different numerical schemes described before in nonhomogeneous subsurface flow processes. One of the conditions is a heterogeneous flow field where the saturated hydraulic conductivity is 1.0 m/day at $x = 0$ and decreases exponentially as described by the following equation:

$$K = \exp\left(-6.0\frac{x}{L}\right) \tag{39}$$

The other heterogeneous field considered was a randomly distributed saturated hydraulic conductivity derived from a parent log-normal distribution given by

$$K = \exp\left[2.3\left(\mu_{\ln(K)} + P_{\ln}\sigma_{\ln(K)}\right)\right], \tag{40}$$

where $\mu_{\ln(K)}$ is the mean of the natural logarithms of K, and $\sigma_{\ln(K)}$ is the standard deviation of $\ln(K)$. $P_{\ln}$ is a random number obtained from the standardized normal

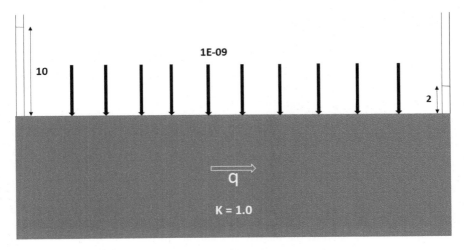

Fig. 1 Schematic representation of steady-state linear flow problem

distribution. A value of 0.0 was chosen for $\mu_{\ln(K)}$ and $\sigma_{\ln(K)}$ was assigned a value of 2.0. Simulations have been carried out for the example flow problem without any sources/sinks, with a single line source and with point sources. A schematic diagram showing the steady-state flow conditions is presented in Fig. 1.

5 Results and Discussion

In presenting the results for the steady-state flow simulations, we divide the results obtained using a uniform source/sink (f = constant) over the entire flow domain and those obtained using a nonuniform source/sink such as point sources and/or sinks. The acronyms used for the different numerical schemes are described in Table 1.

Table 1 Acronyms used for the different numerical schemes

Notation	Description
CFEL	Conventional finite element method with linear basis function
CFEH	Conventional finite element method with Hermite polynomial basis function
R0	Mixed finite element method with zero-order R-T approximation
R1	Mixed finite element method with first-order R-T approximation
QL	Flux lumped during residual integration in MFE formulation
QD	Flux distributed during residual integration in MFE formulation

5.1 Uniform Source/Sink

Each of the six numerical schemes (viz., CFEL, CFEH, R0QL, R0QD, R1QL, and R1QD) produced exact solution for hydraulic head and flux for the case with a *homogeneous* hydraulic conductivity field without any sources/sinks ($f = 0$) as compared to the analytical solution shown below.

$$\phi(x) = -\frac{f}{2K}(x^2 - Lx) + \frac{\phi(L) - \phi(0)}{L}x + \phi(0)$$
$$q(x) = fx - K\frac{\phi(L) - \phi(0)}{L} - \frac{fL}{2} \quad (41)$$

However, with the introduction of a source of constant strength throughout the length of a homogeneous aquifer, the distribution of pressure was inaccurately computed (as compared to the analytical solution presented in Eq. 41) when the MFE method was used. For example, an error of 0.1% or more was observed in the prediction of hydraulic head using the MFE schemes when the flow domain was discretized using nine or less elements with a line source of 10^{-9} m^3/day/m. A maximum error of 17.24, 5.738, and 2.896% was observed respectively from the solutions obtained using R0QL, R0QD, and R1 schemes when the domain was discretized using one element. The degree of the error in the prediction of hydraulic head was directly proportional to the strength of the source used.

Another aspect observed from the simulation of hydraulic potentials using the MFE method was that the R0QD scheme had less error in the solution as compared to the R0QL scheme for any level of discretization while demonstrating similar rate of convergence to the analytical solution. The R1 schemes fell in between the R0QD and the R0QL schemes in converging to the analytical solution when grids are refined. These convergence properties of the different MFE schemes are evident from Fig. 2. Simulation of the Darcy flux was accurate for any strength of source used for any of the CFE or MFE schemes, for all levels of domain discretization.

Results were obtained for a flow domain with four *heterogeneous* zones with zero strength for the source/sink terms. The hydraulic conductivity of each zone was determined using either Eq. 39 (Fig. 3) or Eq. 40 (Fig. 4). All numerical methods with the exception of the CFEH scheme describe the hydraulic head and flux accurately, i.e., the numerical solutions converge for finer meshes to the analytic solution (Eq. 41). Figures 5 and 6 show the Darcy flux computed using the CFEH scheme for the two heterogeneous flow domains. Two distinctive characteristics of the computed flux distribution are observed as the grids are refined. The first one is that the boundary fluxes show higher error for coarser discretization and converge to the analytical solution as the grid size is refined within each heterogeneous zone. In addition, the boundary with a higher conductivity shows greater deviation from the analytical solution for any discretization of the heterogeneous zones. The second characteristic observed from the figures is that it is apparent the CFEH scheme does not describe the Darcy flux at heterogeneities of conductivity in

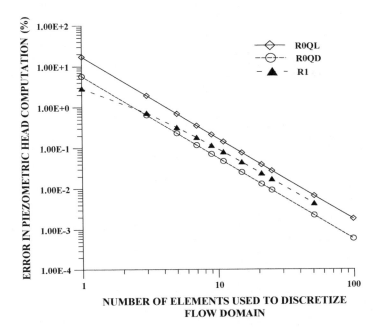

Fig. 2 Rate of convergence of solution in different MFE schemes for piezometric head at $x = 50{,}000$ m and in the presence of a line source of strength equal to 10^{-9} m^3 s^{-1} m^{-1}

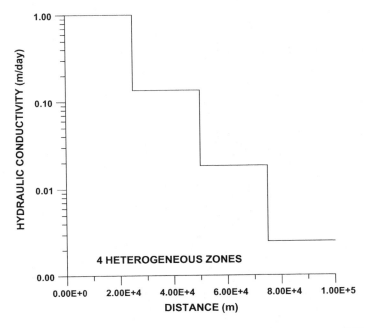

Fig. 3 Hydraulic conductivities of each heterogeneous zone in an exponentially varying conductivity field

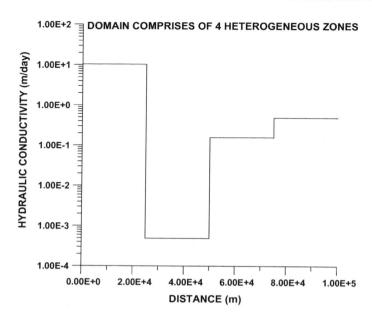

Fig. 4 Hydraulic conductivities of each heterogeneous zone in a randomly varying conductivity field

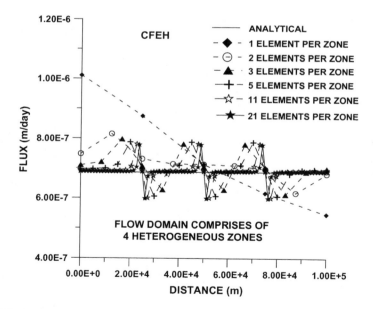

Fig. 5 Distribution of flux computed using the CFEH scheme in an exponentially varying conductivity field when the flow domain

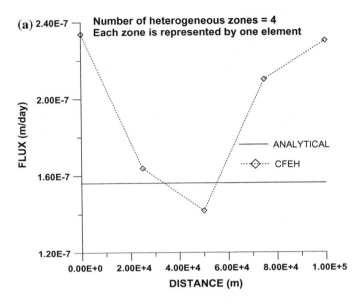
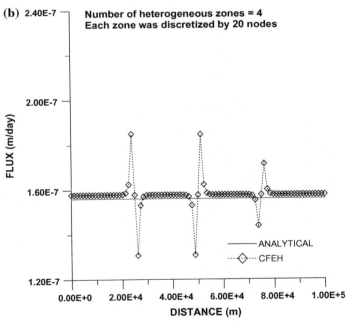

Fig. 6 Distribution of flux computed using the CFEH scheme in a randomly varying conductivity field with four heterogeneous zones and without any uniform sources present in the flow domain when **a** each zone was represented by one element, and **b** each zone was discretized by 20 nodes

a correct manner. With finer grids, the numerical solution does not converge to the analytic solution. This is because at discontinuities of hydraulic conductivity the numerical scheme must be able to describe the gradient of the hydraulic head discontinuously to have a continuous flux. The CFEH scheme is only able to describe the physical process for homogeneous hydraulic conductivity fields. It fails in the case of heterogeneities.

The errors obtained in the computed flux values close to the element boundaries with the CFEH method are due to the discontinuity of the hydraulic conductivity across these interfaces despite the continuity of the pressure gradient. Considering the two boundaries alone, the flux is higher at one end than the other. Using this distribution to simulate contaminant transport can lead to significant storage of contaminant within the aquifer. Alternately, if the computed flux at the boundary was erroneous but of equal magnitude, then the above problem would be minimized because the rate of movement of contaminant into and out of the aquifer would be similar in magnitude. With finer discretization of each heterogeneous zone, the solution at intermediate node points converges to the analytical solution. However, the flux computed at the zone interfaces might lead to significant error in local mass balance of fluid or contaminant transport in the aquifer if the rate of movement of the fluid out of a zone is higher (or lower) than the rate of movement into the neighboring zone for the problem being studied. It was observed that the error in the computed flux near the interface of two heterogeneous zones was higher toward the zone with higher conductivity for refined grids. This resulted in a substantial increase in the hydraulic head at the interface, which is not shown in this paper.

In the simulation of the flow problem with a constant source throughout the length of the aquifer, the domain was divided into three heterogeneous zones. A fine grid solution was used to compare the results obtained from the different numerical schemes being tested. The fine grid solution was obtained by discretizing each zone with 331 elements. As a result, the entire flow domain comprised of 994 nodes. The hydraulic conductivity for each heterogeneous zone as computed using Eq. 39 or Eq. 40 is shown respectively in Figs. 7 and 8. The fine grid solutions were obtained using the CFEL scheme. The choice of the CFE method to obtain the fine grid solution is due to the fact that this method has been in use for a long time and the MFE method is being tested against the CFE method. A line source of 10^{-9} m^3/s/m was used for the simulations.

The distribution of hydraulic head as obtained from the different numerical schemes is presented in Fig. 9a–c for exponential heterogeneity and Fig. 10a–c for random heterogeneity, when each heterogeneous zone was discretized using one element. The schemes used to simulate the pressure distribution that are compared to the fine grid solution are described in the legend of each figure. The figures are plotted on a dimensionless X–Y scale where the pressure heads are scaled down by dividing with the maximum head obtained from the fine grid solution and represented as P/P(max, fine grid) on the y-axis. The location at which these pressure head solution were sought has been scaled down by dividing with the length of the aquifer and is represented as x/L on the x-axis.

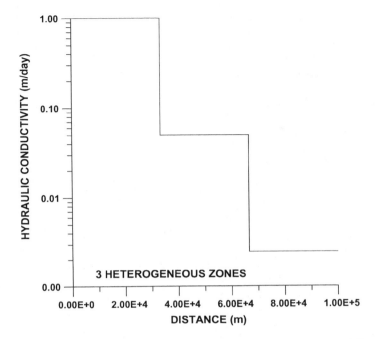

Fig. 7 Hydraulic conductivities of each heterogeneous zone in an exponentially varying conductivity field

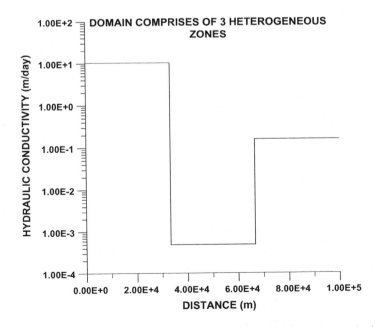

Fig. 8 Hydraulic conductivities of each heterogeneous zone in a randomly varying conductivity field

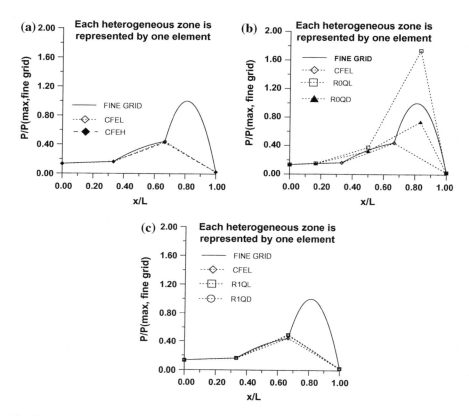

Fig. 9 Pressure head distribution obtained from the simulation of flow in a three-zone exponentially varying conductivity field using **a** CFE schemes, **b** $r = 0$ schemes and **c** $r = 1$ schemes, when each zone is discretized using a single element and in the presence of a uniform source

The choice of a single element discretization of each heterogeneous zone may not sound very practical. The reason for such a choice is to show convergence properties clearly. For the single element representation of the heterogeneous zones, the CFEL and CFEH schemes produced almost identical values of pressure at the nodes (see Figs. 9a and 10a). However, both the schemes missed the peak hydraulic head in the lowest conductivity zone because the head was computed at the node points placed at the beginning and the end of the zone. A single node point placed at the center of each element would have made a big difference in the computed hydraulic head distribution with the CFE methods. This handicap is slightly overcome in the $r = 0$ methods where the hydraulic head is computed on an element basis, helping to reveal the peak not observed in the CFE methods. Considering a single heterogeneous zone, the hydraulic head distribution obtained from the $r = 0$ methods is constant over the element that represents the zone. This distribution results in significant errors in the hydraulic head values over the entire

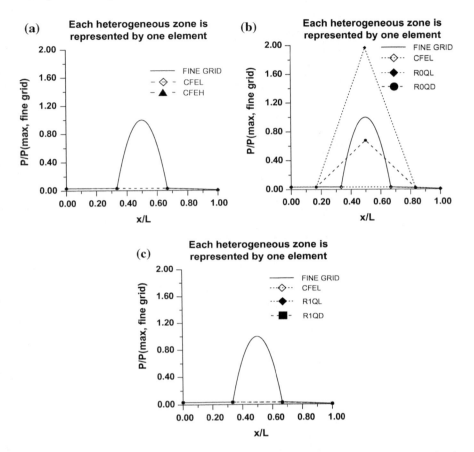

Fig. 10 Pressure head distribution obtained from the simulation of flow in a three-zone randomly varying conductivity field using **a** CFE schemes, **b** $r = 0$ schemes and **c** $r = 1$ schemes, when each zone is discretized using a single element and in the presence of a uniform source

heterogeneous zone as compared to the fine grid solution except at the center of the element where the error in the hydraulic head is smaller than that obtained from the CFE methods.

The property that the $r = 0$ methods have minimum error in the numerical value of pressure head at the center of the element has been demonstrated by Srinivas et al. (1992). The point at the center of the element is a Gauss point and they have shown that the pressure head distribution obtained from the $r = 0$ methods exhibits superconvergence property at these Gauss points. Besides this qualitative analysis, a comparison of the error in the hydraulic head value at the Gauss point within the element representing the least conductive zone (see Figs. 9b and 10b) revealed that the R0QL method produced an error of 74 and 96% and the R0QD method had an error of 26 and 31% in the pressure head value as compared to the fine grid solution, while the CFEL method had an error of 76 and 96%.

The $r = 1$ schemes suffer from the same problem as the conventional finite element methods in the sense that the heads are computed at the nodes. Thus, the hydraulic head distributions obtained from these schemes are almost identical to those obtained from the CFEL scheme. Similar to the $r = 0$ method, the $r = 1$ method shows superconvergence in the pressure head distribution at the two Gauss points in the element that are at a distance of $\pm\sqrt{3^{-1}}$ from the center of the element in a ξ-coordinate system (see Ewing and Wheeler 1983). A comparison of the error in the numerical value of the pressure head at these points in all the zones of the flow domain revealed that the $r = 1$ methods have a smaller error in the hydraulic head value compared to the CFEL method. The CFEL method showed the least error at the nodes in comparison to all the MFE schemes. Hence, the CFEL method demonstrates superconvergence in hydraulic head distribution at the nodes. The convergence properties of the various numerical schemes are illustrated in Figs. 11, 12, 13, 14, 15, and 16.

Using a three-element discretization of each heterogeneous zone improved the computed hydraulic distribution for all the schemes as shown in Fig. 17a–c for exponential heterogeneity and Fig. 18a–c for random heterogeneity. The superconvergence properties of each numerical scheme are visually distinguished. Despite the use of a Hermite cubic interpolant for ϕ in the CFEH scheme, the computed hydraulic heads show error at the nodes as compared to the CFEL

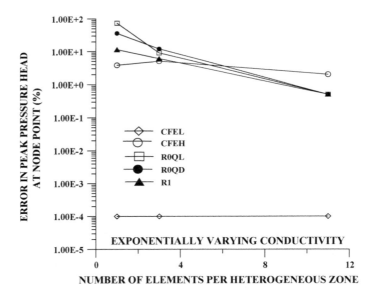

Fig. 11 Convergence of pressure head solutions at node points in an element using different numerical schemes

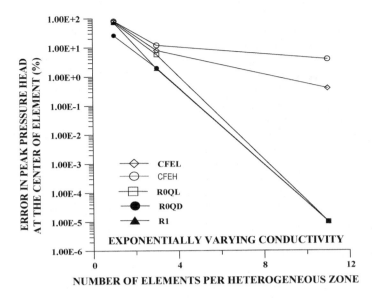

Fig. 12 Convergence of pressure head solutions at the center of an element using different numerical schemes

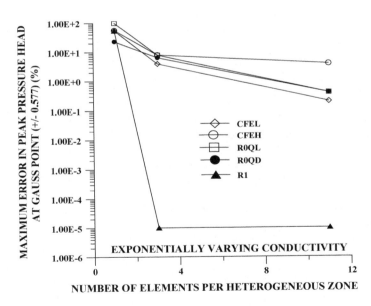

Fig. 13 Convergence of pressure head solutions at Gauss points ($\zeta = \pm 0.577$) in an element using different numerical schemes

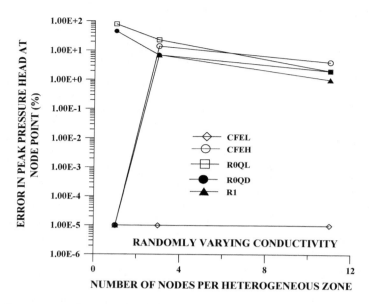

Fig. 14 Convergence of pressure head solutions at node points in an element using different numerical schemes

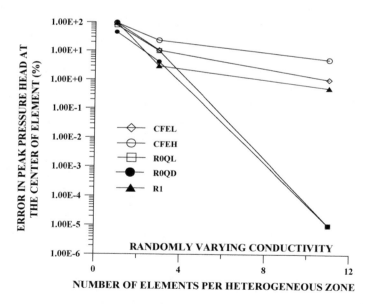

Fig. 15 Convergence of pressure head solutions at the center of an element using different numerical schemes

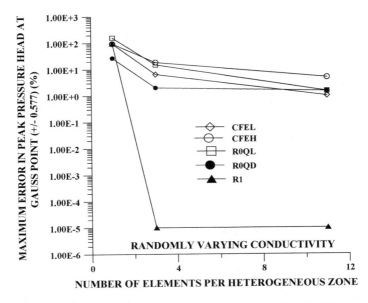

Fig. 16 Convergence of pressure head solutions at Gauss points ($\zeta = \pm 0.577$) in an element using different numerical schemes

scheme where a linear interpolant was used (see Figs. 17a and 18a). The nodal hydraulic head values are very accurate (error of less than 0.001%) for the CFEL scheme. The MFE methods show high accuracy at the corresponding Gauss points. In Figs. 17b and 18b, the error in the hydraulic head value obtained from the CFEL scheme at the center of the element is higher in comparison to those obtained from the $r = 0$ method at this point. The MFE methods did better for prediction of the peak hydraulic head. For example, the peak head obtained from the $r = 0$ methods shows an error of 8.3 and 10.8% with the R0QL method and 2.5 and 3.4% with the R0QD method. The CFEL scheme produced an error of 10 and 11.7% in the prediction of this peak value. Nevertheless, the nodal hydraulic heads computed with the CFEL method have less error than the $r = 0$ methods. The $r = 1$ methods show close convergence to the fine grid solution at the two Gauss points mentioned in the previous paragraph (see Figs. 17c and 18c). The errors associated with the hydraulic heads computed at these Gauss points are less than 0.001% with the $r = 1$ methods. However, the errors at the node points are 7 and 9.2% for the R1QL method and 5.2 and 7.7% for the R1QD method at the peak pressure obtained from these schemes. A visual observation of Figs. 17c and 18c clearly shows the interception of the line connected between the two nodal points of the pressure heads obtained from the $r = 1$ methods with the fine grid solution at the Gauss points near the peak value of the pressure head distribution obtained from this discretization of the domain thus demonstrating the superconvergence properties of the $r = 1$ scheme.

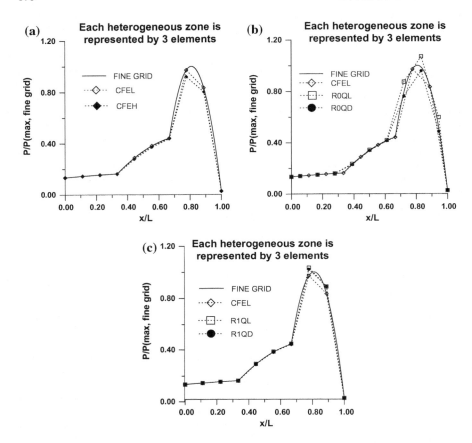

Fig. 17 Pressure head distribution obtained from the simulation of flow in a three-zone exponentially varying conductivity field using **a** CFE schemes, **b** $r = 0$ schemes and **c** $r = 1$ schemes, when each zone is discretized using three elements and in the presence of a uniform source

The flux distributions computed by the different schemes when each heterogeneous zone was represented by one element are shown in Fig. 19 for exponential heterogeneity and Fig. 20 for random heterogeneity. The flux distribution obtained from the CFEL method and the $r = 0$ methods were in exact correspondence with the fine grid solution. The fluxes other than the boundary nodes deviated from the fine grid solution for the $r = 1$ methods. The CFEH scheme produced erroneous velocity values at all the nodes including the boundary nodes. This result was also observed for the heterogeneous flow problem without a line source. It will be shown later that the error in the flux distribution with refinement of the grids in each heterogeneous zone will produce errors that are similar in behavior as observed in the simulation of flow without source/sink terms. The maximum error was observed at $x = 0$ which was a boundary node with the highest hydraulic conductivity.

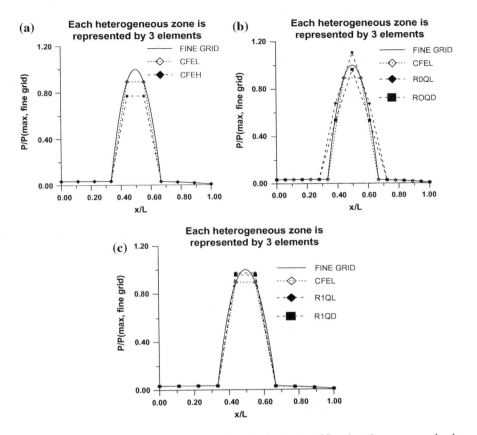

Fig. 18 Pressure head distribution obtained from the simulation of flow in a three-zone randomly varying conductivity field using **a** CFE schemes, **b** $r = 0$ schemes and **c** $r = 1$ schemes, when each zone is discretized using three elements and in the presence of a uniform source

The error in the flux as compared to the fine grid solution varied between 3 and 51%. Thus, we can safely conclude that *the nonlinear interpolation of q in the finite element formulation produces nonuniform flux distribution*.

The flux distribution computed using the three-element discretization of each heterogeneous zone is shown in Fig. 21a, b for exponential heterogeneity and Fig. 22 for random heterogeneity. The computed flux distributions with the CFEL scheme and the $r = 0$ schemes fall exactly on the fine grid solution. Once again the $r = 1$ schemes produced inaccurate flux distributions only at the interface of adjacent heterogeneous zones. Within each zone, the nodal velocities obtained from the $r = 1$ method were very close to the fine grid solution. The maximum errors in the numerical value of the flux with the three-element discretization of each heterogeneous zone were 9.5% (R1QD method) and 6.5% (R1QL method). The concentration of the errors near the interface of adjacent heterogeneous zones lead

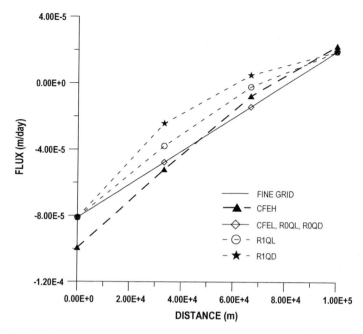

Fig. 19 Distribution of flux obtained from all the numerical schemes in the presence of a uniform source for a one-element discretization of each heterogeneous zone when the domain comprised of three heterogeneous zones

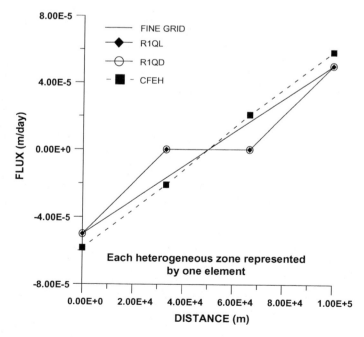

Fig. 20 Distribution of flux in a three-zone randomly varying conductivity field (in the presence of a uniform source) obtained from the CFEH and the $r = 1$ schemes when each zone is discretized using one element

to a *second conclusion* that *the higher order interpolants cause greater fluctuation in velocity when there is a sudden change in the conductivity.*

The CFEH scheme demonstrated a similar behavior as the $r = 1$ schemes except that the fluctuations are oscillatory around the fine grid solution as shown in Figs. 21b and 22. Furthermore, the degree of error in the flux distribution near the interface of two heterogeneous zones was larger when the change in conductivity was higher at the interface. In fact, the nature of oscillation of the computed values of fluxes and the degree of error with respect to the fine grid solution near the interface of two heterogeneous zones are exactly the same as observed in the flow simulation without any source/sink terms. A comparison of the flux distribution obtained from the $r = 1$ methods and the CFEH method reveals that the former method is more robust since the convergence of the solution to the fine grid solution is smoother with refinement of the grid.

The question is what did we learn from the uniformly loaded, steady-state, saturated fluid flow field by computing the hydraulic head and the flux using various numerical schemes? We started with the idea that the strength of the MFE schemes lies in the accurate description of the flux distribution. Similarly, the CFE schemes had their strength in accurately simulating the hydraulic head distribution over the domain. It seems that the results obtained did not indicate that these ideas are entirely true. Furthermore, the higher order schemes like the CFEH and the R1 schemes produced inaccuracy in flux distribution in heterogeneous flow conditions when presumably these schemes should be more robust than the lower order schemes. No dramatic difference was observed in the computed fluxes with the CFEL scheme and the R0 schemes even in heterogeneous conditions. One important aspect was observed that the R0 schemes described the peak hydraulic head better in coarse grid discretization of a heterogeneous flow problem. The Hermite cubic interpolation did not produce accurate hydraulic head and flux distributions in heterogeneous flow domains even with finer discretization. So where does the strength of the MFE methods lie? Durlofsky (1994) observed that for flow domains which are highly variable, particularly those with very abrupt variations in permeability or systems that are more discontinuous, the MFE method provides robust, reliable, and physically realistic solutions. This leads us to investigate flow simulations with nonuniform source/sink terms.

5.2 Nonuniform Source/Sink

For groundwater flow problems, it is more common to have nonuniform sources/sinks. An example of a nonuniform source is a point source or sink such as the presence of a well in an aquifer. Rapidly varying velocity fields are generated near the location of the well point, and for this case the MFE scheme might demonstrate some advantages. The following results and discussion are based on the incorporation of nonuniform sources/sinks in the example one-dimensional flow field described in Sect. 4 except $\phi(0)$ and $\phi(L)$ are now both set to 2.0 m.

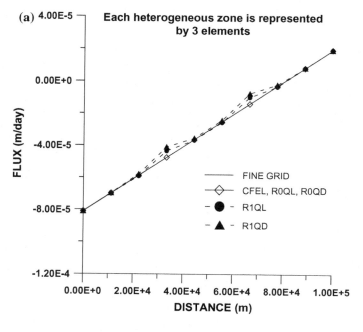
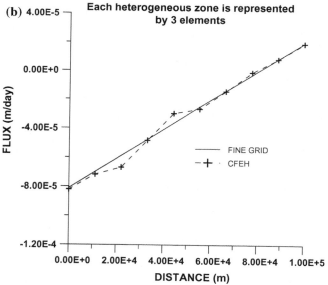

Fig. 21 a Distribution of flux obtained from all the numerical schemes except CFEH in the presence of a uniform source for three-element discretization of each heterogeneous zone when the domain comprised of three heterogenous zones. **b** Distribution of flux obtained from the CFEH scheme in the presence of a uniform source for three-element discretization of each heterogeneous zone when the domain comprised of three heterogeneous zones

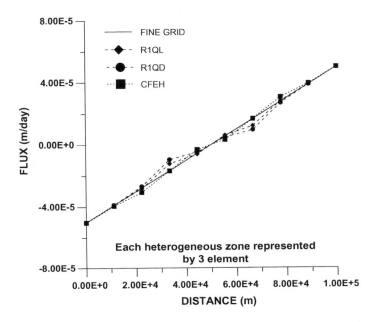

Fig. 22 Distribution of flux in a three-zone randomly varying conductivity field (in the presence of a uniform source) obtained from the CFEH and the $r = 1$ schemes when each zone is discretized using three elements

A numerical analysis was performed using a point source of strength equal to 1.0 m^3/day at $x = 0.3L$ and a point sink of strength equal to -0.4 m^3/day at $x = 0.7L$ in a homogeneous domain. A fine grid solution was obtained using a 1000 node point discretization of the domain using the CFEL scheme. The computed hydraulic heads using the different numerical schemes are presented in Fig. 23a, b for a 5-element discretization of the domain in the case of R0 schemes and a 10-element discretization in the case of CFE and R1 schemes. It is observed from these plots that the R0QD scheme produces inaccurate solution for the hydraulic heads at the location of the source and the sink. The error in the hydraulic heads calculated at the point source and the point sink locations for the different numerical schemes, except the CFEL and the R0QL schemes (which produced exact solution), are shown in Table 2.

The distribution of flux using the spatial discretizations mentioned above is plotted in Fig. 24. The R0 schemes produced the exact distribution of flux at all nodes as compared to the fine grid solution. The errors in the other numerical schemes are maximum at the location of the point source and the point sink. These errors in the calculated flux values are summarized in Table 3. Note that the R0QD scheme produces accurate flux distribution despite the high error in the hydraulic head computation.

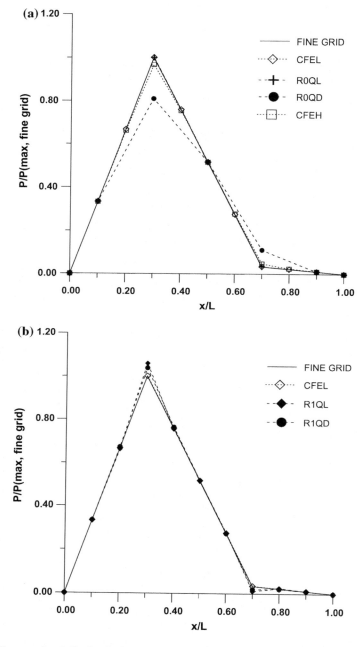

Fig. 23 Pressure head distribution obtained from **a** CFE (10-element discretization) and $r = 0$ (5-element discretization) schemes, and **b** CFEL and $r = 1$ (10-element discretization) schemes, for flow in a homogeneous domain with a point source and a point sink

Table 2 Error in the hydraulic head computation at the location of a point source/sink in a homogeneous aquifer using the CFEH, R0QD and the R1 schemes using a coarse spatial discretization

Numerical method	Error (%) in ϕ at the point source location	Error (%) in ϕ at the point sink location
CFEH	3.4	39.1
R0QD	19.2	221.5
R1QL	5.9	67.8
R1QD	3.7	42.9

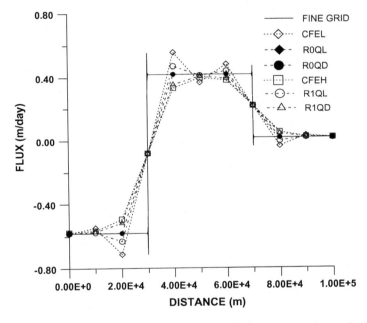

Fig. 24 Flux distribution obtained from CFE (10-element discretization), $r = 0$ (5-element discretization), and $r = 1$ (10-element discretization) schemes, for flow in a homogeneous domain with a point source and a point sink

Table 3 Error in the computed flux values at the closest nodes to the location of a point source/sink in a homogeneous flow condition using the CFE and the R1 schemes for a coarse spatial discretization of the flow domain

Numerical method	Error (%) in q near the point source location	Error (%) in q near the point sink location
CFEL	11.7	5.9
CFEH	8.8	3.9
R1QL	4.9	2.2
R1QD	5.9	2.9

A second analysis was performed using an exponentially decreasing conductivity field with the same parameters and boundary conditions as used for the homogeneous flow domain. A point source of strength equal to 1.0 m^3/day was placed at $x = 0.7L$ and a sink of strength equal to -0.5 m^3/day at $x = 0.3L$. The aquifer was divided into five heterogeneous zones with the hydraulic conductivities as shown in Fig. 25. The fine grid solution was obtained using the CFEL scheme by discretizing each heterogeneous zone into 199 elements which provided a 996 node discretization for the entire flow domain. Hydraulic heads and fluxes were computed using the different numerical schemes for a one-element discretization of each heterogeneous zone (five elements in the entire flow domain) for the R0 schemes and a corresponding two-element discretization of each heterogeneous zone (ten elements in the entire flow domain) for all the other schemes. *For convenience, we will refer to the discretization of R0 schemes only in this section and remind the reader that the corresponding discretization of the other schemes that are compared to the R0 schemes is different.*

The results obtained from the different schemes for the hydraulic head distribution are shown in Fig. 26. The results show similar attributes to those observed in the homogeneous flow with nonuniform sources/sinks. Without repeating any points that have been already mentioned in this section, we present the errors at the locations of the source and the sink for different schemes, except the CFEL and the R0QL scheme which produced exact solution, in Table 4.

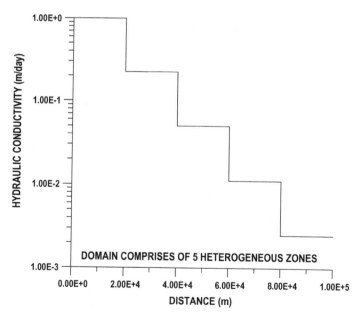

Fig. 25 Hydraulic conductivity of each zone in a five-zone exponentially varying conductivity field used for the simulation of flow with point sources/sinks present in the domain

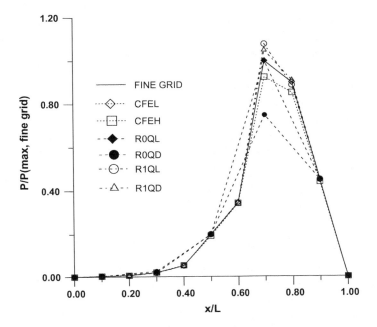

Fig. 26 Pressure head distribution obtained from all the numerical schemes for flow in a five-zone exponentially varying conductivity field with a point source and a point sink when each zone is discretized by one element ($r = 0$) and two elements (CFE and $r = 1$)

Table 4 Error in the hydraulic head computation at the location of a point source and a point sink present in an exponentially varying heterogeneous flow domain using CFEH, R0QD, and R1 schemes

Numerical method	Location of error	Error (%) in ϕ
CFEH	$x = 0.3L$	3.1
	$x = 0.7L$	7.6
R0QD	$x = 0.3L$	31.4
	$x = 0.7L$	25.2
R1QL	$x = 0.3L$	9.6
	$x = 0.7L$	7.7
R1QD	$x = 0.3L$	6.1
	$x = 0.7L$	4.9

The flux distributions obtained from the simulation are shown in Fig. 27. The computed fluxes showed that the R0 scheme is robust and non-oscillatory with no error in the numerical values of the computed fluxes. The CFEH scheme manifested oscillatory behavior in fluxes even when the discretization was made finer (see Misra[26]) due to the heterogeneity in the porous medium. The R1 schemes showed considerable error in simulating the flux, particularly in the vicinity of the point source/sink as did the CFEL scheme. The maximum error in the distribution of flux which was near the vicinity of the source/sink was found to be produced by the CFEL scheme.

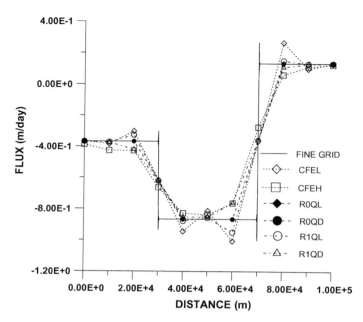

Fig. 27 Flux distribution obtained from all the numerical schemes for flow in a five-zone exponentially varying conductivity field with a point source and a point sink when each zone is discretized by one element ($r = 0$) and two elements (CFE and $r = 1$)

A third analysis was performed using a randomly varying conductivity field where the flow domain was divided into five heterogeneous zones with the hydraulic conductivity of each zone plotted in Fig. 28. A point source of strength equal to 1.0 m^3/day was placed at $x = 0.3L$ and a point sink of strength equal to -0.5 m^3/day was placed at $x = 0.7L$. The fine grid solution was obtained from the CFEL scheme using a 996 node discretization with each heterogeneous zone being divided into 199 elements. Hydraulic head and flux distributions over the domain for both the examples were simulated using the different numerical schemes with the same spatial discretization used for the simulations in the exponentially varying conductivity distribution described above.

The simulated hydraulic head distributions are shown in Fig. 29. The results obtained are similar to those obtained from the previous example in this section. The CFEL scheme and the R0QL scheme described the hydraulic head distribution exactly in the entire domain. The errors in the hydraulic head values for the other schemes at the location of the point source and the point sink are presented in Table 5.

The distribution of flux in the domain obtained from the different numerical schemes is presented in Fig. 30. Similar to the previous example in this section, the CFEH scheme produced an oscillatory solution. The CFEL scheme produced erroneous flux values at the vicinity of the source/sink which were higher than those

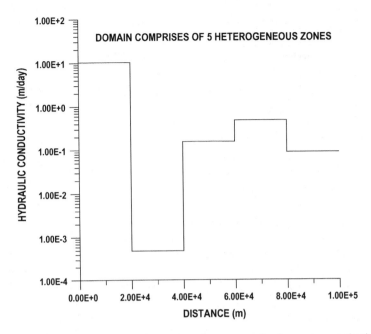

Fig. 28 Hydraulic conductivity of each zone in a five-zone randomly varying conductivity field used for the simulation of flow with point sources/sinks present in the domain

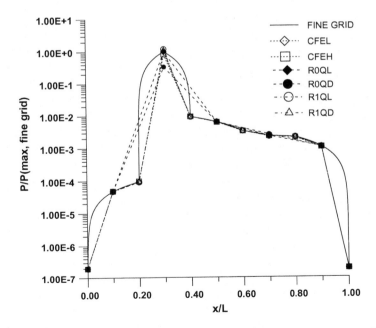

Fig. 29 Pressure head distribution obtained from all the numerical schemes for flow in a five-zone randomly varying conductivity field with a point source and a point sink when each zone is discretized using one element ($r = 0$) and two elements (CFE and $r = 1$)

Table 5 Error in the hydraulic head value computed at the location of a point source and a point sink in a randomly heterogeneous flow domain using the CFEH, the R0QD, and the R1 schemes

Numerical method	Location of error	Error (%) in ϕ
CFEH	$x = 0.3L$	16.6
	$x = 0.7L$	0.6
R0QD	$x = 0.3L$	66.0
	$x = 0.7L$	11.5
R1QL	$x = 0.3L$	19.9
	$x = 0.7L$	3.6
R1QD	$x = 0.3L$	12.4
	$x = 0.7L$	2.3

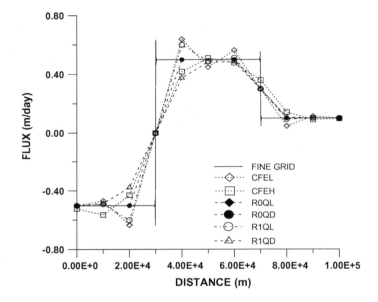

Fig. 30 Flux distribution obtained from all the numerical schemes for flow in a five-zone randomly varying conductivity field with a point source and a point sink when each zone is discretized using one element ($r = 0$) and two elements (CFE and $r = 1$)

obtained from other schemes at these locations. The R0 schemes were error free in the entire domain and the R1 schemes produced erroneous fluxes particularly near the vicinity of the source/sink. The R1 scheme errors were less than those obtained from the CFEL scheme near the source/sink locations.

5.3 Discussion on Accuracy and Computational Efficiency

We have shown the results that are obtained using a very coarse grid. The reason for using coarse grids was to show the level of accuracy of the solutions that are obtained using the numerical schemes. When we used fine grids (see Misra 1994), the errors in the solutions obtained using the CFE and MFE schemes for hydraulic head distribution were very small. However, even with fine grid, the flux distributions obtained using CFE schemes in the presence of point sources/sinks were erroneous. From the results obtained using nonuniform sources/sinks in the flow domain, it can be concluded that the MFE methods are in general superior in simulation of both the hydraulic head and the flux as compared to the CFE methods (similar statement has been made by both Mosé et al. 1990 and Durlofsky 1994). The accuracy that is obtained for the flux distribution using the $r = 0$ method is well suited for particle tracking models to simulate contaminant transport (see Mosé et al. 1990, 1994) as well as other transport models. Our intention is to extend the MFE method to solve unsaturated flow and transport problems.

Cordes and Kinzelbach (1992, 1996) and Srivastava and Brusseau (1995) have explicitly commented about the loss in computational efficiency of the solution while using MFE schemes as compared to other numerical schemes. The reason for the loss is due to the larger number of unknowns required to be solved in the MFE method. We do not agree about the requirement to solve a greater number of unknowns in the MFE schemes because in CFE methods, the fluxes are eventually computed once the values of the hydraulic heads are obtained. This makes the number of unknowns more or less the same in both the MFE and the CFE schemes. We also do not agree about the loss in computational efficiency because we have shown in this paper that highly accurate solutions can be obtained using MFE methods even when extremely coarse grids are used, whereas to obtain similar level of accuracy, a much finer grid is necessary in case of CFE schemes which results in a higher computational cost in the latter case. Several methods have been proposed in literature that could be used to improve the computational efficiency of the MFE methods. The Uzzawa-PCG method is an attractive candidate which has been used by Ewing and Wheeler (1983) and Dougherty (1990). Among the other methods are the penalty method (Carey and Oden 1983, 1986) and the mixed-hybrid method (Kaasschieter 1990, 1995), which have been shown to improve the computational efficiency of the MFE methods significantly.

6 Conclusions and Future Directions

a. In one-dimensional steady-state flow problems without any line sources/sinks, the MFE schemes do not possess any advantage over the CFE schemes in computing either pressure or flux. All the schemes produce identical solutions.

However, in two- or three-dimensional flow problems, the CFE schemes can introduce errors making the MFE schemes advantageous.

b. With the introduction of a line source, the MFE schemes produce errors in simulation of the pressure heads for very coarse discretization of the domain. The solutions converge smoothly to the analytical solution with the refinement of grids in the domain. The R0QD scheme has the fastest convergence rate among all the MFE schemes. The distribution of flux is accurate for any level of discretization. The CFE schemes, however, yield accurate results for both flux and hydraulic head for all levels of discretization.

c. The distribution of hydraulic head obtained from the different schemes are superconvergent at the nodes for the CFE schemes whereas these are superconvergent at the Gauss points within the element in the case of the MFE schemes.

d. In heterogeneous flow domains, the higher order schemes produce spatially oscillatory flux distributions. The oscillations are predominantly located at the interface of adjacent heterogeneous zones. The CFEH scheme retains its oscillatory behavior in simulating the flux even with finer discretization (see Misra 1994). These oscillations are high in higher conductivity areas.

e. The fluxes computed by the R0 schemes are not superior to the CFEL scheme in a heterogeneous porous media with a line source/sink.

f. In a flow field with point sources/sinks, the R0 scheme is robust and has no error in the computed flux values. In a heterogeneous flow domain with drastically changing hydraulic conductivity, these schemes describe the peak hydraulic heads better than any other scheme with the presence of nonuniform sources/sinks.

g. All the schemes except the R0 schemes deviate considerably from the true value of the flux near a point source/sink.

h. The R0QD scheme produces highly inaccurate hydraulic head distribution in the presence of nonuniform sources/sinks. However, the distribution of flux is not affected by this inaccuracy in describing the hydraulic head distribution.

We have presented results from the simulation of a very simple case study in one-dimensional saturated flow domain. However, we strongly believe that the conclusions and the observations are applicable to more complex flow situations. Also, the conclusions drawn from the cases of the heterogeneous flow domains should be applicable to the case of MFE methods applied to unsaturated flow.

Acknowledgements The authors acknowledge the support of the Minnesota Agricultural Experiment Station of the University of Minnesota, the Army High Performance Computing Research Center under the auspices of the Department of the Army, Army Research Laboratory, and EPSCoR Alaska. Additionally, we express our sincere gratitude to the anonymous reviewer of our paper for the critical comments and suggestions that have improved its quality, immensely.

References

Ackerer P, Mosé R, Siegel P, Chavent G (1996) Reply. Water Resour Res 32(6):1911–1913
Allen MB, Ewing RE, Koebbe JV (1985) Mixed finite element methods for computing groundwater velocities. Numer Methods Partial Differ Equ 3:195–207
Allen MB, Ewing RE, Lu P (1992) Well-conditioned iterative schemes for mixed finite-element models of porous-media flows. Siam J Sci Stat Comput 13(3):794–814
Becker EB, Carey GF, Oden JT (1981) Finite elements: an introduction, vol. I. Prentice-Hall, Inc. Englewood Cliffs, New Jersey
Beckie R, Wood EF, Aldama AA (1993) Mixed finite element simulation of saturated groundwater flow using a multigrid accelerated domain decomposition technique. Water Resour Res 29(9):3145–3157
Carey GF, Oden JT (1983) Finite elements: a second course, vol II. Prentice-Hall Inc, Englewood Cliffs, NJ
Carey GF, Oden JT (1986) Finite elements: fluid mechanics, vol VI. Prentice-Hall Inc, Englewood Cliffs, NJ
Chavent G, Roberts JE (1991) A unified physical presentation of mixed, mixed-hybrid finite elements and standard finite difference approximations for the determination of velocities in waterflow problems. Adv Water Resour 14(6):329–348
Chavent G, Cohen G, Jaffre J (1987) A finite element simulator for incompressible two-phase flow. Transp Porous Media 2:465–478
Chiang CY, Wheeler MF, Bedient PB (1989) A modified method of characteristics technique and mixed finite elements method for simulation of groundwater solute transport. Water Resour Res 25(7):1541–1549
Cordes C, Kinzelbach W (1992) Continuous groundwater velocity fields and path lines in linear, bilinear, and trilinear finite elements. Water Resour Res 28(11):2903–2911
Cordes C, Kinzelbach W (1996) Comment on "application of the mixed hybrid finite element approximation in a groundwater flow model: Luxury or necessity?" by R. Mosé, P. Siegel, P. Ackerer, and G. Chavent. Water Resour Res 32(6):1905–1909
Dougherty DE (1990) PCG solutions to flow problems in random porous media using mixed finite elements. Adv Water Resour 13(1):2–11
Durlofsky LJ (1994) Accuracy of mixed and control volume finite element approximations to Darcy velocity and related quantities. Water Resour Res 30(4):965–973
Ewing RE, Wheeler MF (1983) Computational aspects of mixed finite element methods. In: Stepleman R et al (eds) Scientific computing. IMACS/North-Holland Publishing Company, pp 163–172
Hartmann F (1986) Ing Arch 56:221. https://doi.org/10.1007/BF00535887
Huyakorn PS, Pinder GF (1983) Computational methods in subsurface flow. Academic Press Inc, New York
Javandel I, Witherspoon PA (1968) Application of the finite element method to transient flow in porous media. Soc Pet Eng J 8:241–252
Kaasschieter EF (1990) Mixed-hybrid finite elements for saturated groundwater flow. In: Gambolati G, Rinaldo A, Brebbia CA, Gray WG, Pinder GF (eds) Computational methods in subsurface hydrology. Proceedings of the eighth international conference on computational methods in water resources. Computational Mechanics Publication, USA, pp 17–22
Kaasschieter EF (1995) Mixed finite elements for accurate particle tracking in saturated groundwater flow. Adv Water Resour 18(5):277–294
Kaluarachchi JJ, Parker JC (1987) Finite element analysis of water flow in variably saturated soil. J Hydrol 90:269–291
Kolar RL (1992) Environmental conservation laws: formulation, numerical solution, and application. Ph.D. dissertation, Submitted to the Graduate School of the University of Notre Dame, Department of Civil Engineering and Geological Sciences, Notre Dame, Indiana, USA

Lapidus L, Pinder GF (1982) Numerical solution of partial differential equations in science and engineering. Wiley, USA, p 677

Lynch DR (1984) Mass conservation in finite element groundwater models. Adv Water Resour 7:67–75

Meissner U (1973) A mixed finite element model for use in potential flow problems. Int J Numer Methods Eng 6:467–473

Meyling RHJG, Mulder WA, Schmidt GH (1990) Porous media flow on locally refined grids. In: Verheggen T (ed) Proceedings of the workshop in numerical methods for the simulation of multi-phase and complex flow. Springer (in press)

Misra D (1994) Mixed finite element analysis of one-dimensional heterogeneous Darcy flow. Ph.D. dissertation, Submitted to the Graduate School of the University of Minnesota, Department of Biosystems and Agricultural Engineering, St. Paul, MN, USA

Mosé R, Ackerer PH, Chavent G (1990) An application of the mixed hybrid finite element approximation in a three-dimensional model for groundwater flow and quality modeling. In: Gambolati G, Rinaldo A, Brebbia CA, Gray WG, Pinder GF (eds) Computational methods in subsurface hydrology. Proceedings of the eighth international conference on computational methods in water resources, Venice, Italy, June 11–15. Springer, USA, pp 349–355

Mosé R, Siegel P, Ackerer P, Chavent G (1994) Application of the mixed hybrid finite element approximation in a groundwater flow model: luxury or necessity? Water Resour Res 30 (11):3001–3012

Mulder WA, Meyling RHJG (1991) Numerical simulation of two-phase flow using locally refined grids in three-space dimensions. In: Proceedings of the 11th SPE symposium on reservoir simulation, Anaheim, CA, pp 299–306

Oden JT, Carey GF (1983) Finite elements: mathematical aspects, vol IV. Prentice-Hall Inc, Englewood Cliffs, NJ

Pollock DW (1988) Semianalytical computation of pathlines for finite difference models. Ground Water 26(6):743–750

Poceski A (1992) Mixed finite element method. In: Brebbia CA, Orszag SA (eds) Lecture notes in engineering. Springer, Berlin, p 345

Prenter PM (1975) Splines and variational methods. Wiley, USA

Price HS, Cavendish JC, Varga RA (1968) Numerical methods of higher order accuracy for diffusion convection equations. Soc Pet Eng J 8(3):293–303

Raviart PA, Thomas JM (1977) A mixed finite element method for 2nd order elliptic problems. In: Galligani I, Magenes E (eds) Mathematical aspects of finite element methods, vol 606. Lecture notes in mathematics. Springer, Berlin, pp 292–315

Reissner E (1950) On a variational theorem in elasticity. J Math Phys 29:90–95

Russell TF, Wheeler MF (1983) Finite element and finite difference methods for continuous flows in porous media. In: Ewing RE (ed) The mathematics of reservoir simulation (Chapter II). Frontiers in applied mathematics series. SIAM, Philadelphia, pp 35–105

Segerlind LJ (1984) Applied finite element analysis. Wiley, USA

Segol G, Pinder GE, Gray WG (1975) A Galerkin finite element technique for calculating the transient position of the salt water front. Water Resour Res 11(2):343–347

Sovich TJ (1988) Solute transport in heterogeneous reactive aquifers using particle and mixed finite element methods. Master of Science Thesis, Submitted to the Department of Civil Engineering, University of California, Irvine

Srinivas C, Ramaswamy B, Wheeler MF (1992) Mixed finite element methods for flow through unsaturated porous media, In: Russell TF, Ewing RE, Brebbia CA, Gray WG, Pinder GF (eds) Numerical methods in water resources. Computational methods in water resources, vol 1, Computational Mechanics Publication and Elsevier Applied Science, USA, pp 239–246

Srivastava R, Yeh T-CJ (1992) A three-dimensional numerical model for water flow and transport of chemically reactive solute through porous media under variably saturated conditions. Adv Water Resour 15:275–287

Srivastava R, Brusseau ML (1995) Darcy velocity computations in the finite element method for multidimensional randomly heterogeneous porous media. Adv Water Resour 18(4):191–201

van Genuchten MT (1982) A comparison of numerical solutions of the one-dimensional unsaturated-saturated flow and mass transport equations. Adv Water Resour 5:47–55

van Genuchten MT (1983) A Hermitian finite element solution of the two-dimensional saturated-unsaturated flow equation. Adv Water Resour 6:106–111

Yeh G-T (1981) On the computation of Darcian velocity and mass balance in the finite element modeling of groundwater flow. Water Resour Res 17(5):1529–1534

Yeh T-CJ, Srivastava R, Guzman A, Harter T (1993) A numerical model for water flow and chemical transport in variably saturated porous media. Ground Water 31(4):634–644

Zienkiewicz OC, Parekh CJ (1970) Transient field problems: two-dimensional and three-dimensional analysis by isoparametric finite elements. Int J Numer Meth Eng 2:61–71

Subsurface Acid Sulphate Pollution and Salinity Intrusion in Coastal Groundwater Environments

Gurudeo Anand Tularam and Rajibur Reza

1 Introduction

Water is a critical resource for humans and now it is a source of major concern (Hassan and Tularam, in press). It is probably much more important than oil and continues to grow in importance. Climate change studies have predicted rather unfavourable conditions and outcomes for water in our world generally (Tularam and Krishna 2009; Tularam and Marchisella 2014; Tularam and Hassan 2016a, b). The extreme nature of temperature variations and consequent rainfall patterns have been attributed to the new changing climate conditions—leading to many droughts or floods conditions around the world as noted in recent times (Hassan and Tularam 2017). As the temperature level increases, there is much more evaporation of stored water and also increased transpiration levels throughout; both changing conditions have serious implications for human search for sustainable potable water for consumption; desalination and purification of the groundwater, for example. Although not preferred, it is true that humans are becoming used to the consumption of recycled water and many are increasingly accepting of potable water sourced from groundwater, which is indeed a natural storage tank of water. What is well known is that some coastal aquifers and groundwater stores are often almost pure water that could be used for consumption with little treatment. The coastal groundwater is indeed the

---

G. A. Tularam (✉)
Mathematics and Statistics, Griffith Sciences [ENV], Griffith University,
Nathan, QLD 4111, Australia
e-mail: a.tularam@griffith.edu.au

G. A. Tularam · R. Reza
Environmental Futures Research Institute, Griffith University,
Nathan, QLD 4111, Australia

R. Reza
Department of Accounting, Finance and Economics, Griffith Business School,
Griffith University, Nathan, QLD 4111, Australia

© Springer Nature Singapore Pte Ltd. 2018
N. Narayanan et al. (eds.), *Flow and Transport in Subsurface Environment*,
Springer Transactions in Civil and Environmental Engineering,
https://doi.org/10.1007/978-981-10-8773-8_6

result of the rainfall-led infiltration of water's long path from its start on hills to reach of the sea. The rather long travel and the long length of time taken for such waters to reach the sea have helped in the filtering of the waters to a high degree, almost suitable for drinking. That is, long travel way from the mountains to the sea allows the water to pass through many natural cleaning processes such as sand and rocky filled soils underground. However, most of the human potable water needs are presently sourced from either surface stores such as lakes or rivers that are then blocked and the water stored in large dams. The dams are increasingly becoming difficult to manage due to the variation in climate, which means they are expensive to maintain in terms of costs to infrastructure—dangers of failing in dams, particularly for the population that live downstream of the dams such as Brisbane in Australia, for example.

The main aim of this chapter is to review several studies that have been conducted by the author and colleagues over the years in their attempts to deal with both groundwater pollution caused by the existence of acid sulphate soils in coastal regions and seawater intrusion in coastal aquifers leading to salinity pollution in subsurface environment. Some solutions to the problems that have been posed and suggested in such studies have now been used successfully in many areas to solve pollution problems. Both types of pollution are equally damaging to the groundwater. For example, coastal groundwater is often used for irrigation of various crops, but they mostly lead to rather poor production levels in coastal farms (sugar cane, for example). The use of such water at times leads to irreversible destruction of the coastal grazing lands. This chapter reviews the main findings of the above research over the past 15 or so years.

The nature of the subsurface flows and causes of both acid sulphate soils and salinity based on flow and transport equations that underlie the groundwater flow processes are presented first. Then, a number of studies conducted in Australia are reviewed to examine the nature of work that has been done regarding pollution of coastal areas by acidity and salinity. Next, some problems concerning the modelling tools are examined in relation to acidity- and salinity-related processes. An alternative to rainfall drinking water is desalination of seawater and this is also briefly discussed as one of the alternatives but desalination has its own problems. A brief summary of the work that is presented concludes this chapter.

2 Background of Groundwater

There is no substitute for fresh drinking water and water is becoming a scarce resource in our world all around. This shortage has been noted more acutely in recent times in Africa and Asia (Tularam and Krishna 2009; Tularam and Marchisella 2014; Tularam and Murali 2015; Tularam and Hassan 2016a, b; Tularam and Properjohn 2011). About 75% of the earth is covered with water but this is not the amount available for human use. The scarcity had led to water being now a major investing option—the need for supply of water and the costs of its related infrastructure (Roca et al. 2015; Tularam and Reza 2016). In budgeting drinking water, most of the world's water is in the oceans as saltwater with around 2% of this water is frozen fresh water. Around 1% is considered to be useful fresh water for human use (Roca and Tularam 2012; Roca et al. 2015) with groundwater being around 98% of the fresh water available on the planet. Clearly, the amount of water in the earth's environment never changes—for it is as a liquid (fresh water, seawater, rain and tiny droplets in clouds), gas (water vapour) and in a solid state (snow, ice or hail). The sun heats water evaporating much of it making it gaseous–vapour via evaporation. The currents of air carry the water vapour upwards due to the warming of the earth's surface. Rising higher and higher, the water vapour becomes cooler whence it condenses into tiny drops—forming clouds. These drops are then attracted to each other and thus join to form a heavier group of drops; and later due to changing temperature, pressure and other climatic conditions, they fracture to fall back to the earth as rain, hail or snow.

Rainfall may be evaporated directly from water, land or vegetation; the runoff water from rainfall moves into streams and wetlands, and some infiltrate and soak into the surface ground to groundwater, or are absorbed by plant roots. The water trapped in plant and trees can also return to the water vapour and flow into the air by a process known as evapotranspiration—from the leaves of plants. Some of the rainfall water do indeed infiltrate deeper into the ground and thus add to the existing groundwater. This volume of water then slowly flows along the direction of high to low elevation levels—that is, the groundwater flows towards the rivers, wetlands or the sea (Tularam and Singh 2009; Water and Rivers Commission 1998a, b). It is noted that any groundwater is then an important source of drinking water and irrigation water. The management of this particularly important resource and commodity is critically important if we are to fulfil the present human water demand and in addition, sustain it for the future generations.

Figure 1 shows that the water is constantly flowing and moving and forming an important system—after rainfall, water is flowing from an area of recharge (rainfall on the hills) to an aquifer before moving towards a discharge area—a spring, wetland or creek. The water flows in a pattern that follows the surface topography but in a downhill direction. There is a difference in the amount of rain that falls in regions around the world—the fall amount differs from country to country as some parts experience high to very high rainfall levels, while other areas experience drought conditions known as a fresh water crisis (Tularam and Krishna 2009; Tularam and Hassan 2016a, b).

3 Groundwater Flow Modelling

A flow model of groundwater can be developed by studying the differences in energy. The water flows from high-energy areas to low-energy areas. The total energy content (unit volume of water) is determined by the sum of gravitational potential energy, pressure energy and kinetic energy and this may be written mathematically as

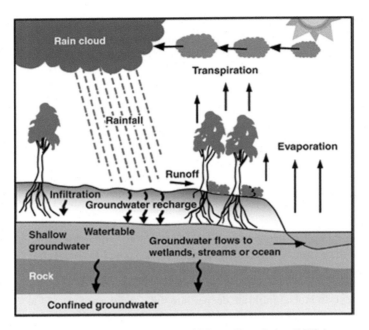

Fig. 1 The natural water cycle. *Source* Water and Rivers Commission (1998a)

$$\text{Total Energy (TE)} = \rho g z + P + \frac{\rho V^2}{2},$$

where ρ is the fluid density, g is the gravitational acceleration, z is the elevation of the measuring point relative to a datum, P is the fluid pressure at the measurement point and V is the fluid velocity.

The groundwater volume flows slowly (on the order of 1 m/day or less), and this implies that the volume of water has a nearly zero level of kinetic energy (KE) relative to the other terms in the TE expression above. Ignoring the KE leads to an expression for the mechanical energy per unit weight (ρg); thus, the concept of hydraulic head is developed (h): Hydraulic head$(h) = z + \frac{P}{\rho g}$. The flow is from a high hydraulic head to low hydraulic head. The porous media through which the water flows causes some resistance and this is taken into account by considering soil properties—the rate of flow depends on soil properties such as the permeability, k, which is a measure of the ease with which a fluid flows through a soil matrix. Since the Darcy's law states that the flow rate is linearly proportional to the hydraulic gradient we may write $q = -\frac{\rho g k}{\mu}(\nabla h)$, where q is the Darcy flux, or flow rate per unit surface area, and μ is the fluid viscosity. Darcy's law can also be written as $q = -\frac{k}{\mu}(\nabla P + \rho g \nabla z)$. Assuming a constant density of groundwater, the Darcy form is reduced to a simpler Darcy's law but when the assumption is not appropriate another form is used as given above (Mulligan and Charette 2008).

4 Coastal Aquifers

In coastal aquifers, the groundwater travels from upland (hills) passing through inland aquifers. The flow is variable in which it is affected by a number of forces that are acting and this causes the great variability in the flow regime concerning space and time (Fig. 2). As noted earlier, the hydraulic gradient seems to be the major driver of the flow. As the fresh water flux reaches the coastal boundary, a density difference mixing occurs making the water circulate simply under the influence of the force of gravity (Li et al. 1999; Li and Jiao 2003); there is an influence from the oceanic waves and tides as well (Taniguchi et al. 2002; Burnett et al. 2003). This mixing process together leads to a dispersive circulation particularly along the fresh water–saltwater boundary within the aquifer; this is because of the density difference of fresh water with the salty water (Kohout 1960; Michael et al. 2005). In fact, there are also other forcing mechanisms that affect the rate of fluid flow for both fresh and saline groundwater at the coastal boundary.

For the sake of simplicity, it is assumed that there is a sharp transition, or interface between the fresh water from upstream and saline water from the seaward side. In fact, the interface is much more complex given the many number of forces that are acting on the zone. An estimate of the position of the fresh water–saltwater interface can be determined using some assumptions: a sharp interface, horizontal groundwater flow only, and lack of flow in the saline water. This simplifies the analysis in which pressures at adjacent points along the interface are equal. In this manner, the depth of the interface can be calculated $h = \frac{\rho_1}{\rho_2 - \rho_1} = 40z$, where ρ_1 is the density of fresh water (1000 kg/m^3) and ρ_2 is the density of seawater (1025 kg/m^3). In this manner of calculation, the depth of the interface is 40 times the elevation of the water table relative to the mean sea level. While this is an approximation of the interface location, it has been used as a benchmark for when

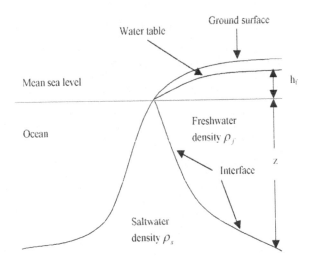

Fig. 2 SWI in coastal lowlands (Tularam et al. 2006)

recharge increases, groundwater levels increase, which in turn forces the interface downwards. At the times of little or no recharge, the reverse process occurs as correctly predicted by the rule.

4.1 Groundwater Salinity and Saltwater Intrusion (SWI)

Groundwater salinity and saltwater intrusion (SWI) are important concerns for all environmentalists (Tularam and Glamore 2004a; Tularam and Singh 2007, 2009). Any signs of pollution can be devastating, particularly for farmers but also our future generations. It is said that one day water will be worth more than price of gold and this underground resource will then become highly sought after and valuable.

The landward encroachment of seawater into fresh coastal aquifers is a problem that has been noted in many parts of the world due to rising sea levels or decreasing upland coastal groundwater levels (Tularam and Glamore 2004a; Tularam and Singh 2007, 2009) and such a process is known as SWI. Seawater intrusion (SWI) is the landward migration of seawater into fresher coastal aquifers, occurring underground. So this process is not visible to the population living in the coastlands. Australia is a country that is particularly susceptible to SWI since most of the population live around the coastal lands and much farming occurs in the coastal lowlands such as sugar cane, which is often irrigated using coastal aquifers (Ivkovic et al. 2012; see Fig. 2).

The coastal population is on the increase, and this is turn will lead to much industrial and farming demands, which will make the coastal aquifers susceptible to SWI. In most cases, however, SWI is caused by borewater extraction in coastal lands for irrigation purposes, or drinking; but the intrusion inland could also occur due to sea level rise (ice melts or tsunami type surges occur, etc.) causing the flow gradient to reverse or when the upland groundwater recharge variations occur due to rainfall extremes. The climate change appears to be the main culprit in this regard for the changes in the earth's climate that has been attributed to it. Climate change appears to be the cause of the extreme conditions such as tsunami events, droughts and/or rather low/high rainfall frequency and levels. Each of these impacts can further enhance the effects of "normal" or natural SWI in coastal lowlands.

More specifically, the meeting of the seawater and the upland fresher groundwater flowing when mixes actually forms an underground wedge (diffusive rather than a sharp wedge). It is referred to as an interface of the two, where the less dense fresh water is above; that is, the flow of the groundwater is above the interface because it is less dense. The separation position is often called saltwater wedge (Fig. 2), which clearly cannot be seen from aboveground. However, drilling can provide evidence of the same, for example, the evidence from boreholes confirms this interface but as a diffusive rather than a sharp as often in diagrams drawn for convenience. The difference in density means that saltwater also flows towards the landward side from the seaward side but stays below the fresher groundwater; as noted earlier, the upland groundwater is flowing from the landward side to the sea; thus, there is saltwater wedge beneath the fresh water. This wedge often causes saline water upland of the coastline and is sometimes many metres beneath fresh water in coastal aquifers towards the landward side.

Ghyben–Herzberg was one of the first to study this particular wedge and interestingly, this model is in use even today. If z is the depth of the interface below sea level then we may estimate the depth to the saltwater interface using the relation: $z = 40$ h, where h is the height of the groundwater above the sea level. So this may mean that 1 m decline in fresh groundwater level could potentially result in a 40 m rise in the position of the fresh water–saltwater interface.

Figure 2 also shows a mixing region at the interface and this is mixing between the ground fresh water and saltwater. There are a number of processes that drive the mixing that lead to the creation of the transition zone but in the main there are two: the atomic–molecular diffusion driven by concentration gradients and mixing propagated by mechanical movement processes—referred to as mechanical dispersion. The transition zone and the saltwater wedge are variable and dependent on the climate conditions.

A number of authors have studied the above relationship and the extent of seawater penetration into the aquifer (Banasiak and Indraratna 2012; Tularam and Singh 2009). The properties of the aquifer largely determine what happens and this has been well described by Custodio and Bruggeman (1987) in their approximation of L the penetration inland of the so-called 'toe' of the interface $L = \frac{kb^2}{2aq_o}$, where a is the approximately 1.025; k is the hydraulic permeability; b is the width of the aquifer and q_0 is the fresh water discharge per unit coastal length. From the equation, the so-called toe penetration is related to the permeability of the aquifer; also, it is related to the square of aquifer thickness. L is inversely proportional to fresh water discharge.

Daily tidal features, rainfall recharge variations or variable irrigation patterns may all influence the interface position (diffusive). As will be noted later, tsunami effects may also affect the position. If there are other aquifers across the depth of the coastal lands, drilling and irrigation pumping may cause "upconing" of groundwater; and this, in turn, may force the saline seawater into any of the existing fresh water aquifers, thus polluting them. While there are other sources of salinity pollution of coastal groundwater such as waste from agricultural features, in the end essentially there are two main concerns in coastal areas: Acid Sulphate Soils (ASS) pollution and SWI. The pollution effects of these are often irreversible and therefore any pollution is not sustainable for the longer term.

To summarize, the main aim of this chapter is to review some of the studies by the author and colleagues in the area of ASS and SWI. Both types of pollutions are equally damaging to the groundwater and often lead to poor production levels in coastal farms—at times total destruction of the coastal lands may occur. In the following, we first review water flow, ASS and salinity in the context of groundwater pollution. A number of studies that were conducted in Australia are then examined in terms what led to pollution. Also, some models of flow and transport of polluted waters in mainly coastal areas in terms of any difficulties experienced with models are discussed. Finally, a summary of the key/main points related to groundwater and its pollution completes the chapter.

5 Acid Sulphate Soils (ASS)

ASS affects more than 3 million ha of coastal Australia, with up to 0.6 million ha in New South Wales (NSW) (Indraratna et al. 2001a, b; Indraratna et al. 2002; White et al. 1997). In this case, iron pyrite (FeS_2) is the main culprit and when this exists, the relevant soils are labelled as potential acid sulphate soils (PASS). However, these soils may indeed contain other sulphidic materials such as iron monosulphide (FeS), greigite (Fe_3S_4) including many organic sulphides (Bush and Sullivan 1999). If the PASS is undisturbed and is in non-oxidizing conditions then the pyritic materials are relatively inert. However, when oxidized (the soils are now ASS) they can cause highly acidic conditions in the soil matrix. If this condition of ASS is followed by rainfall events and thus re-flooding the soil waters are acidified and later flow into the coastal waterways causing severe environmental, economic and social damage to the flora and fauna; this ASS problem is not only a problem in Australia but indeed in many coastal areas worldwide.

The pyrites found in Australia (FeS_2) were formed many thousands of years ago during saline inundation of sea conditions that included soils rich in iron. As noted above, when PASS is submerged by the water table (thus preventing atmospheric oxygen reacting with the pyritic layer), the pyrite is inert but much of the Australian coastline has oxidized sulphide minerals (ASS) due to various reasons. An engineering solution to lowering risks of coastal flooding in low-lying lands led to the installation of rather a large number of deep surface drains in coastal lowlands, but this, in the end, has had a rather negative effect—a general lowering of the water table elevation, where ever such deep drains were constructed. As a result, the existing pyritic layers (PASS) were exposed to oxidizing conditions—via atmospheric oxygen diffusing to drain soil matrix profiles. Once the soil matrix has ASS, any amount of flood type rainfall events would then cause the acid stored in the matrix to dissolve into soil water—this ultimately led to the pollution of the surrounding flood mitigation drains, creeks and river systems.

Indraratna et al. (2001a, b, 2002) and Banasiak and Indraratna (2012) have completed many projects over the years to help control this ASS problem in coastal Illawarra. His team not only investigated the installation of weirs in the flood mitigation drains in order to raise the water table elevation but also via tidal intrusion they neutralized some of the highly acidic conditions in the drains and creeks in Berry (NSW, Australia). The uncovering of the many relationships in the soil conditions has led to the development of a number of mathematical models that can simulate acid production and transport in Illawarra.

5.1 Chemical Equations

Acid sulphate soils possess chemical and biological constituents, which when exposed to oxygen will generate acid (Indraratna et al. 2001a, b). PASS are generally under anaerobic sub-aqueous conditions. Sulphate-reducing bacteria break down the organic matter and reduce dissolved sulphate ions to sulphides, and iron III oxides into iron II. The main source of sulphate is seawater (Dent 1986) and the reaction is

$$Fe_2O_{3(s)} + 4SO_{4(aq)}^{2-} + 8CH_2O + 1/2O_{2(aq)} \rightarrow 2FeS_{2(s)} + 8HCO_{3(aq)} + 4H_2O. \quad (1)$$

A number of environmental impacts may be noted as mentioned earlier such as acid leaching, groundwater contamination and salinity intrusion. PASS poses chemical and biological constituents, which when exposed to oxygen will generate acid (Indraratna et al. 2001a; Tularam and Indraratna 2001). Sulphate-reducing bacteria break down the organic matter and reduce dissolved sulphate ions to sulphides, and iron III oxides into iron II. When exposed to oxygen, the naturally occurring iron sulphide minerals (pyrite) are oxidized forming sulfuric acid as shown:

$$FeS_2 + \frac{15}{4}O_2 + \frac{7}{2}H_2O \rightarrow Fe(OH)_3 + 2SO_4^{2-} + 4H^+ \quad (2)$$

If the water table drops below the PASS strata, oxygen reacts with the pyrite producing acid in the soil matrix that later pollutes groundwater. The contaminants seep via groundwater to creeks and rivers often killing fish.

5.2 Tularam and Indraratna (2001) Model

As noted earlier, the oxidation of pyrites and acid pollution of surface waters close to estuaries can be attributed to pyrite oxidation (Indraratna et al. 2011, 2012; Tularam and Indraratna 2001). The leaching of acid has caused changes in pH of waters and in turn caused a number of environmental problems—fish kills in rivers, affected many oyster farms and influenced the local flora and fauna—marine life and habitats. The local infrastructure such as bridge walls have been affected by the acid conditions. The farmers report lower levels of agricultural production due to acid sulphate soils conditions (Indraratna et al. 2001a, b, 2011, 2012; Blunden et al. 1997). The above authors have applied a number of remedial strategies such as manipulation of the water table particularly in the Illawarra (New South Wales, Australia). The idea to submerge the pyritic layer through the raising of the existing water table was considered to prevent as much oxygen from reaching and oxidizing the PASS (Indraratna et al. 2001a, b, 2002). Blunden and Indraratna (2000) among others (Indraratna et al. 2001a, b, 2002) studied the groundwater manipulation method—and in the end developed a three-dimensional model to predict the amount of acid produced during droughts in order to manage the raising of the water table. A new tidal buffering strategy using gated drains was also examined (Indraratna et al. 2002). The drain water has a high level of acidity that is to be neutralized and it was suggested that the bicarbonates and carbonates that can be transported to the acid conditions may help neutralize the acid (Broughton creek, New South Wales). For such remediation to be effective, an estimate of the total acid within the pyritic soil matrix is needed. There are some models in the literature, but a cylindrical coordinate model was developed and used by Tularam and Indraratna (2001). The new model compared well with other models when calculated for the 120 ha field site (South Coast of Illawarra, NSW).

5.3 Mathematical Model

Detailed analysis of soils at the Berry site in the South Coast of Illawarra clearly show the existence of extant root channels caused by previous organic activity. The pyritic material mostly exists close to the root channels. In the past, a two-way linear modelling has been considered: The distribution of oxygen has been determined in the channel, and then another model has been considered for the linear oxygenation of soil. The current cylindrical approach allows these two processes to occur simultaneously. Tularam and Indraratna (2001) shown a possible radial approach that can model such root channels as they pass through the pyritic layer. Even though the root channels are not exactly parallel in real life, the parallel assumption nevertheless is the first approximation of a much more complex issue of angular determination. The root channels are surrounded by acidic soil, which is assumed to contain homogeneously distributed sulphides (mainly FeS_2). This is also a reasonable assumption because the radius considered in this modelling is very small indeed and is close to the channels. The gaps in the cylindrical model provide an improvement area for future work. These gaps have been ignored in this article because of the rather small volume lost. At the Berry site, pyrites are found in clay soils about a metre below the ground surface. The chemical equation representing the pyrite oxidation process shows that the amount of pyrite oxidized is directly proportional to the amount of oxygen consumed (Blunden and Indraratna 2000).

There were approximately 300 root channels per square metre at the berry site (Blunden and Indraratna 2000). In actual site, only a few root channels were noted to be "exactly" parallel in the soil matrix but such an assumption simplified the solution approximation. This approach provides a geometry that may be modelled in cylindrical coordinates (r is the vertical coordinate and θ is the angle based on the radial direction). The oxygen concentration within the cylinder is ideally dependent on these three variables and is denoted by $c(r, z, \theta)$, while the time variable is donated by t. Assuming oxygen transport into soil profile is solely based on diffusive process, both within the pore channels and in the soil matrix, the oxygen distribution may be given as

$$\frac{\partial c}{\partial t} = \frac{1}{r}\frac{\partial}{\partial r}\left(rD_r\frac{\partial c}{\partial t}\right) + \frac{\partial}{\partial z}\left(D_z\frac{\partial c}{\partial z}\right) + \frac{1}{r^2}\frac{\partial}{\partial \theta}\left(D_\theta\frac{\partial c}{\partial \theta}\right) - k(c), \quad (3)$$

where k is the sink term dependent on concentration of oxygen. A cylindrical process allowed for the rotational symmetry in the matrix using a vertical inward directed root channel axis. In this manner, any of the derivatives with respect to the angular variable did not exist so Eq. 3 was changed to

$$\frac{\partial c}{\partial t} = \frac{1}{r}\frac{\partial}{\partial r}\left(rD_r\frac{\partial c}{\partial t}\right) + \frac{\partial}{\partial z}\left(D_z\frac{\partial c}{\partial z}\right) - k(c). \quad (4)$$

To solve this problem, a number of boundary conditions need to be determined. At $r = 0$, there is symmetry hence no boundary condition. Assuming oxygen is totally consumed as it reaches the maximum radius of influence (r_s), the extreme boundary condition is given by $\frac{\partial c}{\partial r}(r_s, z, t) = 0$. Also, assuming that all oxygen centres the soil matrix only from the channel walls, the annular ends may be bounded by the following conditions: $\frac{\partial c}{\partial z}(r, 0, t) = \frac{\partial c}{\partial z}(r, z_c, t) = 0$. As far as the root channel and wall interface are concerned, the oxygen concentration is assumed to be continuous across the interface, and given by $c_c(r_c, z, t) = c_s(r_c, z, t)$. The subscripts c and s refer to root channel and soil matrix, respectively, while the transfer rates must also match across the interface, the solubilities and diffusion coefficients in the different mediums may differ. Hence, the mass balance condition may be written $S_c D_r^c \frac{\partial c_c}{\partial r}(r_c, z, t) = S_s D_r^c \frac{\partial c_c}{\partial r}(r_c, z, t)$, where S_c and S_s are the respective solubility coefficients in the root channel and soil matrix surrounding the channel, respectively (in this case, $S_c = 1$).

5.4 Consumption Kinetics or Sink Term

The consumption term, $k(c)$, earlier can be approximated by a number of models. Davis and Ritchie (1986) developed a model that is commonly used in the literature on acid soils. According to the authors, pyrite crystals devour oxygen at a rate proportional to the surface area of the crystals given as $\frac{\partial m}{\partial t} = kA$, where m is the concentration of oxygen, A is the surface area of the crystal given (m^2), k is the rate constant $(kg/m^2/day)$, m is the mass (kg) and t is the time (day). The authors use $0.052\sqrt{c}$ as an approximation for k attributed to Mckibben and Barens (1986). Simplifying the expression and portioning the oxygen in the soil solution into that consumed by pyrite and organic matter (that is, $c_s = c_p + c_{com}$), the rate of pyrite oxidation using binominal theorem may be simplified as

$$\frac{\partial m}{\partial t} = \frac{0.311 C\,\text{FeS}_2 \sqrt{C}}{\rho d}\frac{1}{r} + \frac{0.311 C\,\text{FeS}_2\, C_{\text{om}} \sqrt{C}}{2\rho dc}, \qquad (5)$$

where C_{FeS_2} the concentration of pyrite in the soil matrix, c is the oxygen consumed by pyrite, ρ is the bulk density of matrix and d is the diameter of crystals. The second on the right represents a small amount of oxygen consumed by organic materials, usually assumed to be a constant. The right-hand side of equation is directly related to the total amount of oxygen consumed in the soil matrix per unit time and can be modelled more simply as $\frac{0.31126 C\,\text{FeS}_2 \sqrt{C}}{f} + \text{om}$. At similar distance from the root channel wall, the consumption of oxygen by both factors will have equal supply of oxygen laterally, thus justifying one of the boundary conditions presented earlier. The model based on the above consumption equation can be written as

$$\frac{\partial c}{\partial t} = \frac{1}{r}\frac{\partial}{\partial r}\left(rD_r \frac{\partial c}{\partial t}\right) + \frac{\partial}{\partial z}\left(D_z \frac{\partial c}{\partial z}\right) - \frac{0.311 C\,\text{FeS}_2 \sqrt{C}}{\rho d}. \qquad (6)$$

The Michaelis–Menten (MM) uptake kinetics model usually written as $k(c) = \frac{k_1 c}{c^* + c}$, where k_1 is the maximum rate of consumption and c^* is the concentration at half the maximum rate. When oxygen concentration is low the linear rate is dominant, and when the concentration increases (that is, c^* is negligible when compared with c in the denominator), a constant rate trends to describe the oxygen uptake.

When compared with the various simplifications performed on the complex Davis and Ritchie (1986) model, the MM equation appears to better describe the relationship between the rate of consumption and dissolved oxygen concentration (for example, when $k_1 = 6.5$ and $c^* = 0.0022$ (Tularam and Indraratna 2001). The MM model is not only simpler but seems to be a more elegant equation that takes into account the fundamentals of pyritic and matter consumption. Based on the MM uptake kinetics, the oxygen distribution may be written as

$$\frac{\partial c}{\partial t} = \frac{1}{r}\frac{\partial}{\partial r}\left(rD_r \frac{\partial c}{\partial t}\right) + \frac{\partial}{\partial z}\left(D_z \frac{\partial c}{\partial z}\right) - \frac{k_{1c}}{c^* + c}, \qquad (7)$$

where k and c^* may be estimated. Other relevant field measurements are presented in Tularam and Indraratna (2001).

Salinity and ASS are two rather important polluters of groundwater and in the earlier sections, a brief overview has been presented for readers so that they are familiar with the nature of the models involved in modelling flow and transport of acidity and salinity in complex domains. Groundwater pumping is considered next in that pumping from aquifers can lead to rather damaging after effects and longer term issues of land, creeks, rivers and the ecosystem of the region.

6 Impacts of Groundwater Pumping

The pumping of groundwater solves many problems that are faced by those living in coastal areas and low rainfall regions. Yet, there is a negative side to long-term pumping as well, such as those noted earlier—pyrite oxidation, seawater intrusion, aquifer overdraft and groundwater depletion that may also lead to some damage to infrastructure, as well as in some cases land subsidence problems.

There are a number of reasons why pumping is critical, for irrigation, for example. But acid pollution and salinity intrusion may result from this process. While there is much literature concerning pumping problems, less exist on issues such as land subsidence, loss of flora and fauna. In some countries, however, groundwater resources are vital for the public use and domestic supply but it is to be noted that excessive pumping levels do lead to decrease in groundwater levels that in turn lead to overdraft problems.

Overdraft is pumping of water from an aquifer in an excess of the supply flowing into the basin. This results in the depletion of groundwater that then can greatly harm the environment. This is only a problem if the pumping is above the safe yield. The lakes, ponds wetlands and aquatic species can all be affected. The reduction of water tables reduction is something that has been noted around the world and it has impacted agricultural production in lowlands in particular. The flora and fauna depend on the water levels in rivers, lakes, ponds and streams. In this manner, the aquatic and wildlife habitats are affected.

Surawski et al. (2005), Tularam and Singh (2007) and Tularam and Krishna (2009) among others have all explored the effects of longer term pumping in north and south of the Brisbane region. In the main the affected areas are coastal lowlands. SWI, aquifer overdraft, groundwater depletion, and interesting land subsidence have all been noted in such conditions. In the low-lying coastal areas, it is important to note that acid pollution and SWI have been also reported.

Tularam and Krishna (2009) found that land subsidence was associated with the pumping. Land subsidence can be observed when there is a 'downward movement or sinking of the earth's surface caused by removal of underlying support'. Increased groundwater borehole pumping often leads to land subsidence—the groundwater withdrawal lessens the soil water movement within the soil matrix; and thus the support provided by the soil water in the soil matrix; essentially, the soil water fills void and supports. This appears to be one of the least reported aspects but a 'potentially destructive impact associated with continued groundwater pumping' (Tularam and Krishna 2009). So without adequate recharge a compression occurs in the sediments causing some settling or subsidence. All such processes occur underground, so it is difficult to appreciate by those who live around such low-lying areas. This is where the importance of modelling of groundwater flow and transport becomes such an important tool.

A number of places in US are affected by land subsidence due to high groundwater withdrawals. The flow patterns in the groundwater and surface are also affected by subsidence; there is thus a decline in storage capacity. There has been damage to roads, railways, canals, levees public and private buildings that have been attributed to overdraft conditions. Due to the lack of infiltration, there is therefore also an increased flood risk and infrastructure settlement.

Although land subsidence and other sources of pollution are important, in this paper, only pyrite oxidation and saline water intrusion are examined since excessive pumping lower water levels which then facilitate the oxidation of materials through exposure to air via the latent vertical root systems. Later re-flooding of aquifers causes the dissolving of the oxidized materials into groundwater making it often highly acidic in the case of acid sulphate, for example.

7 Acid Sulphate Soils and Salinity

In their paper on canal estates, Tularam and Dobos (2005a, b) study housing developments in coastal lowlands. Such developments almost always cause the lowering of the water table for some period of time, as required during construction periods. Many of the Australian coastal lowlands contain pyrites that cause acid pollution upon oxidation. The lowering of the water table to allow for dry excavation constructions may lead to the oxidation of the pyritic soils producing acid in the soil matrix. Moreover, SWI may also occur due to the imbalance in the flows of the subsurface fresh water from upstream and salty waters from the seaward side causing, that is, changes in the flow dynamics as noted earlier. The results showed that the water level decreased during dewatering activities and increased after re-flooding. The dissolved oxygen, titrated acidity and salinity levels rose during periods of increased groundwater depth, while pH levels fell. The increase in oxygen availability and the associated increase in acidity may lead to the dissolution of heavy metals, but it was noted that the monitoring data was lacking. Since salinity may increase due to other close by saline surface water bodies such as a tidal affected river, further studies are required to understand subsurface flow patterns during dewatering operations.

The dry excavation of three separate cells led to an increase in groundwater depth. The subsequent return to the original depth was noted after re-flooding (time lag noted). Dissolved oxygen levels in groundwater rose, the pH levels decreased and the groundwater acidity content increased. The levels of groundwater pH and acidity did not recover after re-flooding of the dry cells as rapidly as dissolved oxygen; in some piezometers, the acidity levels continued to rise after re-flooding. This may be due to rainfall events that flush residual acidity from the soil pores into the groundwater.

Salinity increased during periods of dewatering and dry excavation when the groundwater table depth increased. There was a return to lower salinity levels when the dry cells were re-flooded and the groundwater table rose. Essentially, lowering the water table in ASS material will increase dissolved oxygen levels, generate acidity and lower pH levels; and salinity intrusion may occur depending on the location of the canal estate.

Tularam and Singh (2009) studied salinity intrusion in many coastal estuarine and waterways in various parts of Australia. Brisbane is a river city close to the coast and as such the condition of the river influences the surrounding groundwater quality. The FEMWATER finite element package is used to model the subsurface flow and transport and the simulations showed that salinity intrudes as far as 15–25 km from the coast. The model simulations show tidal effects on Brisbane River and surrounding groundwater and water quality deterioration appear to be related to coastal seawater intrusion over time. The model describes the flow dynamics well and also captures the influence of a number of factors such as the tidal inundation process in the river. The salinity that is present in the river compounds SWI close to the river boundary. This is why some high levels of salinity were noted upstream. Higher salinity was also noted in the groundwater that was adjacent to the river bank boundary, than when compared to subsurface waters in the regions away from the bank.

A time series-based coastal study by the authors showed similar inland intrusion in Brisbane coastline region. Tularam and Keeler (2006a, b) modelled tidal intrusion to study the influence of tidal behaviour on groundwater height. The effect on salinity intrusion was also included in the study using spectral time-series methodology. Tidal behaviour and groundwater depth were related and the salinity levels varied due to tidal influence for the Brisbane coastal lowlands. The process can be incorporated into new models in order to accurately allow for the seaward boundary conditions. The time-series approach was appropriate to study the effect of tidal behaviour on coastal groundwater dynamics including salinity variability caused by the process. In fact, the coastal weather conditions will also be an important factor. A 3D flow and transport model may be included to model such a process and this process will allow for the salinity varying due to the changes in the groundwater depth inland. Tularam and Keeler (2006a, b) further examined the effect of tidal behaviour on groundwater level and salinity intrusion on the Port of Brisbane. The tidal wave dampens the amplitude of the groundwater head affecting the energy or pressure of the system.

8 Surface Flooding and Leakage into Groundwater

Tularam and Singh (2007) in their paper examined groundwater pollution following a tsunami event. Tsunami type event was simulated on the Pine River's coastal aquifer system. A 3D density-dependent flow and solute transport model finite element methods was used to develop the model. The longer term influence on the aquifer water quality was noted. Further, little variations in the soil water permeability may significantly affect the water system.

A physical conceptual model of the Pine River's shire aquifer was developed using the geological features of the coastal region. This model was used to develop a three-dimensional finite element model of the same using the FEMWATER3D package (Hsin-Chi et al. 1990). Attaching appropriate soil water and hydrology parameters to the model allowed a 'tsunami' type event to be simulated utilizing a special capability of FEMWATER. Importantly, FEMWATER package involves a 'cold' start as well as a 'hot' option for initial conditions. An estimate of heads and concentrations of the system is initiated in the model till the flow converges based on the governing equations (Hsin-Chi et al. 1990). This output of heads and concentrations is then initiated into the FEMWATER model to simulate the process efficiently and economically in terms of time.

The contamination of groundwater due to a tsunami type event is particularly dependent upon the nature of the permeability of the soil matrix and the water table height. From the simulations, it was observed that the surface sediments of the Pine Rivers Shire aquifer are not permeable enough to allow much of the seawater to contaminate the groundwater during the first two hours of the event. The 1–2 m AHD depth of the water table was also responsible for groundwater not being contaminated in that period. However, in areas nearby where the surface sediments are more permeable, the aquifer will be much more affected by infiltration increasing the risk of contamination of the soils water system and groundwater.

It is true that some drainage occurs, some seawater often remains on land inland as surface water bodies in local low-lying areas disperse slowly. This suggests that seawater will be retained in the surface zone affecting the soil and groundwater conditions over a longer period of time. This contamination can be simulated as point sources of salt in soils and allowed to diffuse through the aquifer over longer periods of time using FEMWATER. While the simulation shows little change in the salinity levels for the longer simulation period, more research is required to study more closely the different soil types and soil–water systems.

9 Tidal Intrusion—Solution to Acidity Problems

9.1 Buffering Capacity

Tidal buffering relies on carbonates and bicarbonates that are transported up the estuarine system via the tidal system. For example, if a neutralizing capacity of 0.625 mol H^+/m^3 (1/4 neutralizing capacity of seawater) was discharged into a buffering zone, then 625 mol of H^+ would be neutralized. Depending on the rate of acid flow and volume, the buffering capacity should help increase somewhat the pH of the drain water more than two units. Equation 2 represents the buffering reaction of sulphuric acid $(pK_a - 3)$ to carbonic acid $(pK_a - 6.3)$ a weaker, slightly ionized acid (Tularam and Glamore 2004a).

To simulate intrusion in the creeks or drains in lowlands, a new type of model is developed. Some estuarial models exist but a significant difference exists between this case and earlier models. Early models did not concern smaller creeks or drain conditions, where the flow condition is minimal. Due to difficulties involved, two simplifications were considered in modelling, namely (i) a constant cross section of the creek/drain and (ii) well-mixed intruding conditions.

$$\text{Seawater} \quad \text{Sulphate Oxidation} \quad \text{Carbonic Acid}$$
$$HCO_3^- + H^+ + SO_4^{2-} \rightarrow H_2CO_3 + SO_4^{2-} \tag{8}$$

Indraratna et al. (2001a, 2002, 2011) designed and developed one-way floodgates for flood mitigation drains in NSW acid soils (Illawarra coastal region, Australia). This was mainly to allow some carbonate/bicarbonate buffering of initial acidic cations (pH < 4:5) that normally discharge during the ebb tide. Improvements in the carbonate/bicarbonate buffering processes have led to much improved water quality upstream of the one-way floodgate; this also has high concentrations of aluminium and iron in addition to high acidity. Essentially, the alkaline water intrudes upstream and reacts with H^+ ions and the study showed that tidal buffering via modified floodgates was not only possible but led to better water conditions all around. A rainfall event of 131.8 mm later showed evidence of a strong recovery time as well as the average pH being markedly improved.

It seems now that the process has included tidal buffering via two-way floodgates that seems to improve water quality without the addition of anthropogenic chemicals. Indraratna et al.'s (2002) results from a 15-month field trial near the town of Berry show that 'Short wet periods, triggered by rainfall, were characterised by neutral pH and low EC readings' (p. 1). The pyrite-induced acidic conditions pollute the groundwater table that in turn leaches the pyrite oxidation products into the flood mitigation canal (pH < 4:5); also, there were increased concentrations of aluminium and iron in the drains. Their results provided field evidence to confirm the effectiveness of tidal buffering. Thus, the installation of a two-way, vertically lifting floodgate 10 months into the study permitted tidal intrusion and buffering thus decreasing Al and Fe concentrations, even in the extended dry periods. There are still conditions that need further investigation particularly the extreme acid leakage and high rainfall conditions because the saline intrusion into the soil matrix tends to be dependent on the rainfall. Here is more evidence that confirms the previous ASS management strategies need to be modified to avoid further deterioration of the subsurface environment.

9.2 Mathematical Modelling

Existing salinity intrusion models, both stationary and non-stationary (Tularam and Glamore 2004b; Abarca et al. 2002; Bear 1979; Bear et al. 1999), deal with estuaries, rivers or creeks, where the width of the estuary is considered to change over distance. Savenije (1988) used exponential equations to model the width of an estuary using examples of river mouth estuaries. Savenije's model only deals with funnel type estuarial geometry. Such models may be useful but there are no current models specifically relating to creek/drain SWI. It was decided that the creek/drain intrusion is a simpler case of saline intrusion. The width of the flood mitigation creeks/drains can be assumed to be a constant over some distance and the small creek or drain water is assumed to be well-mixed during intrusion so a salt wedge and diffusion zone are not modelled. A 1D advection–diffusion equation of saline intrusion of a tidal creek or drain is $\frac{\partial c}{\partial t} = \frac{1}{A}\frac{\partial}{\partial x}(AD\frac{\partial c}{\partial x}) - v\frac{\partial c}{\partial x}$, where c is the salinity concentration, A is the cross section, D is the total dispersion coefficient, v is the velocity of flow [small] and x is the distance along the drain from the mouth of a tidal creek. To account for the tidal process, an averaging over a tidal cycle was considered and the averaged model is $A\frac{\partial \bar{c}}{\partial t} = \frac{\partial}{\partial x}(AD\frac{\partial \bar{c}}{\partial x}) + Q_f\frac{\partial \bar{c}}{\partial x}$. Over a tidal cycle, the fresh upstream discharge (Q_f) remains effective, thus changing the directional flow of salinity. In this case, $\bar{c}$ is the time-averaged salinity concentration of the small creek or drain water upland of the floodgates. When salinity intrusion has reached equilibrium (between downward and upward flow) the steady-state situation of Eq. 4 may be found, that is, when $\frac{\partial c}{\partial t} = 0$. The solution with the following boundary conditions $x = 0$, $\bar{c} = \bar{c}_0$; $x = L$, $\bar{c} = 0$ is

$$\bar{c} = \bar{c}_0 - \frac{\bar{c}_0}{1 - e^{-Q_f L/AD}} + \frac{\bar{c}_0}{1 - e^{-Q_f L/AD}} e^{\frac{Q_f x}{AD}}. \tag{9}$$

In equation above, $\bar{c}_0$ is the salinity concentration at the river end of the creek or drain and L represents a distance (m) along the creek or drain, where the saline concentration is negligible (zero or very close to zero).

To simulate salinity intrusion in coastal waterways in South Eastern Coastal lowlands of Gold Coast (Australia), Tularam and Glamore (2004a) developed a mathematical model. This process was to neutralize acidity caused by pyrite oxidation. The model results compared favourably with the field conditions. It was noted that SWI could in fact reduce pH of the upland water from 2.4 to about 6.0. The salinity left in the drains is of concern but controlled conditions allow the levels to remain low enough so that it is not of a concern to the landholders. The Nerang river flood mitigation system ought to be redesigned to allow controlled tidal water intrusion to neutralize acid leached into the waterways. An existing one-way floodgate can allow some saline water upland during tidal upland inflow periods. So we note that even some saline creek water may be useful to neutralize some of the acidic water stored in drains and creeks, thus lowering the impact on local flora and fauna—fish and other estuarine communities including their habitats downstream following heavy rainfalls.

Tularam and Indraratna (2001) developed a simplified mathematical model for tidal creek/drain water conditions. An analytical solution was developed to determine the amount of chloride levels in terms of EC that would intrude into the upper side of the flood-gated creek/drain over an average tidal cycle. The analytical solution compared well with field results obtained from two field sites. Sensitivity analyses showed that the model was appropriate for such intrusion modelling being computationally economical. The mathematical model could predict with reasonable accuracy the total amount of chloride level at any distance upland. Further simulations showed that, given the chloride conditions, very little if any intrusion would occur into the surrounding land. This is mainly due to the low concentrations of salt in the upland creek water and low diffusion levels against the normal flow conditions of the groundwater. The low flow conditions made diffusion an important factor in chloride transport upstream thus explaining the low salt intrusion levels upstream. It seems therefore that the landholders have little to be concerned about regarding chloride intruding their land from the smaller creeks/drains. During floods, the low levels would be diluted even further with the large amounts of surface water and therefore would also be of little concern as well.

This investigation involving saline intrusion is the first step of a major study that attempts to remediate the high levels of acid leakage due to the oxidation of ASS in the Nerang River. Much of the acid is first drawn into the deep flood mitigation drains and hence into the creeks and subsequently into the Nerang River causing a number of environmental problems. In this case, a weir-based groundwater management strategy involving higher water tables that submerges the ASS layer is not appropriate because the acid sulphate is only centimetres from the surface. For this reason, oxidation of pyrite has already occurred in large amounts and thus high levels of acid are already stored in the soil (Indraratna 2001a, b, 2002; Tularam and Indraratna 2002). Much of the leached acid is stored on the upland side of the flood-gated creek from previous high rainfall events, which remains a problem in many locations Australia wide. However, as shown earlier, the acid storage in the creeks can be buffered using a saline buffering strategy (Indraratna et al. 2002). It is recommended that a strategy be devised to allow restricted saline intrusion from Nerang Creek into the upland section of the creek beyond the floodgates through redesigned or modified floodgates similar to that applied in Shoalhaven (NSW) by Glamore (2003) and Indraratna et al. (2001a, b, 2002). Also, further studies involving chemical processes such as cation exchange buffering and a detailed study of the actual neutralization capacity of brackish water need to be conducted. Nonetheless, the results of this study and other studies done in Shoalhaven by Indraratna et al. (2002) suggest that the buffering strategy would be appropriate for such sites Australia wide. Therefore, the tidal buffering strategy ought to be a major part of a comprehensive approach to the management of ASS pollution in coastal regions.

10 Groundwater Modelling Tools

Tularam et al. (2006) and others have studied saline intrusion and clearly it has the potential to cause a number of difficulties for coastal lowlands around Australia but the negative impacts of salt intrusion are not observed in the short- to medium-term calculations. Models such as SALTFLOW and PDE2D are two such packages that may be used here to study the effects of the longer term. A 2D mathematical-coupled salinity intrusion model is developed to examine to compare the results of the models. The performances of the model packages are analysed in terms of ease of use, accuracy of simulations and capacity to represent physical processes related to the boundary conditions.

When compared to standard problems, SALTFLOW was favoured in modelling Australian lowlands over SALTFLOW. The SALTFLOW and PDE2D solutions were similar in many respects but SALTFLOW was favoured in the Henry problem for it was more accurate generally. The PDE2D required more user input but allowed more options in solving. Frind has used a dispersion coefficient different to that of Henry. The steady-state transport equation used by Henry was $D\nabla^2 c - q\nabla c = 0$. Both Pinder and Cooper (1970), Frind (1982) used $D\nabla^2 c - \frac{q}{n}\nabla c = 0$. They divided the equation by the porosity, whereas Henry did not (Voss and Souza 1987). Multiplying the equation simplifies to $nD\nabla^2 c - q\nabla c = 0$. Henry and Frind both used a dispersion coefficient (D) of 6.6×10^{-6} (m^2 s^{-1}) but the total dispersivity is given by the product of the porosity (n) and D. In this way, Frind solution used dispersivity of 2.31×10^{-6} m^2 s^{-1} (D^*). The different dispersion coefficients used by Frind may explain the poor agreement with the simulations performed. Moreover, the presence of numerical dispersion appears in Frind's original solution. This would cause the isochlor to migrate coastward (Volker and Rushton 1982).

10.1 SALTFLOW Modelling of Pollution

Surawski et al. (2005) studied acid pollution in the Pimpama (QLD, Australia) coastal plain using SALTFLOW. Their model included factors such as subsurface upland flow and groundwater extraction as well as the impact of any sea level rise. Pimpama is located between Brisbane and the Gold Coast in South East Queensland. The region is experiencing population growth and therefore became an area of interest for the study. There is ASS in the region in which sugar cane farms require irrigation; and thus there is some groundwater pumping needed leading to a possibility of sea levels rising due to very low coastal lands. Whether pumping from the aquifer is sustainable will be an important supply question. The simulations conducted represent the first attempt to assess the sustainability allowing for the stresses such as sea level rise and SWI.

Some preliminary work tested the performance of SALTFLOW package against some existing benchmark solutions from the literature. The ease of use of the SALTFLOW package meant that it was selected over codes such as PDE2D and SUTRA. Also, SALTFLOW allowed for time-varying boundary conditions for groundwater flow.

Use of time-varying boundary conditions was used to consider the effect of groundwater pumping chloride concentrations and the results showed significant SWI near the coastal boundary over the next two decades. The effect of the sea level rise was significant. It causes seawater intrusion into the coastal aquifer. Oude Essink (2001) found that the influence of the sea level rise would be less than that of the groundwater extraction when salinity intrusion is considered. However, Oude Essink (2001) had neglected to mention that relatively small groundwater yields (such as that of the Pimpama coastal plain), rising sea levels appeared to be a more important factor in of SWI than groundwater extraction than pumping. But the results do show some sensitivity to groundwater extraction. Since groundwater extraction is already limited throughout the Pimpama coastal floodplain (Harbison and Cox 2002), the above result was surprising when allowing for the conditions. Harbison and Cox (2002) argued that the maximum rate at which groundwater can be pumped from such estuarine deposits ought to be around $1\ L\ s^{-1}$ per pumping well. So even small rates of groundwater extraction may dangerously affect the plains.

Surawski et al. (2005) noted that the breakthrough curves were not at steady-state conditions. The other stresses placed on the aquifer may have been responsible for this. The seawater boundary condition is time varying, which could lead to a steady-state solution being impossible. The sensitivity runs conducted include groundwater extraction as well and a steady-state solution is hindered by the fact that a maximum rate of groundwater pumping may be exceeded, (Dagan and Bear 1968) thus suggesting that it has exceeded for the Pimpama coastal plains.

But the SALTFLOW was at a disadvantage since it had only the pre-conditioned conjugate gradient method. So a possible convergent solution cannot be found at times for problems with violated matrix symmetry during the running of the iterative solution. However, this an iterative matrix solver that provides solutions for both symmetric and non-symmetric matrices that could solve this problem. For example, when and if the pre-conditioned conjugate gradient method fails, the package may switch to a Jacobi or Gauss–Seidel iterative scheme.

The process led to results such as sensitivity to sea level rise, moderately sensitive to groundwater extraction and somewhat insensitivity to fresh water inflow. Climate change tells us sea levels will rise by about a few centimetres and so the Pimpama aquifer will be influenced about time particularly, the next 20 years. The present pumping levels do not seem to be sustainable and some remedial actions ought to be taken so that a fresher supply of groundwater is available for future domestic and agricultural supply conditions (Oude Essink 2001; Indraratna et al. 2011). Others ideas such as artificially injecting water into the aquifer or the introduction of physical coastal flow reduction barriers may also be considered.

11 Desalination as a Solution—Case of Surface and Subsurface Waters

One important and widely used method to deal with groundwater pollution problems in terms of potable drinking is by the desalination, and since vast amounts of seawater are available on earth, it is a great solution it would seem. However, desalination has its own problems and these were highlighted in Tularam and Ilahee

(2007). It was noted that planning and monitoring stages are critical to the successful management and operation of desalination plants. The site ought to be selected carefully away from residential areas because the brine pollution can occur in seas close to shores or in deep groundwater aquifers. The authors mentioned that Reverse Osmosis (RO) method is preferred when other methods are rather expensive. The RO method has higher efficiency (30–50%) when compared with distillation type plants (10–30%). However, RO membranes are easily polluted, with fouling and scaling. The cleaning of the filters is very costly and chemicals may be toxic to receiving waters.

As noted the byproduct is mainly brine concentration that is almost twice that of seawater. A deep ingrained long pipe that is usually installed far into the sea or the coastline to discharge this product. It is noted that the long-term effects of concentrated discharge have not been studied much. The cleaning of filters discharges traces of various chemicals used in cleaning including any anticorrosion products used in the plant. Even small traces of toxic substances can be harmful to the flora and fauna, including those in the marine ecosystem. Such deep pipes have been installed underground in Tugun (QLD). The deep piping structures more often than not pass below the pristine coastal aquifers.

There are also many other issues of concern such as when, where, how and which plant to consider. Any decision should be made by considering economic, environmental and socio-economical concerns together; the plant must be considered in light of global climatic changes, sea level rise and possible expansions in later water demand. In addition, the piping structure, discharge of toxic substances and brine concentrations at discharge points should all be monitored and were some of the many recommendations of this study (Tularam and Ilahee 2007).

12 Summary and Conclusion

The main aim of this chapter is to review studies on groundwater pollution related to ASS and salinity in subsurface environment. These pollution types are equally damaging to the groundwater and indeed lead to poor production levels in coastal farms and at times even irreversible destruction of the coastal lands. The paper first reviewed the nature of the subsurface flows and causes of both ASS and salinity based on flow and transport equations that underlie the groundwater flow processes. Then, a number of studies conducted in Australia are reviewed to examine the nature of work that has been done in recent times regarding pollution of coastal areas by acidity and salinity. A few problems with the modelling tools were examined in relation to flow and transport of acidity and salinity. An alternative to groundwater for potable water is desalination, which is briefly discussed and the problems related to the osmosis process are also analysed.

In sum, the nature of the subsurface flows and causes of both salinity and ASS based on flow and transport equations that underlie the groundwater flow processes highlighted the problems that exist. A number of studies conducted in Australia show the nature of work that has been done in recent times regarding pollution of coastal areas by salinity and acidity by Indraratna et al. team from the University of Wollongong over a number of years in Australia. They have also examined in some detail the modelling of oxidation, leaching into matrix and flow into drains and creeks and ultimately to rivers damaging the local system. Clearly, some problems exist with the modelling tools particularly with the boundaries, etc. and the modelling representation process. However, the models provide insights with not only how pollution/flow occurs but also in fact how to remediate using tidal intrusion into groundwater to neutralize some of the acidity. Further, the desalination plant alternative showed that it has a number of groundwater issues as well. The pollution of groundwater is a major area of research and various studies have examined subsurface environments—nature of pollution and possible remediation of the coastal groundwaters but clearly there is much yet to be done. This review shows that acidity, salinity and flow of them can be modelled successfully using partial differential equations but the analysis can also be conducted using time series.

In conclusion, groundwater is an important and in fact a critical resource for future generations and indeed an alternative to methods of storing large amounts of water could be used instead of the existing man-made water dams that require much maintenance and are susceptible to significant levels of evaporation. Groundwater pumping for irrigation and drinking in coastal areas can be also a problem but this process has been curtailed in many parts of Australia or indeed regulated. The placement of essential control on the pumping levels in the coastal aquifers lowers the susceptibility of coastal aquifers to salinity intrusion problems that can lead to large amounts of coastal low lands being useless for agriculture, for example, this has already occurred in many regions around the globe. However, the tidal intrusion is in return a powerful method that has been shown to be particularly useful in neutralizing much of the acidity that has leached into drains, creeks and rivers due to ASS. The acid in drains and creeks being leached after the oxidation of pyrites due to the oxygenation of the soil matrix. Much ASS exist in Australia and around the world and have been particularly problematic for developing countries such as coastal Bangladesh and South Vietnam. The work done by the team in the University of Wollongong (Australia) led by Professor Indraratna, including Dr. Blunden and Dr. Glamore, can be used as a guide for the remediation of pyrite caused acidity in the soil matrix and resulting acidity leaching to irrigation drains, creeks and rivers in coastal lowlands. Also, the team has developed sophisticated tidal intrusion methods to deal with acidic leakage into rivers, creeks and drains. Although desalination is a possible alternative to potable water, given that much saltwater is available on earth, the methodologies used in desalination are particularly power-hungry and rather expensive to maintain; but more importantly, there are indeed a number of issues regarding possible leakage into groundwater aquifers, thus pollution of groundwater; and vastly increasing the salt concentration in local seawater environments caused by dumping of concentrated salt into marine ecosystems.

References

Abarca E, Carrera J, Voss CI et al (2002) Effect of aquifer bottom morphology on seawater intrusion. In: Proceedings of 17th salt water intrusion meeting (SWIM-17), Delft University of Technology, Delft, The Netherlands, pp 116–126

Banasiak LJ, Indraratna B (2012) Key strategies for managing acid sulphate soil (ASS) problems on the southeastern coast of New South Wales, Australia. In: Narsilio GA, Arulraja A, Kodikara J (eds) Proceedings 11th Australia-New Zealand conference on geomechanics: ground engineering in a changing world, pp 1–6

Bear J (1979) Hydraulics of groundwater. McGraw-Hill, New York

Bear J, Cheng AHD, Sorek S et al (1999) Sea water intrusion in coastal aquifers-concepts, methods and practices. Kluwer Academic Publishers, New York

Blunden B, Indraratna B (2000) Evaluation of surface and ground-water management strategies for drained acid sulfate soil using numerical simulation models. Aust J Soil Res 38:569–590

Blunden B, Indraratna B (2001) Pyrite oxidation model for assessing groundwater management strategies in acid sulfate soils. J of Geotech and Geo-Env 127(2):146–157

Blunden B, Indraratna B, Nethery A (1997) Effect of groundwater table on acid sulafte Soil remediation. In: Bouazza A, Kodikar I, Parker B (eds) Proceedings of geoenvironment 97, Melbourne, Balkema, Rotterdam, pp 549–554

Burnett WC, Bokuniewicz H, Huettel M et al (2003) Groundwater and pore water inputs to the coastal zone. Biogeochem 66:3–33

Bush R, Sullivan L (1999) Pyrite morphology in three Australian Holocene sediments. Aust J Soil Res 37:637–653

Custodio E, Bruggeman GA (eds) (1987) Groundwater problems in coastal areas: a contribution to the international hydrological programme. In: Studies and reports in hydrology, p 45. Paris, UNESCO

Dagan G, Bear J (1968) Solving the problem of local interface upconing in a coastal aquifer by the method of small perturbations. J Hydraul Res 6:15–44

Davis GB, Ritchie AIM (1986) A model of oxidation in pyritic mine wastes; part 1: equations and approximate solution. App Math Model 10(5):314–322

Dent D (1986) Acid sulphate soils: a baseline for research and development. Wageningen, International Institute for Land Reclamation and Improvement

Frind EO (1982) Simulation of long-term transient density-dependent transport in groundwater. Adv Water Resour 5:73–88

Glamore W (2003) The effects of tidal buffering on acid sulphate soil environments in coastal areas. Unpublished Doctoral Thesis, University of Wollongong, Australia

Harbison J, Cox M (2002) Hydrological characteristics of groundwater in a subtropical coastal plain with large variations in salinity: Pimpama, Queensland, Australia. Hydrol Sci 47(4):651–665

Hassan OM, Tularam GA (2017) Impact of rainfall fluctuations and temperature variations on people movement in Sub-Saharan: a time series analysis of data from Somalia and Ethiopia. In: Proceedings of MODSIM, international congress on modelling and simulation, modelling and simulation society of Australia and New Zealand, 03–08 Dec 2017, Hobart, Australia

Hassan OM, Tularam GA (in press) The effects of climate change on rural-urban migration in Sub-Saharan Africa (SSA): The cases of Democratic Republic of Congo, Kenya and Niger. In: Waterbody hydrodynamic studies. INTECH Open Access Publisher, ISBN 978-953-51-5673-4

Hsin-Chi JL, David RR, Cary AT, George Y et al (1990) FEMWATER—a three-dimensional finite element computer model for simulating density dependent flow and transport in variably saturated media. Version 3.0 (Reference manual)

Indraratna B, Tularam GA, Blunden B (2001a) Reducing the impact of acid sulphate soils at a site in Shoalhaven Floodplain of New South Wales Australia. Q J Eng Geol Hydrogeol 34(4):333–346

Indraratna B, Tularam GA, Glamore W, Downey J (2001b) Engineering strategies for controlling problems of acid sulphate soils in low-lying coastal areas. Austral Geomech 1(1):133–146

Indraratna B, Tularam GA, Blunden B (2002) The effects of tidal buffering on acid sulphate soil environments in coastal areas of New South Wales. Geotech Geol Eng 20(3):181–199

Indraratna B, Regmi G, Nghiem LD, Golab AN (2011) Geo-environmental approaches for the remediation of acid sulphate soil in low lying floodplains. In: Han J, Alzamora DE (eds) Geo-Frontiers. ASCE, USA, pp 856–865

Indraratna B, Regmi G, Nghiem LD, Golab A (2012) Performance of a PRB for the remediation of acidic groundwater in acid sulfate soil terrain. J Geotech Geoenviron Eng 136(7):897–906

Ivkovic KM, Marshall SM, Morgan LK et al (2012) National-scale vulnerability assessment of seawater intrusion: summary report. National Water Commission, Canberra

Kohout F (1960) Cyclic flow of salt water in the Biscayne aquifer of southeastern Florida. J Geophys Res 65:2133–2141

Li HL, Jiao JJ (2003) Tide-induced seawater-groundwater circulation in a multi-layered coastal leaky aquifer system. J Hydrol 274:211–224

Li L, Barry DA, Stagnitti F, Parlange JY (1999) Submarine groundwater discharge and associated chemical input to a coastal sea. Water Resour Res 35:3253–3259

McKibben M, Barnes H (1986) Oxidation of pyrite in low temperature acidic solutions: rate laws and surface textures. Geochimica Cosmoc Acta 50(7):1509–1520

Michael HA, Mulligan AE, Harvey CF (2005) Seasonal water exchange between aquifers and the coastal ocean. Nature 436:1145–1148

Mulligan AE, Charette MA (2008) Groundwater flow to the coastal ocean. Encyclopedia of ocean sciences, in press. http://www.whoi.edu/science/MCG/groundwater/pubs/PDF/Nov06/Encyc_OS_SGD.pdf Accessed 18 Jan 2016

Oude Essink GHP (2001) Improving fresh groundwater supply-problems and solutions. Ocean Coast Manag 44:429–449

Pinder GF, Cooper H (1970) A numerical technique for calculating the transient position of the saltwater front. Water Resour Res 6(3):875–882

Roca E, Tularam GA (2012) Which way does water flow? An econometric analysis of the global price integration of water stocks. App Econ 44(23):2935–2944

Roca E, Tularam GA, Reza R (2015) Fundamental signals of investment profitability in the global water industry. Int J water 9(4):395–424

Savenije HHG (1988) Influence of rain and evaporation on salt intrusion in estuaries. J Hydr Eng 114(12):1509–1524

Surawski N, Tularam GA, Braddock R (2005) Sustainability of groundwater extraction for the Pimpama coastal-plain, Queensland, Australia. MODSIM, pp 2366–2372 http://www.mssanz.org.au/modsim05/papers/surawski_2.pdf Accessed 29 Nov 2015

Taniguchi M, Burnett WC, Cable JE, Turner JV (2002) Investigation of submarine groundwater discharge. Hydrol Process 16(11):2115–2129

Tularam GA, Dobos J (2005a) Coastal canal estate constructions and related environmental issues. In Walker D (ed), pp 571–576 http://search.informit.com.au/fullText;dn=490307412901440;res=IELENG Accessed 22 Nov 2015

Tularam GA, Dobos J (2005b) Changes in subsurfrace water quality during coastal canal estate constructions. Aust Geomech 40(3):85–94

Tularam GA, Glamore W (2004a) Salinity intrusion in coastal and creeks. Aust Geomech 39(2):73–78

Tularam GA, Glamore W (2004b) Salinity intrusion in coastal lowlands. Aust Geomech J 39(2):73–76

Tularam G, Hassan OM (2016a) The vulnerable nature of water security in Sub-Saharan Africa (SSA)—a country by country analysis. In: Sherman W (ed) Handbook on Africa: challenges and Issues of the 21st century. Nova Science, NY, pp 47–83

Tularam G, Hassan OM (2016b) Water availability and food security: implication of peoples movement and migration in Sub-Saharan Africa (SSA). In: Thangaranjan M, Singh V (eds) Groundwater assessment, modeling, and management. Taylor and Francis, FL, pp 405–426

Tularam GA, Ilahee M (2007) Environmental concerns of desalinating seawater using reverse osmosis. J Env Monit 9:805–813

Tularam GA, Indraratna B (2001) A cylindrical model of pyrite oxidation in coastal acidic soils. Aust Geomech 5–12

Tularam GA, Indraratna B (2002) A cylindrical model of pyrite oxidation in coastal acidic soils with Michaelis-Menten uptake kinetics. Environ Eng Geosci 8(4):329–334

Tularam GA, Keeler HP (2006a) The study of coastal groundwater depth and salinity variation using time-series analysis. Environ Imp Assess Rev 26(7):633–642

Tularam GA, Keeler HP (2006b) A time series analysis of tridal effects on ground water salinity. Aust Geomech 41(4):97–104

Tularam GA, Krishna M (2009) Long term consequences of groundwater pumping in Australia: a review of impacts around the globe. J App Sci Env Sanitation 4(2):151–166

Tularam GA, Marchisella P (2014) Water scarcity in Asia and its long-term water and border security implications for Australia. In: Securing water wastewater systems, pp 189–211

Tularam GA, Murali KK (2015) Water security problems in Asia and longer term implications for Australia. Sustainable water use and management. Springer, New York, pp 119–149

Tularam GA, Properjohn M (2011) An investigation into water distribution network security: risk and implications. Security J 4:1057–1066

Tularam GA, Properjohn M (2013) An investigation into modern water distribution network security: risk and implications. Secur J 24(4):283–301

Tularam GA, Reza R (2016) Water exchange traded funds: a study on idiosyncratic risk using Markov switching analysis. Cogent Econ Fin 4:1–12

Tularam GA, Singh R (2007) Simulating a Tsunami event in Pine Rivers shire (Brisbane) Australia. Aust Geomech 42(4):31–34

Tularam GA, Singh R (2009) Estuary, river and surrounding groundwater quality deterioration associated with tidal intrusion. J App Sci Env Sanitation 4:141–150

Tularam GA, Surawski N, Braddock R (2006) Capabilities of saltflow and Pde2d in modelling saltwater intrusion in coastal Australian lowlands. Aust Geomech 41(1):75–80

Volker R, Rushton K (1982) An assessment of the importance of some parameters for seawater intrusion in aquifers and a comparison of dispersive and sharp-interface modelling approaches. J Hydrol 56:239–250

Voss CI, Souza WR (1987) Variable density flow and solute transport simulation of regional aquifers containing a narrow freshwater-saltwater transition zone. Water Resour Res 23(10):1851–1866

Water and Rivers Commission (1998a) The water cycle, Water Facts 7

Water and Rivers Commission (1998b) What is groundwater? Water Facts 8

White I, Melville M, Sammut J et al (1997) Reducing acidic discharges from coastal wetlands in eastern Australia. Wetlands Ecol Manag 5(1):55–72

Notes

http://www.worstpolluted.org/projects_reports/display/92
http://umccc.org.au/node/25

Part III
Synergetic Effects of Heat and Mass Transfer in Porous Media

Fully Developed Magnetoconvective Heat Transfer in Vertical Double-Passage Porous Annuli

M. Sankar, N. Girish and Z. Siri

Nomenclature

- B_0 Magnetic field strength
- Br Brinkmann number
- Da Darcy number
- g Acceleration due to gravity
- Gr Grashof number
- GR Modified Grashof number
- Ha Hartman number
- K Permeability of porous medium
- k Thermal conductivity
- N Baffle position
- Nu Nusselt number
- T Temperature
- Re Reynolds number
- u,v Velocity components
- V Dimensionless axial velocity

Greek Symbols

- β Volumetric coefficient of thermal expansion
- γ Pressure gradient

M. Sankar (✉)
Department of Mathematics, School of Engineering, Presidency University, Bangalore, India
e-mail: manisankarir@yahoo.com

N. Girish
Department of Mathematics, JSS Academy of Technical Education, Bangalore, India

Z. Siri
Institute of Mathematical Sciences, University of Malaya, 50603 Kuala Lumpur, Malaysia

© Springer Nature Singapore Pte Ltd. 2018
N. Narayanan et al. (eds.), *Flow and Transport in Subsurface Environment*,
Springer Transactions in Civil and Environmental Engineering,
https://doi.org/10.1007/978-981-10-8773-8_7

λ Radius ratio
θ Dimensionless temperature
ρ Density
Λ Viscosity ratio

Subscripts

1   Inner passage
2   Outer passage
c   Cold wall
h   Hot wall

1 Introduction

The understanding of mixed convective or combined forced and free convective flows, where the combined forces of buoyancy and external mechanisms are of comparable order of magnitude, are important in many industrial applications. In particular, the analysis of laminar fully developed mixed convection in vertical channels has been extensively investigated and is evident in the reviews conducted by Incropera (1988), Aung (1987) and Gebhart et al. (1988). Among the vertical passages, the channels formed by the concentric annulus, tubes, and parallel plates are prominent geometries as they aptly represent the physical configurations of important practical applications. Hence, a detailed literature survey is carried out on these three types of vertical channels in the subsequent sections.

Combined free and forced convection flow in a differentially or uniformly heated channel formed by vertical parallel plates has been extensively investigated due to its direct relevance in many practical applications such as heat exchangers, cooling of modern electronic systems, solar energy collectors, chemical processing equipment, etc. Numerous theoretical and experimental investigations on the mixed convective flow in vertical channels can be found in the literature with a prime objective of acquiring a quantitative understanding of this configuration in the important engineering applications. The pioneering analysis on fully developed mixed convection in a vertical parallel-plate channel with linearly varying temperature boundary condition is due to Tao (1960). He has proposed a new method based on the complex function which decouples the velocity and temperature equations by combining them into a single second-order differential equation, namely the Helmholtz wave equation. Later, Bodoia and Osterle (1962) developed the finite difference method to examine the developing free convective flow in a vertical parallel-plate channel. Aung (1972) obtained the closed-form solutions for the fully developed flow in a vertical, parallel-plate channel with the channel walls that are asymmetrically heated by either uniform heat fluxes (UHF) or uniform wall temperatures. Further, Aung and Worku (1986a, b) systematically investigated the conditions for flow reversal for both developing and fully developed flows in a vertical parallel-plate channel by considering symmetric as well as asymmetric heating conditions. Cheng et al. (1990) discussed the phenomenon of flow reversal and heat transfer for the fully developed mixed convection in a vertical channel. Hamadah and Wirtz (1991) analyzed fully developed mixed convective flow in a vertical channel formed by two parallel plates. The exact solutions are obtained for the opposing buoyancy flow by considering three sets of thermal boundary conditions. In terms of Fourier series and polynomials, Mcbain (1999) obtained the exact solutions for the fully developed flow in a vertical cavity or duct of rectangular and elliptic sections.

In the above investigations of mixed convection in a parallel-plate vertical channel, the effect of viscous dissipation has not been taken into consideration. However, the influence of viscous dissipation cannot be neglected for fluids with high values of the dynamic viscosity as well as for high velocity flows. Also, since the momentum and energy equations are coupled for mixed convection flows, the viscous dissipation influences the velocity and temperature fields significantly. Hence, the inclusion of viscous dissipation gives rise to a nonlinear velocity term in the energy equation and this coupling term would be absent when viscous dissipation is neglected. Realizing this important fact, many investigations are carried out to understand the influence of viscous dissipation on the mixed convective heat transfer in vertical channels. Employing the perturbation method, Barletta and his co-workers (Barletta 1998, 1999a; Barletta and Zanchini 1999) examined the fully developed mixed convection flow in a vertical parallel-plate channel in the presence of viscous dissipation. The effects of viscous dissipation on the fully developed flow and thermal fields and the corresponding heat transfer rate have been examined for different thermal boundary conditions. The combined influences of cavity inclination and viscous dissipation on the mixed convection in an inclined vertical parallel-plate channel have been analytically investigated by Barletta and Zanchini (2001).

Further, the influence of an external magnetic field on the mixed convection flows in vertical channels filled with porous media has gained increasing attention and a comprehensive review on this topic can be found in Nield and Bejan (2013). The flow characteristics of electrically conductive fluid in vertical channels are significantly altered when the magnetic field is applied perpendicular to the flow direction. In practical applications, such as magnetohydrodynamic (MHD) pumps and generators, this effect is useful to control the unwanted movement of the fluids. Chamkha (2002) made a detailed analysis on the hydromagnetic fully developed mixed convection flow in a vertical channel with different thermal boundary conditions. Also, the investigation examined the effects of viscous dissipation, magnetic field, heat generation or absorption on the velocity and thermal fields, and the heat transfer rate. The effect of viscous and ohmic dissipations on magnetohydrodynamic mixed convective flow in a vertical channel is analyzed by Umavathi and Malashetty (2005). Using the perturbation method, the analytical solutions are obtained in the absence of viscous dissipation and the complete governing equations are solved by finite difference method. Barletta and Celli (2008) discussed the effects of magnetic field and viscous dissipation on fully developed mixed convection in a vertical channel. Kumar et al. (2009) examined the influence of viscous dissipation on fully developed laminar mixed convection flow in an infinite vertical porous channel using a two-region model. Using the perturbation series method, analytical solutions are obtained for three types of thermal boundary conditions. Saleh and Hashim (2010) reported the flow reversal phenomena of the fully developed mixed convection in a vertical parallel-plate channel in the presence of an applied magnetic field and viscous dissipation. Chen et al. (2011) used the differential transformation method to analyze fully developed, mixed convection flow in a parallel-plate vertical channel. From the known velocity and temperature fields, the entropy generation equation is also solved to obtain the entropy generation number and irreversibility distribution ratio. Kumar et al. (2011, 2012) used a two-fluid continuum model to analyze the fully developed mixed convective flow in a vertical channel in the presence of viscous dissipation and magnetic field. The other notable studies on the effects of Joule heating and/or viscous dissipation on the fully developed magnetohydrodynamic convective flow in a parallel-plate vertical channel are due to Liu and Lo (2012) and Sarveshanand and Singh (2015).

Fully developed mixed convection in a vertical tube has also received considerable attention in the literature, as the knowledge of fluid flow and heat transfer processes in this geometry is relevant in the modeling process of many important physical situations. One of the earliest investigations on laminar convective along a vertical pipe with uniformly heated or cooled boundaries is due to Mortan (1960). The exact solution is obtained by assuming fully developed flow condition. Iqbal et al. (1970) reported the fully developed mixed convection in a vertical circular tube with a uniform boundary heat flux using three different mathematical techniques. Their investigations reveal the fully developed convective flow and heat transfer processes in the presence and absence of viscous dissipation. Later, the effect of viscous dissipation on the combined free and forced convection of non-Newtonian fluid in a vertical circular tube with uniform wall heat flux is investigated by Marner and Hovland (1973) by assuming the fully developed upflow. Other notable contributions on the investigation of fully developed convective flow and heat transfer in a vertical tube with and without viscous dissipation can be found in (Barletta 1999b; Barletta and Rossi 2001; Orfi et al. 1993). The combined influence of viscous dissipation and porosity on the developing forced convective flow in an isothermal tube has been investigated by Ranjbar-Kani and Hooman (2004). Later, Aydin (2005) analyzed thermally developing forced convective flow in a pipe with viscous dissipation and by considering two different thermal boundary conditions, namely the constant heat flux (CHF) and the constant wall temperature (CWT).

Among the different vertical passages, an annular enclosure formed by vertical concentric cylinders is a commonly employed geometry in variety of heat transfer equipment. Owing to the industrial applications of this geometry, such as tube extrusion of high viscosity fluids, cooling of electronic components and transmission cables, numerous investigations have been carried out on developing and fully developed convection in the vertical annular passages. Fully developed mixed convection in a vertical annulus has been investigated by Rokerya and Iqbal (1971). They obtained the analytical solutions in terms of Kelvins functions in the absence of viscous dissipation and in the presence of viscous dissipation, numerical solutions are obtained. Using an implicit finite difference technique, Coney and El-Shaarawi (1975) made a detailed analysis of the development of laminar convective flow and heat transfer in vertical concentric annuli for three radius ratios with an isothermal inner wall and the adiabatic outer wall kept and vice versa. The influence of radial fins on the fully developed convective flow in a vertical finned annulus with the fins attached to the outer surface of the inner wall has been analyzed by Prakash and Renzoni (1985). Joshi (1987) obtained the closed-form solutions for the fully developed flow and heat transfer in a differentially heated vertical annular duct with isothermal annular walls. The analytical results are validated against the numerical results for developing flow and also derived a condition for the fully developed flow assumptions. El-Shaarawi and Al-Nimr (1990) derived analytical expressions for fully developed flow and heat transfer rates in an open-ended vertical annular passage. The closed-form solutions are obtained for four fundamental boundary conditions of temperature at the annular walls. Later, Al-Nimr (1993) presented analytical solutions for transient fully developed flow in an open-ended vertical annulus by considering four thermal conditions, which are termed as the basic fundamental boundary conditions of four kinds. Coelho and Pinho (2006) obtained the closed-form solutions for fully developed laminar convective flow in a concentric annulus with viscous dissipation for uniform, wall heat fluxes as well as uniform wall temperatures. Later, Zanchini (2008) analyzed the effect of variable viscosity on mixed convection in an isothermally heated vertical annulus.

In many important applications such as packed-bed catalytic reactors, geological disposal of high-level nuclear waste and petroleum resources, a deep understanding and rigorous analysis of convective heat transfer in porous annuli essential. Using fully developed flow assumptions, Al-Nimr (1995a) examined the natural convective flow and heat transfer in an open-ended vertical porous annulus. The velocity and temperature distributions and heat transfer rate are obtained for four fundamental boundary conditions which consist of uniform heat flux and uniform wall temperature boundary conditions. Using non-Darcy model, Kou and Huang (1997) analytically investigated the fully developed convection in a vertical annulus filled with porous media by considering three different thermal boundary conditions. Free convective flow of an electrically conducting fluid in a porous annulus has important applications in geothermal applications, where the electrically conducting gases are influenced by the existing magnetic field. Al-Nimr (1995b) obtained analytical solutions for the magnetohydrodynamic free convective flow in a vertical porous annulus under fully developed flow conditions. Using the Darcy's law, Barletta et al. (2008) examined the combined influences of magnetic field and viscous dissipation on the fully developed convection in a vertical porous annulus and found that the magnetic field strongly restrains the fluid flow and heat transfer. Recently, Dawood et al. (2015) made a comprehensive review on fluid flow and convective heat transfer in a vertical porous and nonporous annulus. The review consists of the analysis of free, forced, and mixed convection in concentric and eccentric annulus and the possible areas in which this geometry has potential applications have also been discussed in detail. Using perturbation method, Jha et al. (2016a) reported the effect of time-periodic boundary condition on fully developed mixed convective flow in a vertical porous annulus. Jha et al. (2016b) investigated analytically fully developed convection in a vertical micro-concentric annulus filled with heat generating/absorbing fluid by taking into account the velocity slip and temperature jump. Recently, Oni (2017) examines the effects of heat source and thermal radiation on mixed convection flow in a vertical porous annulus.

Due to various possible applications, such as cooling of electronic equipment and cooling of turbine blades and nuclear reactors, investigations on fully developed mixed convection heat transfer in vertical channels in which a thin baffle is placed, termed as "double-passage channels", have become a subject of increased interest. El-Din (2002) made a combined analytical and numerical investigation of fully developed mixed convection in a vertical double-passage channel. The left and right walls of the channel are isothermally heated with different temperatures, and the baffle is placed in between the end walls. The effects of viscous dissipation and baffle position on the velocity and temperature profiles; heat transfer rate has been discussed for different values of the parameters. Analytical solutions are obtained in the absence of viscous dissipation, while numerical solution is obtained in the presence of viscous dissipation. Later, El-Din (2007) extended his earlier work (El-Din 2002) for uniform wall heat flux in the same configuration. The analysis has been performed for three different heat flux ratios, namely, symmetric, nonsymmetric, and adiabatic thermal boundary conditions. In the similar kind of configuration, other noted investigations with additional constraints, such as porosity and magnetic field geometries are due to (Kumar et al. 2009, 2011, 2012). A thorough survey of the literature reveals the lack of information on fluid flow and heat transfer processes in vertical double-passage channel formed by three vertical, concentric cylinders, in which the middle cylinder is considered as a perfectly heat conducting baffle. The existing investigations are focused only on the double-passage channels formed by parallel plates (El-Din 2002, 2007; Kumar et al. 2009, 2011, 2012). However, many important applications involve a cylindrical annular geometry, where the curvature effects are important, and in many studies, the double-passage annuli are overlooked. Although the flow and heat transfer analysis of fully developed mixed convection flows in the vertical double-passage annuli has not received attention in the literature, it is known that the flow behavior in double-passage annuli can exhibit interesting dynamical phenomena, and with this contribution we aim to fill the gap existing in the current literature. In particular, our main emphasis is to investigate the various effects, such as magnetic field, porosity, curvature ratio, baffle position, and viscous dissipation on the fully developed mixed convection in the vertical double-passage annuli formed by three vertical, concentric cylinders.

2 Mathematical Formulation

The physical configuration for the present study, as shown in Fig. 1, is the double-passage annuli formed by three vertical, concentric cylinders of which middle cylinder is a thin and perfectly conductive. The radii of inner, middle, and outer cylinders are r_i, r_m, and r_o, respectively, and the annuli passages are filled with fluid-saturated porous material. We considered two-dimensional, laminar, incompressible steady flow in the annular passages. It is assumed that the fully developed fluid enters the annuli with constant temperature and uniform upward velocity. Also, the viscous dissipation effect is being taken into consideration in the present study. The sidewalls of the annuli are asymmetrically heated, where temperatures of outer and inner cylinders are more compared to the fluid temperature. Apart from the buoyancy term present in momentum equation, all fluid properties are assumed to be constant. Further, a uniform magnetic field has been applied in the radial direction and we assumed that the induced magnetic field can be neglected when compared with the magnitude of applied magnetic field. Employing the above mentioned assumptions, the governing equations for present analysis are as follows.

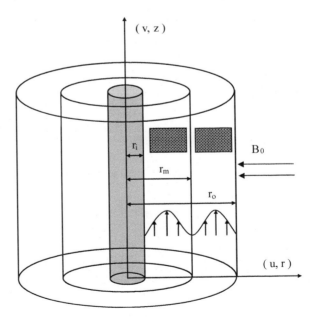

Fig. 1 Physical configuration and coordinate system

2.1 Nondimensional Variables and Governing Equations

Continuity equation:

$$\frac{\partial u}{\partial r} + \frac{u}{r} + \frac{\partial v}{\partial z} = 0 \tag{1}$$

Momentum equation in radial direction:

$$u\frac{\partial u}{\partial r} + v\frac{\partial u}{\partial z} = -\frac{1}{\rho_o}\frac{\partial p}{\partial r} + v\left[\frac{\partial^2 u}{\partial r^2} + \frac{1}{r}\frac{\partial u}{\partial r} + \frac{\partial^2 u}{\partial z^2} - \frac{u}{r^2}\right] \tag{2}$$

Momentum equation in axial direction:

$$u\frac{\partial v}{\partial r} + v\frac{\partial v}{\partial z} = -\frac{1}{\rho_o}\frac{\partial p}{\partial z} + v\left[\frac{\partial^2 v}{\partial r^2} + \frac{1}{r}\frac{\partial v}{\partial r} + \frac{\partial^2 v}{\partial z^2}\right] + g\beta(T - T_r) - \frac{1}{\rho_o}\left(\frac{\mu}{K}v + \sigma B_o^2 v\right) \tag{3}$$

Energy equation:

$$u\frac{\partial T}{\partial r} + v\frac{\partial T}{\partial z} = k\left[\frac{\partial^2 T}{\partial r^2} + \frac{1}{r}\frac{\partial T}{\partial r} + \frac{\partial^2 T}{\partial z^2}\right] + \mu\left(\frac{\partial v}{\partial r}\right)^2 \tag{4}$$

Transverse velocity and derivative of temperature in z-direction are assumed to be zero due to the fully developed flow assumption. That is, for the assumption of fully developed flow, we have $u = 0$, $\frac{\partial}{\partial z}() = 0$ and $\frac{\partial^2}{\partial z^2}() = 0$. Using these conditions, Eqs. (1)–(4) reduce to

$$\frac{\partial p}{\partial r} = 0 \Rightarrow p \neq p(r) \text{ but } p = p(z) \tag{5}$$

$$v\left[\frac{\partial^2 v_j}{\partial r^2} + \frac{1}{r}\frac{\partial v_j}{\partial r}\right] = \frac{1}{\rho_o}\frac{\partial p_j}{\partial z} - g\beta(T_j - T_r) + \frac{\mu}{\rho_o K}v_j + \frac{\sigma B_o^2 v_j}{\rho_o} \tag{6}$$

$$k\left[\frac{\partial^2 T_j}{\partial r^2} + \frac{1}{r}\frac{\partial T_j}{\partial r}\right] + \mu\left(\frac{\partial v_j}{\partial r}\right)^2 = 0 \tag{7}$$

In the above equations, $j = 1$ and 2 represent the passage 1 and 2, respectively. The following are nondimensional parameters:

$$R = \frac{r}{r_o}, \quad V = \frac{v}{v_r}, \quad Z = \frac{z}{r_o Re}, \quad T_r = \frac{(T_c + T_h)}{2}, \quad \theta_i = \frac{T_j - T_r}{T_h - T_c}, \quad P = \frac{p}{\rho_o v_r^2},$$

$$\Lambda = \frac{\mu}{\mu_e}, \quad \nu = \frac{\mu_e}{\rho_o}, \quad Re = \frac{v_r r_o}{\nu}, \quad Gr = \frac{g\beta(T_h - T_c)r_o^3}{\nu^2}, \quad Br = \frac{\mu v_r^2}{k(T_h - T_c)},$$

$$Da = \frac{K}{r_o^2}, \quad Ha^2 = \frac{\sigma B_o^2 r_o^2}{\mu_e}, \quad \lambda = \frac{r_i}{r_o}, \quad N = \frac{r_m}{r_o}.$$

Using the above transformations, the nondimensional governing equations can be written as

$$\left[\frac{\partial^2 V_j}{\partial R^2} + \frac{1}{R}\frac{\partial V_j}{\partial R}\right] - \left(\frac{\Lambda}{Da} + Ha^2\right)V_j = \frac{\partial P_j}{\partial Z} - \frac{Gr}{Re}\theta_j \quad (8)$$

$$\left[\frac{\partial^2 \theta_j}{\partial R^2} + \frac{1}{R}\frac{\partial \theta_j}{\partial R}\right] + Br\left(\frac{\partial V_j}{\partial R}\right)^2 = 0 \quad (9)$$

Since V_j and θ_j are the functions of R only and $P_j = P_j(Z)$, the above partial differential equations can be written as the following ordinary differential equations:

$$\left[\frac{d^2 V_j}{dR^2} + \frac{1}{R}\frac{dV_j}{dR}\right] - \left(\frac{\Lambda}{Da} + Ha^2\right)V_j = \frac{dP_j}{dZ} - \frac{Gr}{Re}\theta_j \quad (10)$$

$$\left[\frac{d^2 \theta_j}{dR^2} + \frac{1}{R}\frac{d\theta_j}{dR}\right] + Br\left(\frac{dV_j}{dR}\right)^2 = 0. \quad (11)$$

2.2 Boundary Conditions and Heat Transfer Rate

The dimensional boundary conditions in passages 1 and 2 are considered as

$$r = r_i, \quad v_1 = 0, \quad T_1 = T_h$$
$$r = r_m, \quad v_1 = 0 = v_2, \quad T_1 = T_2$$
$$r = r_o, \quad v_2 = 0, \quad T_2 = T_c$$

Using the dimensionless variables defined earlier, the corresponding dimensionless boundary conditions can be written as

$$\begin{aligned} R &= \lambda, \quad V_1 = 0, \quad \theta_1 = 0.5 \\ R &= N, \quad V_1 = 0 = V_2, \quad \theta_1 = \theta_2 \\ R &= 1, \quad V_2 = 0, \quad \theta_2 = -0.5 \end{aligned} \quad (12)$$

The conservation of mass at any cross section of the annuli is given in nondimensional form as

$$\int_\lambda^N RV_1 dR = \frac{1}{2}(N^2 - \lambda^2) \quad \text{and} \quad \int_N^1 RV_2 dR = \frac{1}{2}(1 - N^2) \tag{13}$$

The heat transfer rate in both passages can be obtained by evaluating the Nusselt numbers on hot and cold walls which are defined as

$$Nu_h = \frac{4(1-\lambda)}{(2\theta_{b1} - 1)\lambda \ln \lambda} \quad \text{and} \quad Nu_c = \frac{-4(1-\lambda)}{(1 + 2\theta_{b2}) \ln \lambda} \tag{14}$$

In the above equations, θ_{b1} and θ_{b2} are the dimensionless bulk temperatures in the passages 1 and 2 and are defined as

$$\theta_{b1} = \frac{\int_\lambda^N \theta_1 V_1 R dR}{\int_\lambda^N V_1 R dR} \quad \text{and} \quad \theta_{b2} = \frac{\int_N^1 \theta_2 V_2 R dR}{\int_N^1 V_2 R dR}. \tag{15}$$

3 Method of Solution and Validation

In this study, the mathematical model equations (10) and (11) along with the boundary conditions (12) are solved both numerically and analytically. The analytical solutions of the problem are obtained in the absence of viscous dissipation. However, when the effect of viscous dissipation is taken into account, the governing differential equations are coupled and nonlinear. Therefore, an implicit finite difference method along with successive over-relaxation technique has been used to obtain the solutions when the viscous dissipation is present. Further, the numerical results have been successfully validated with the analytical solution in the absence of viscous dissipation and found an excellent agreement.

3.1 Analytical Solution

The governing differential equations (10) and (11) are solved analytically when the viscous dissipation effects are absent, i.e., $(Br = 0)$. In this case, the governing ordinary differential equations are solved using Cauchy's linear differential equation method. The details of the method are not provided for brevity, but the solutions are given below:

$$\theta = \frac{\ln R}{\ln \lambda} - 0.5 \tag{16}$$

$$V_1 = \left(\frac{N^2 - \lambda^2}{2\alpha^2 C_1} + \frac{C_2 Gr}{C_1 \alpha^2 Re \ln \lambda}\right) \left(\frac{\begin{array}{c}I_0(\alpha R)(K_0(\alpha\lambda) - K_0(\alpha N)) \\ + K_0(\alpha R)(I_0(\alpha N) - I_0(\alpha\lambda)) \end{array}}{K_0(\alpha\lambda)I_0(\alpha N) - K_0(\alpha N)I_0(\alpha\lambda)} - 1\right)$$

$$- \left(\frac{Gr}{Re\alpha^2 \ln \lambda}\right) \left(\frac{\begin{array}{c}I_0(\alpha R)(\ln N K_0(\alpha\lambda) - \ln \lambda K_0(\alpha N)) \\ + K_0(\alpha R)(\ln \lambda I_0(\lambda N) - \ln N I_0(\alpha N))\end{array}}{K_0(\alpha\lambda)I_0(\alpha N) - K_0(\alpha N)I_0(\alpha\lambda)} - \ln R\right) \tag{17}$$

$$V_2 = \left(\frac{1 - N^2}{2\alpha^2 C_3} + \frac{C_4 Gr}{\alpha^2 Re C_3 \ln \lambda}\right) \left(\frac{\begin{array}{c}I_0(\alpha R)(K_0(\alpha) - K_0(\alpha N)) \\ + K_0(\alpha R)(I_0(\alpha N) - I_0(\alpha))\end{array}}{K_0(\alpha)I_0(\alpha N) - K_0(\alpha N)I_0(\alpha)} - 1\right)$$

$$- \left(\frac{Gr}{Re\alpha^2 \ln \lambda}\right) \left(\frac{\ln N(I_0(\alpha R)K_0(\alpha) - K_0(\alpha R)I_0(\alpha))}{K_0(\alpha)I_0(\alpha N) - K_0(\alpha N)I_0(\alpha)} - \ln R\right) \tag{18}$$

where $\alpha^2 = \frac{\Lambda}{Da} + Ha^2$

$$C_1 = \frac{\frac{1}{\alpha^4}\left[\left\{\begin{array}{c}-2 + \alpha N(K_0(\alpha\lambda)I_1(\alpha N) + I_0(\alpha\lambda)K_1(\alpha N)) \\ + \alpha\lambda(K_0(\alpha N)I_1(\alpha\lambda) + I_0(\alpha N)K_1(\alpha\lambda))\end{array}\right\} - \frac{\alpha^2}{2}(N^2 - \lambda^2)\right]}{(K_0(\alpha\lambda)I_0(\alpha N) - K_0(\alpha N)I_0(\alpha\lambda))}$$

$$C_2 = \frac{\frac{1}{\alpha^4}\left[\left\{\begin{array}{c}-\ln(N\lambda) + \alpha N \ln N(I_1(\alpha N)K_0(\alpha\lambda) + I_0(\alpha\lambda)K_1(\alpha N)) \\ + \alpha\lambda \ln \lambda(K_0(\alpha N)I_1(\alpha\lambda) + I_0(\alpha N)K_1(\alpha\lambda))\end{array}\right\} - \frac{\alpha^2}{4}(2(N^2 \ln N - \lambda^2 \ln \lambda) - (N^2 - \lambda^2))\right]}{(K_0(\alpha\lambda)I_0(\alpha N) - K_0(\alpha N)I_0(\alpha\lambda))}$$

$$C_3 = \frac{1}{\alpha^4}\left[\frac{2 - \alpha\{K_0(\alpha N)I_1(\alpha) + I_0(\alpha N)K_1(\alpha) + N(K_0(\alpha)I_1(\alpha N) + I_0(\alpha)K_1(\alpha N))\}}{K_0(\alpha)I_0(\alpha N) - K_0(\alpha N)I_0(\alpha)} - \frac{\alpha^2}{4}(1 - N^2)\right]$$

$$C_4 = \frac{1}{\alpha^4}\left[\frac{\ln N\{1 - \alpha N(K_0(\alpha)I_1(\alpha N) + I_0(\alpha)K_1(\alpha N))\}}{K_0(\alpha)I_0(\alpha N) - K_0(\alpha N)I_0(\alpha)} - \frac{\alpha^2}{4}(N^2 - 1 - 2N^2 \ln N)\right],$$

In the above equations, Eqs. (16)–(18) represent the temperature and velocity profiles in the passages 1 and 2, respectively. Here, N is the baffle position and $\frac{d}{dz}P_j = \gamma_j$ are obtained by considering conservation of mass at any cross section of the annuli as given in (13).

3.2 Numerical Solution

For the case of viscous dissipation effects taken into consideration ($Br \neq 0$), the governing equations (10) and (11) are nonlinear and coupled through the Brinkman term present in the energy equation. Hence, the analytical solution is not possible to obtain and the governing equations are solved using an implicit finite difference technique. The second-order finite difference approximations are used to discretize the derivatives in the governing equations, and the resulting finite difference equations are solved using Successive Point Over-Relaxation (SPOR) method. The grid sizes are varied from 81 grids to 201 grids in the R-direction to check the grid independency test. During the selection of grid sizes, it has been ensured that both passages have equal number of grids. For example, for 101 grids, 51 grids are placed in the first passage and 51 grids on the second passage with one grid on the middle cylinder or baffle. The maximum velocity and temperature and Nusselt number are considered as sensitive measures for grid independency tests. Based on the detailed grid independency tests, we found that 161 grids, with 81 grids on each passage, provide sufficiently accurate results and the solutions are not varying significantly with further increasing the grid size. To evaluate the Nusselt numbers and bulk temperatures in both passages, Simpson's rule has been used to numerically evaluate the integrals. Further, to validate the obtained results, the numerical results are compared with the analytical results in the absence of viscous dissipation ($Br = 0$) and is exhibited in Fig. 2. The velocity and temperature profiles for various values of Darcy and Hartmann numbers are shown in Fig. 2, and an excellent agreement has been found between the numerical and analytical results.

4 Results and Discussion

In this chapter, we examine the fully developed mixed convection in the vertical double-passage annuli, formed by three vertical, concentric cylinders. The top and bottom of the annuli are kept open, and the inner and outer cylinders are uniformly heated with asymmetric temperatures. Using fully developed flow assumptions, the governing partial differential equations are reduced to system of ordinary differential equations and are solved analytically as well as numerically. In the absence of viscous dissipation, closed-form solutions are obtained in terms of Bessel's functions and the numerical simulations are sought when the viscous dissipation effect is considered in the analysis. Our study mainly focuses on the influences of various effects, such as magnetic field, porosity, viscous dissipation, baffle location and radius ratio on the velocity and temperature profiles, and heat transfer rates. Further, the numerical results are compared with the analytical results and obtained excellent agreement between the two solutions. The physical parameters, such as Darcy number (Da), Hartmann number (Ha), Brinkman number (Br), modified Grashof number ($GR = Gr/Re$), and geometrical parameters, such as radius ratio (λ) and

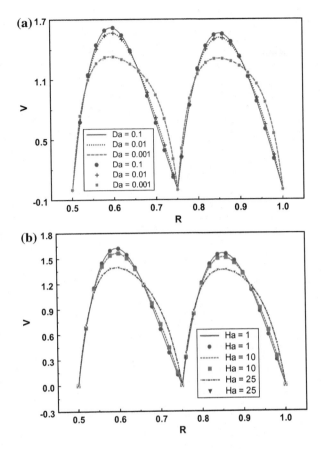

Fig. 2 Comparison of numerical (lines) and analytical (symbols) velocity profiles for different values of **a** Da and **b** Ha for $\lambda = 0.5, N = 0.75, GR = 10^3, Br = 0$

baffle position (N), are varied over wide range of values, and their effects on flow pattern, thermal distribution, and heat transfer rate are analyzed in detail. It is worth to mention a note about the velocity and temperature profiles in two passages. In all the graphs involving velocity and temperature profiles, the passage 1 varies from $R = \lambda$ to $R = N$ and the passage 2 varies from $R = N$ to $R = 1$.

4.1 Effect of Magnetic Field

In this section, the influence of magnetic field on the fluid flow and heat transfer process are investigated by considering the fully developed flow in the vertical double-passage annuli. The combined influence of modified Grashof number, GR, and Hartmann number, Ha, on the velocity and temperature profiles is depicted in Fig. 3 by fixing the values of $\lambda = 0.5$, $N = 0.75$, $Br = 0.01$. The modified Grashof number GR is the ratio of Grashof number to Reynolds number. It is found that the magnitude of velocity increases as GR increases, since an increase in GR indicates the enhancement of mixed convection flow rate. As such, greater flow rates are achieved for higher value of GR, namely $GR = 2 \times 10^3$. Since the baffle is placed in the middle of the annulus, $N = 0.75$, it can be observed that the velocity profiles in both passages are symmetric. As regards to the influence of magnetic field on the

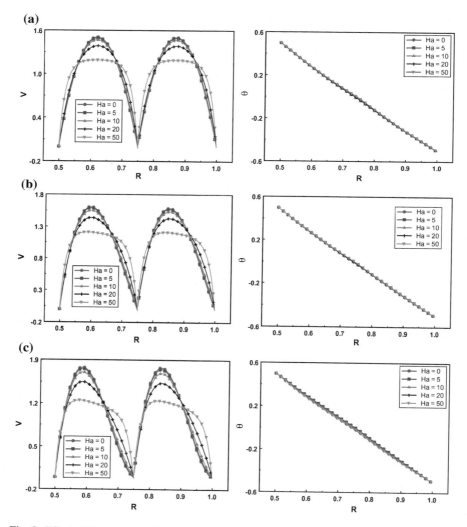

Fig. 3 Effect of Hartmann number on velocity and temperature profiles for different values of GR at $\lambda = 0.5, N = 0.75, Br = 0.01$. **a** $GR = 10$, **b** $GR = 10^3$, **c** $GR = 2 \times 10^3$

fluid velocity, we found that the magnitude of velocity decreases with an increase in Hartmann number, Ha. The impact of magnetic field on velocity is apparent, as the parabolic velocity profile for $Ha = 0$ changes to a flat velocity profile for higher value of $Ha = 50$. A careful observation of temperature profile reveals that the thermal profiles in both passages are unaltered with GR and Ha. This can be expected, as the energy equation does not depend on these two parameters but strongly depends on Br.

Figure 4 exemplifies the combined influences of Hartmann number and baffle position on the velocity and temperature profiles. A change in the baffle position leads to three different passages, namely, wider, narrow, or equal annuli passages. For $N = 0.75$, both passages are of the same width, while for $N = 0.9$ (or $N = 0.6$), we get wider (or narrow) inner passage and narrow (or wider) outer passage. In

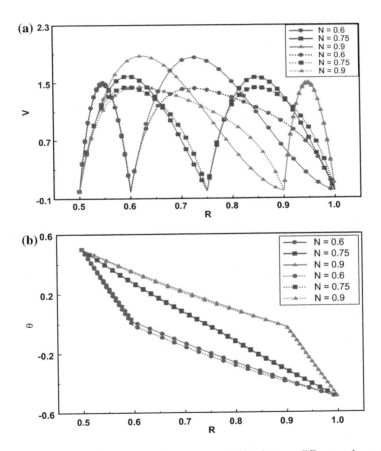

Fig. 4 Effect of N on **a** velocity and **b** temperature profiles for two different values of Ha at $\lambda = 0.5, GR = 10^3, Br = 0.01, Da = 0.1$. Continuous lines correspond to $Ha = 5$ and dotted lines correspond to $Ha = 20$

Fig. 4, the influence of three baffle positions and two values of *Ha* are analyzed on velocity and thermal profiles by fixing other relevant parameters. The velocity profiles reveal that the maximum velocity occurs in wider passage and minimum velocity is observed in narrow passage. Further, the influence of Hartmann number on velocity profiles indicates that the magnitude of velocity decreases with an increase in *Ha*. The thermal profiles are not altered with the Hartmann number; however, the influence of baffle position is significant on thermal profiles. The quantitative measure in any heat transfer analysis is the rate of heat transfer measured by the Nusselt number. The change in heat transfer rate with Hartmann number is displayed in Fig. 5 at both inner and outer walls. The rate of heat transfer increases with modified Grashof number, while an increase in the Hartmann number reduces the heat transfer rate. Also, as *GR* increases, the Nusselt number

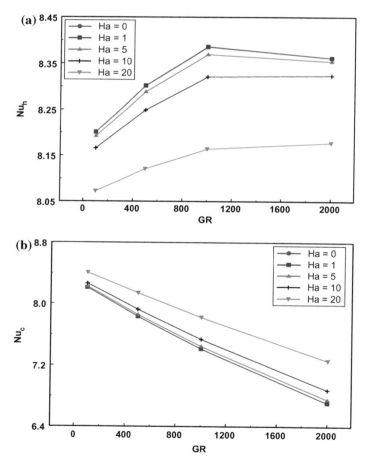

Fig. 5 Effect of *Ha* and *GR* on the hot (Nu_h) and cold (Nu_c) wall Nusselt numbers for $\lambda = 0.5$, $N = 0.75, Br = 0.01$ and $Da = 0.1$

shows an increasing trend along the inner wall, but decreasing trend is found along the outer wall. The variations of heat transfer rate for different baffle positions are of practical importance and are explained in Fig. 6 for different values of Ha. From the Nusselt number profiles, it can be seen that the narrow passage produces higher heat transfer rate, while broader passage produces lower heat transfer rate. The outer wall Nusselt number decreases with N and Ha. Since the magnetic field produces a drag force, the fluid velocities are suppressed and hence the heat transfer is reduced.

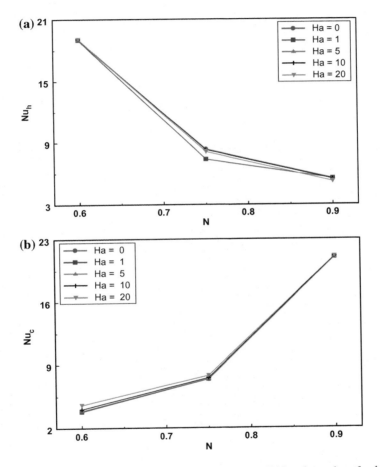

Fig. 6 Effect of Ha and N on the hot (Nu_h) and cold (Nu_c) wall Nusselt numbers for $\lambda = 0.5$, $GR = 10^3$, $Br = 0.01$ and $Da = 0.1$

4.2 Effect of Porosity

The influence of porosity on the flow pattern, thermal distribution, and heat transfer rate is examined in this section against different physical and geometrical parameters. Figure 7 depicts the combined effects of Darcy number and modified Grashof number on the flow and thermal fields for fixed values of $\lambda = 0.5, N = 0.75, Br = 0.01$.

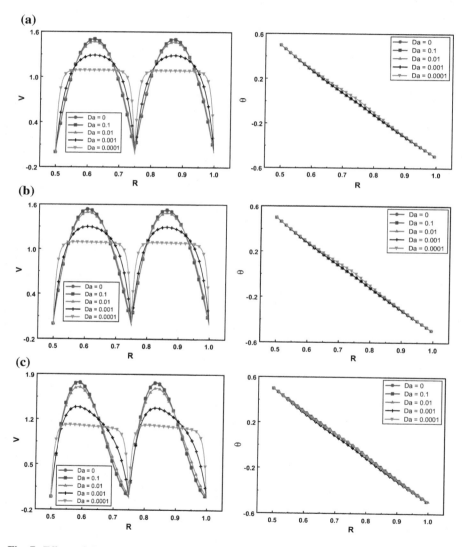

Fig. 7 Effect of Da on velocity and temperature profiles for $\lambda = 0.5, N = 0.75, Br = 0.01$. (Top) $GR = 10^2$, (Middle) $GR = 5 \times 10^2$, (Bottom) $GR = 2 \times 10^3$

The effect of Darcy number on the flow field is visibly apparent for all values of GR. An overview of the figure reveals that the velocity increases with GR but decreases with Da. A decrease in the Darcy number decreases the porosity, and hence the flow penetration reduces considerably. As a result, the flow velocity suppresses significantly, as can be seen from the figure that the velocity profile attains flat shape for the lowest value of Da, $Da = 10^{-4}$. As expected, the change in thermal field is not significant with respect to Da and GR. However, slender variation is observed in the thermal pattern, as the Brinkman number is different from zero, $Br = 0.01$. To understand the effects of baffle position and Darcy number, the velocity and thermal profiles are illustrated in Fig. 8 for three different values of N and two different values of Da. The baffle position and Darcy number significantly alter the flow pattern as can be seen from the figure. In particular, the Darcy number has strong influence on the

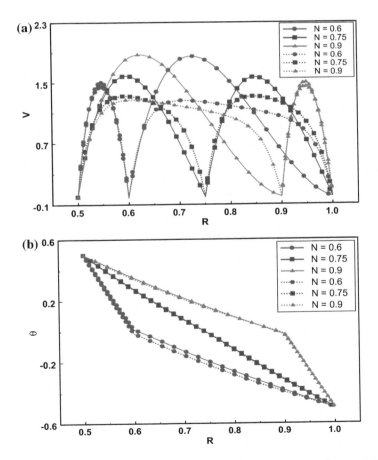

Fig. 8 Effect of N on velocity and temperature profiles for $\lambda = 0.5, GR = 10^3, Br = 0.01, Ha = 5$. Continuous lines correspond to $Da = 0.1$ and dotted lines correspond to $Da = 0.001$

flow pattern in wider annular passage rather than narrow passage. The temperature distribution does not change with Darcy number, but changes with baffle positions.

Figure 9 illustrates the variation of heat transfer with different values of Da and GR for fixed values of $\lambda = 0.5, N = 0.75, Br = 0.01, Ha = 1$. An overview of the figure reveals that the hot wall Nusselt number increases sharply with GR up to $GR = 10^3$ and then remains invariant. However, as the Darcy number is decreased to $Da = 10^{-3}$, the Nusselt number does not vary significantly and remains flat for all values of GR. As the Darcy number decreases, the flow resistance increases, which results in the reduction of heat transfer at low values of Da. The Nusselt number along the cold wall decreases steadily with GR. In this investigation, one of the important parameters is baffle position and its influence on heat transfer is

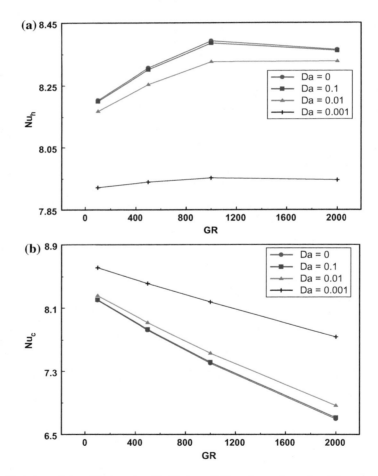

Fig. 9 Variation of hot (Nu_h) and cold (Nu_c) wall Nusselt numbers for different values of GR and Da at $\lambda = 0.5, N = 0.75, Br = 0.01, Ha = 1$

important in terms of application point of view. Figure 10 illustrates the effect of baffle position on heat transfer rate for various values of Da. It has been observed that the hot wall Nusselt number is high for narrow passage and the heat transfer is minimum for wider passage, whereas cold wall Nu is higher for wider passage and lower for narrow passage.

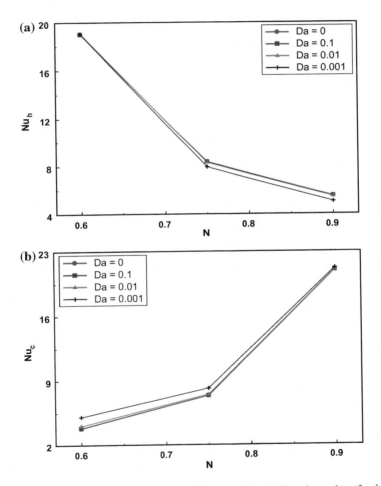

Fig. 10 Effect of Da and N on the hot (Nu_h) and cold (Nu_c) wall Nusselt numbers for $\lambda = 0.5$, $GR = 10^3$, $Br = 0.01$ and $Ha = 1$

4.3 Effect of Viscous Dissipation

In this study, another important parameter of interest is Brinkman number, which is due to the presence of viscous dissipation. This section is devoted to the results pertaining to viscous dissipation effects. Figure 11 depicts the variation of flow and thermal fields for different values of Br and three values of Da. The presence of

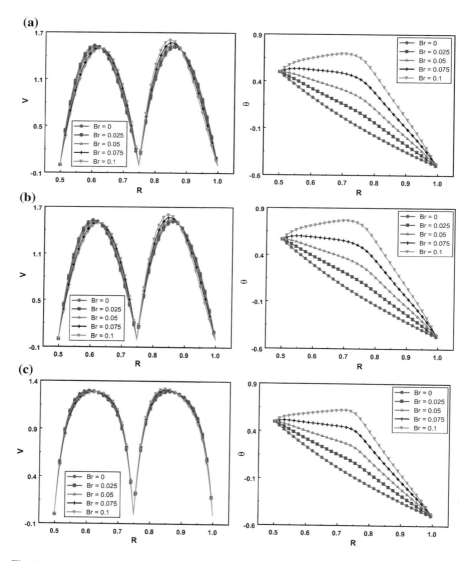

Fig. 11 Effect of Br on velocity and temperature profiles for $\lambda = 0.5, N = 0.75, GR = 5 \times 10^2$ $Ha = 5$, (Top) $Da = 0$, (Middle) $Da = 0.1$ and (Bottom) $Da = 0.001$

viscous dissipation is strongly reflected in thermal profiles rather than velocity distribution. Since the viscous dissipation effect originates from the energy equation, it is expected that the effect of Br is strongly detected in temperature distribution. In the absence of viscous dissipation ($Br = 0$), a linear temperature profile is observed for all values of Da. However, as the Brinkman number is increased, temperature gradients along the radial direction in the inner passage decrease and then sharply decrease in the outer passage. As for the velocity profiles are concerned, the magnitude of velocity decreases with an increase in Br and decrease in Da. The effect of Ha and Br on the velocity and thermal profiles is depicted in Fig. 12 for fixed values of other parameters. The presence of magnetic field reduces the flow circulation rate and does not have significant impact on thermal pattern. However, the inclusion of viscous dissipation significantly alters the temperature profiles.

In this analysis, the main emphasis is to understand the effects of Brinkman number on the heat transfer rate in the presence of magnetic field and porosity. Figures 13 and 14 illustrate the combined influences of Da and Ha for different values of Brinkman number. An overview of the Nusselt number profile reveals that the heat transfer rate decreases with viscous dissipation effect characterized by Brinkman number for all values of Da and Ha. As observed earlier, the hot wall Nu decreases with Br, whereas the cold wall Nu increases with Br. It is interesting to observe from Fig. 13 that the heat transfer at hot wall decreases with a decrease in Da, while the cold wall Nu increases as the Darcy number decreases. Also, for higher value of $Br = 0.05$, heat transfer at hot wall increases as the Darcy number decreases. Figure 14 shows the variation of hot and cold wall Nusselt numbers for different values of Ha and Br. The general observation of the figure reveals that the hot wall Nu decreases and cold wall Nu increases with Br and Ha. As noted in the variation of Nu with Darcy and Brinkman numbers, the hot wall Nu increases with an increase in Ha for higher Br.

4.4 *Effect of Radius Ratio*

The important and unique geometrical parameter in annular passage is the ratio of inner to outer radius, known as radius ratio and its effect on flow pattern, thermal distribution, and heat transfer rate is shown in Figs. 15, 16, and 17 for two different values of Da and Ha. Figure 15 exemplifies the combined effects of radius ratio and Darcy number on velocity and thermal profiles for low and high values of Da and λ. The variation of radius ratio influences the width of the inner annular passage by fixing the width of the outer annular passage. A smaller radius ratio ($\lambda = 0.1$) corresponds to a wider inner passage, and a larger radius ratio ($\lambda = 0.5$) indicates the narrow inner passage. In general, an increase in the radius ratio increases the magnitude of velocity and a decrease in the Darcy number results in a decrease in velocity magnitude. A careful observation of the figure reveals the existence of flow reversal in the inner annular passage for the lower radius ratio ($\lambda = 0.1$) at both

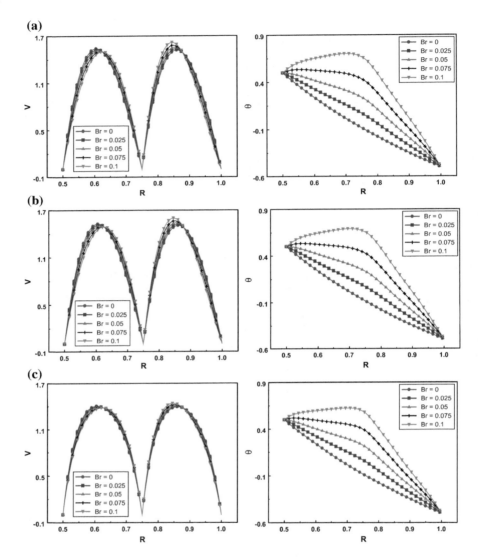

Fig. 12 Effect of Brinkmann number on velocity and temperature profile for $\lambda = 0.5, N = 0.75$, $GR = 5 \times 10^2, Da = 0.1$, (Top) $Ha = 0$, (Middle) $Ha = 5$ and (Bottom) $Ha = 20$

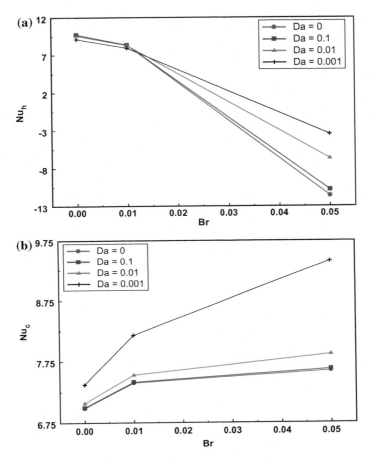

Fig. 13 Effect of *Da* and *Br* on the hot (Nu_h) and cold (Nu_c) wall Nusselt numbers for $\lambda = 0.5$, $N = 0.75, GR = 10^3$ and $Ha = 1$

values of *Da*. The radius ratio also influences the thermal profiles in a significant manner; however, the Darcy number does not change the thermal distribution appreciably. Figure 16 exhibits the combined influence of magnetic field and radius ratio on the velocity and thermal profiles. It can be noted that the magnetic field reduces the flow velocity and radius ratio augments the flow velocity. It is interesting to observe that the effect of *Ha* on velocity field is marginal for larger radius ratio. Also, flow reversal phenomenon near the baffle can be seen for smaller values of λ and *Ha*, and the flow reversal can be controlled by increasing these parameters.

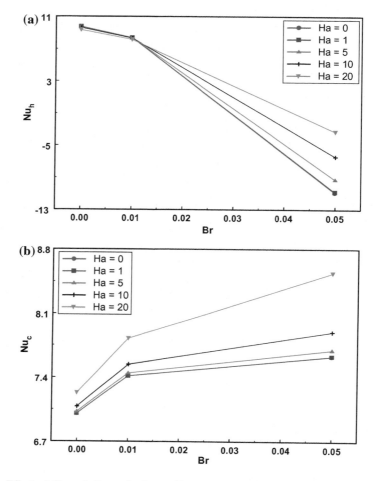

Fig. 14 Effect of Ha and Br on the hot wall (Nu_h) and cold wall (Nu_c) Nusselt numbers for $\lambda = 0.5, N = 0.75, GR = 10^3$ and $Da = 0.1$

The variation of hot and cold wall Nusselt numbers is interesting when both Darcy and Hartmann numbers are varied, and is illustrated in Fig. 17. In general, the hot wall Nu decreases with a decrease in Da and an increase in Ha and for the cold wall Nu, the situation is opposite. In particular, the hot wall Nusselt number decreases sharply for $Ha > 10$ and $Da \leq 10^{-2}$.

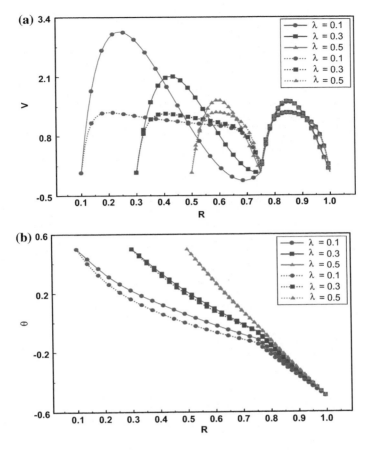

Fig. 15 Effect of radius ratio on velocity and temperature profiles for $N = 0.75, GR = 10^3, Br = 0.01, Ha = 5$. Continuous lines correspond to $Da = 0.1$ and dotted lines correspond to $Da = 0.001$

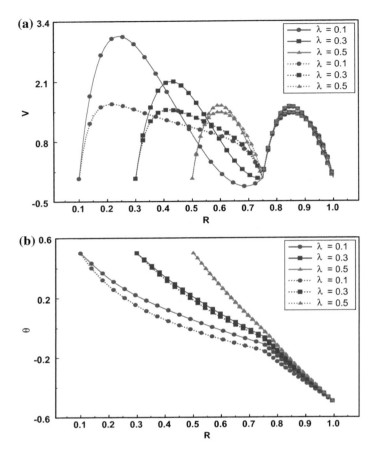

Fig. 16 Effect of radii ratio on velocity and temperature profiles for $N = 0.75, GR = 10^3, Br = 0.01, Da = 0.1$. Continuous lines correspond to $Ha = 5$ and dotted lines correspond to $Ha = 20$

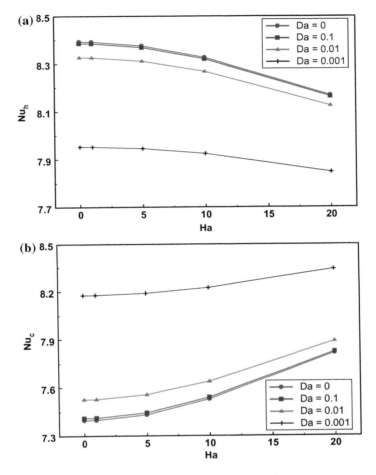

Fig. 17 Effect of Ha and Da on the hot (Nu_h) and cold (Nu_c) wall Nusselt numbers for $\lambda = 0.5$, $N = 0.75, GR = 10^3$ and $Br = 0.01$

5 Conclusions

Fully developed mixed convection in the vertical porous double-passage annuli has been investigated numerically in the presence of viscous dissipation and magnetic field. The analytical expressions are derived for velocity and temperature profiles, and heat transfer rates in the absence of viscous dissipation, while the numerical solutions are obtained by taking the viscous dissipation into account.

Based on the influences of different physical and geometrical parameters on the flow pattern, thermal distribution, and heat transfer rate, the following important conclusions are drawn from the present analysis:

(I) In the absence of viscous dissipation, the velocity profiles are significantly altered with the variation of Darcy and Hartmann numbers, but the temperature profiles are unaltered with the variation of these parameters.
(II) The flow velocity and hot wall Nusselt number are suppressed by the presence of magnetic field and porosity. However, the cold wall Nu increases with these two parameters.
(III) The rate of heat transfer at the hot wall is greatly suppressed as the viscous dissipation effect increases.
(IV) The location of baffle (middle cylinder) substantially changes the flow pattern, thermal distribution, and heat transfer processes in both passages.
(V) As regards to the radius ratio, maximum and minimum velocities occur for smaller and larger radius ratios, respectively.
VI) In general, the viscous dissipation have dominant effect on thermal distribution and the flow patterns are profoundly altered by the presence of porosity, magnetic field, and baffle position.

Acknowledgements The authors MS and GN are, respectively, grateful to the managements of Presidency University, Bengaluru and J S S Academy of Technical Education, Bengaluru, and to VTU, Belgaum, India for their support and encouragement. Also, M. Sankar was supported by the Vision Group of Science and Technology (VGST) K-FIST (L1) grant funded by the Government of Karnataka.

References

Al-Nimr MA (1993) Analytical solution for transient laminar fully developed free convection in vertical concentric annuli. Int J Heat Mass Transf 36(9):2385–2395. https://doi.org/10.1016/S0017-9310(05)80122-X
Al-Nimr MA (1995a) Fully developed free convection in open-ended vertical concentric porous annuli. Int J Heat Mass Transf 38(l):1–12. https://doi.org/10.1016/0017-9310(94)00148-o
Al-Nimr MA (1995b) MHD free-convection flow in open-ended vertical concentric porous annuli. Appl Energy 50(4):293–311. https://doi.org/10.1016/0306-2619(95)98800-H
Aung W (1972) Fully developed laminar free convection between vertical plates heated asymmetrically. Int. J Heat Mass Transf 15(8):1577–1580. https://doi.org/10.1016/0017-9310(72)90012-9
Aung W (1987) Mixed convection in internal flow: handbook of single-phase convective heat transfer. Wiley, New York
Aung W, Worku G (1986a) Developing flow of flow reversal with asymmetric wall temperatures. ASME J Heat Transf 108(2):299–304. https://doi.org/10.1115/1.3246919
Aung W, Worku G (1986b) Theory of fully developed, combined convection including flow reversal. ASME J Heat Transf 108(2):485–488. https://doi.org/10.1115/1.3246958
Aydin O (2005) Effects of viscous dissipation on the heat transfer in a forced pipe flow. Part 2: Thermally developing flow. Energy Convers Manag 46(18–19):3091–3102. https://doi.org/10.1016/j.enconman.2005.03.011
Barletta A (1998) Laminar mixed convection with viscous dissipation in a vertical channel. Int J Heat Mass Transf 41(22):3501–3513. https://doi.org/10.1016/S0017-9310(98)00074-X

Barletta A (1999a) Analysis of combined forced and free flow in a vertical channel with viscous dissipation and isothermal-isoflux boundary conditions. ASME J Heat Transf 121(2):349–356. https://doi.org/10.1115/1.2825987

Barletta A (1999b) Combined forced and free convection with viscous dissipation in a vertical circular duct. Int. J Heat Mass Transf 42(12):2243–2253. https://doi.org/10.1016/S0017-9310(98)00343-3

Barletta A, Celli M (2008) Mixed convection MHD flow in a vertical channel: effects of Joule heating and viscous dissipation. Int J Heat Mass Transf 51(25–26):6110–6117. https://doi.org/10.1016/j.ijheatmasstransfer.2008.04.009

Barletta A, di Rossi SE (2001) Effect of viscous dissipation on mixed convection heat transfer in a vertical tube with uniform wall heat flux. Heat Mass Transf 38(1–2):129–140. https://doi.org/10.1007/s002310100204

Barletta A, Zanchini E (1999) On the choice of the reference temperature for fully-developed mixed convection in a vertical channel. Int J Heat Mass Transf 42(16):3169–3181. https://doi.org/10.1016/S0017-9310(99)00011-3

Barletta A, Zanchini E (2001) Mixed convection with viscous dissipation in an inclined channel with prescribed wall temperatures. Int J Heat Mass Transf 44(22):4267–4275. https://doi.org/10.1016/S0017-9310(01)00071-0

Barletta A, Lazzari S, Magyari E, Pop I (2008) Mixed convection with heating effects in a vertical porous annulus with a radially varying magnetic field. Int J Heat Mass Transf 51(25–26):5777–5784. https://doi.org/10.1016/j.ijheatmasstransfer.2008.05.018

Bodoia JR, Osterle JF (1962) The development of free convection between heated vertical plates. ASME J Heat Transf 84(1):40–43. https://doi.org/10.1115/1.3684288

Chamkha AJ (2002) On laminar hydromagnetic mixed convection flow in a vertical channel with symmetric and asymmetric wall heating conditions. Int J Heat Mass Transf 45(12):2509–2525. https://doi.org/10.1016/S0017-9310(01)00342-8

Chen C-K, Lai H-Y, Liu C-C (2011) Numerical analysis of entropy generation in mixed convection flow with viscous dissipation effects in vertical channel. Int Commun Heat Mass Transf 38(3):285–290. https://doi.org/10.1016/j.icheatmasstransfer.2010.12.016

Cheng C-H, Kou H-S, Huang W-H (1990) Flow reversal and heat transfer of fully developed mixed convection in vertical channels. J Thermophy Heat Transf 4(2):375–383. https://doi.org/10.2514/3.190

Coelho PM, Pinho FT (2006) Fully-developed heat transfer in annuli with viscous dissipation. Int J Heat Mass Transf 49(19–20):3349–3359. https://doi.org/10.1016/j.ijheatmasstransfer.2006.03.017

Coney JER, El-Shaarawi MAI (1975) Finite difference analysis for laminar flow heat transfer in concentric annuli with simultaneously developing hydrodynamic and thermal boundary layers. Int J Numer Meth Eng 9(1):17–38. https://doi.org/10.1002/nme.1620090103

Dawood HK, Mohammed HA, Sidik NAC, Munisamy KM, Wahid MA (2015) Forced, natural and mixed-convection heat transfer and fluid flow in annulus: a review. Int Commun Heat Mass Transf 62:45–57. https://doi.org/10.1016/j.icheatmasstransfer.2015.01.006

El-Din MMS (2002) Effect of viscous dissipation on fully developed laminar mixed convection in a vertical double-passage channel. Int J Therm Sci 41(3):253–259. https://doi.org/10.1016/S1290-0729(01)01313-8

El-Din MMS (2007) Laminar fully developed mixed convection with viscous dissipation in a uniformly heated vertical double-passage channel. Therm Sci 11(1):27–41. https://doi.org/10.2298/TSCI0701027S

El-Shaarawi MAI, Al-Nimr MA (1990) Fully developed laminar natural convection in open-ended vertical concentric annuli. Int J Heat Mass Transf 33(9):1873–1884. https://doi.org/10.1016/0017-9310(90)90219-K

Gebhart B, Jaluria Y, Mahajan R, Sammakia B (1988) Buoyancy-Induced flows and transport. Hemisphere, Washington, DC

Hamadah TT, Wirtz RA (1991) Analysis of laminar fully developed mixed convection in a vertical channel with opposing buoyancy. ASME J Heat Transf 113(2):507–510. https://doi.org/10.1115/1.2910593

Incropera FP (1988) Convection heat transfer in electronic equipment cooling. J Heat Transf 110(4b):1097–1111. https://doi.org/10.1115/1.3250613

Iqbal M, Aggarwala BD, Rokerya MS (1970) Viscous dissipation effects on combined free and forced convection through vertical circular tubes. ASME J Appl Mech 37(4):931–935. https://doi.org/10.1115/1.3408720

Jha BK, Daramola D, Ajibade AO (2016a) Mixed convection in a vertical annulus filled with porous material having time-periodic thermal boundary condition: steady-periodic regime. Meccanica 51(8):1685–1698. https://doi.org/10.1007/s11012-015-0328-4

Jha BK, Oni MO, Aina B (2016b) Steady fully developed mixed convection flow in a vertical micro-concentric-annulus with heat generating/absorbing fluid: an exact solution. Ain Shams Eng J. https://doi.org/10.1016/j.asej.2016.08.005

Joshi HM (1987) Fully developed natural convection in an isothermal vertical annular duct. Int Commun Heat Mass Transf 14(6):657–664. https://doi.org/10.1016/0735-1933(87)90045-5

Kou H-S, Huang D-K (1997) Fully developed laminar mixed convection through a vertical annular duct filled with porous media. Int Commun Heat Mass Transf 24(1):99–110. https://doi.org/10.1016/S0735-1933(96)00109-1

Kumar JP, Umavathi JC, Biradar BM (2011) Mixed convection of magnetohydrodynamic and viscous fluid in a vertical channel. Int J Non-Linear Mech 46(1):278–285. https://doi.org/10.1016/j.ijnonlinmec.2010.09.008

Kumar JP, Umavathi JC, Biradar BM (2012) Two-Fluid mixed magnetoconvection flow in a vertical enclosure. J Appl Fluid Mech 5(3):11–21

Kumar JP, Umavathi JC, Pop I, Biradar BM (2009) Fully developed mixed convection flow in a vertical channel containing porous and fluid layer with isothermal or isoflux boundaries. Transp Porous Med 80(1):117–135. https://doi.org/10.1007/s11242-009-9347-8

Liu C-C, Lo C-Y (2012) Numerical analysis of entropy generation in mixed-convection MHD flow in vertical channel. Int Commun Heat Mass Transf 39(9):1354–1359. https://doi.org/10.1016/j.icheatmasstransfer.2012.08.001

Marner WJ, Hovland H (1973) Viscous dissipation effects on fully developed combined free and forced non-newtonian convection in a vertical tube. Warme-und Stoffubertragung 6(4):199–204. https://doi.org/10.1007/BF02575265

Mcbain GD (1999) Fully developed laminar buoyant flow in vertical cavities and ducts of bounded section. J Fluid Mech 401:365–377. https://doi.org/10.1017/S0022112099006783

Morton BR (1960) Laminar convection in uniformly heated vertical pipes. J Fluid Mech 8(2):227–240. https://doi.org/10.1017/S0022112060000566

Nield DA, Bejan A (2013) Convection in porous media, 4th edn. Springer, New York

Oni MO (2017) Combined effect of heat source, porosity and thermal radiation on mixed convection flow in a vertical annulus: an exact solution. Eng Sci Technol Int J 20(2):518–527. https://doi.org/10.1016/j.jestch.2016.12.009

Orfi J, Galanis N, Nguyen CT (1993) Laminar fully developed incompressible flow with mixed convection in inclined tubes. Int J Numer Meth Heat Fluid Flow 3(4):341–355. https://doi.org/10.1108/eb017535

Prakash C, Renzoni P (1985) Effect of buoyancy on laminar fully developed flow in a vertical annular passage with radial internal fins. Int J Heat Mass Transf 28(5):995–1003. https://doi.org/10.1016/0017-9310(85)90281-9

Ranjbar-Kani AA, Hooman K (2004) Viscous dissipation effects on thermally developing forced convection in a porous medium: circular duct with isothermal wall. Int Commun Heat Mass Transf 31(6):897–907. https://doi.org/10.1016/S0735-1933(04)00076-4

Rokerya MS, Iqbal M (1971) Effects of viscous dissipation on combined free and forced convection through vertical concentric annuli. Int J Heat Mass Transf 14(3):491–495. https://doi.org/10.1016/0017-9310(71)90167-0

Saleh H, Hashim I (2010) Flow reversal of fully-developed mixed MHD convection in vertical channels. Chin Phys Lett 27(2):024401

Sarveshanand, Singh AK (2015) Magnetohydrodynamic free convection between vertical parallel porous plates in the presence of induced magnetic field. SpringerPlus 4:333. https://doi.org/10.1186/s40064-015-1097-1

Tao LN (1960) On combined free and forced convection in channels. ASME J Heat Transf 82(3):233–238. https://doi.org/10.1115/1.3679915

Umavathi JC, Malashetty MS (2005) Magnetohydrodynamic mixed convection in a vertical channel. Int J Non-Linear Mech 40(1):91–101. https://doi.org/10.1016/j.ijnonlinmec.2004.05.018

Zanchini E (2008) Mixed convection with variable viscosity in a vertical annulus with uniform wall temperatures. Int. J Heat Mass Transf 51(1–2):30–40. https://doi.org/10.1016/j.ijheatmasstransfer.2007.04.046

Effect of Nonuniform Heating on Natural Convection in a Vertical Porous Annulus

M. Sankar, S. Kiran and Younghae Do

Nomenclature

Ar	Aspect ratio
A_l	Amplitude at inner wall (m)
A_r	Amplitude at outer wall (m)
D	Width of the annulus (m)
Da	Darcy number
g	Acceleration due to gravity (m/s^2)
H	Height of the annulus (m)
K	Permeability of the porous medium (m^2)
k	Thermal conductivity (W/m K)
$\overline{Nu}$	Average Nusselt number
Nu_l	Nusselt number at inner wall
Nu_r	Nusselt number at outer wall
p	Fluid pressure (Pa)
Pr	Prandtl number
Ra	Rayleigh number
Ra_D	Darcy–Rayleigh number
T_1, T_2	Dimensionless temperatures at the inner and outer walls
t^*	Dimensional time (s)
t	Dimensionless time
(r_i, r_o)	Radius of inner and outer cylinders (m)

M. Sankar (✉) · Y. Do
Department of Mathematics, KNU-Center for Nonlinear Dynamics,
Kyungpook National University, Daegu 41566, Republic of Korea
e-mail: manisankarir@yahoo.com

M. Sankar
Department of Mathematics, School of Engineering, Presidency University,
Bangalore, India

S. Kiran
Department of Mathematics, Sapthagiri College of Engineering,
Bangalore, India

© Springer Nature Singapore Pte Ltd. 2018
N. Narayanan et al. (eds.), *Flow and Transport in Subsurface Environment*,
Springer Transactions in Civil and Environmental Engineering,
https://doi.org/10.1007/978-981-10-8773-8_8

(r, z)	Dimensional radial and axial coordinates (m)
(R, Z)	Dimensionless radial and axial coordinates
(u, w)	Dimensional velocity components in (r, z) direction (m/s)
(U, W)	Dimensionless velocity components in (R, Z) direction

Greek Letters

β	Coefficient of thermal expansion (1/K)
ε	Amplitude ratio
ζ	Dimensionless vorticity
θ_1, θ_2	Dimensional temperatures at inner and outer walls (K)
κ	Thermal diffusivity (m^2/s)
λ	Radius ratio
Λ	Viscosity ratio
ν_e	Effective kinematic viscosity of the porous medium (m^2/s)
ν_f	Fluid kinematic viscosity (m^2/s)
ρ	Fluid density (kg/m^3)
σ	Heat capacity ratio
φ	Porosity
ϕ	Phase deviation
ψ	Dimensionless stream function
ψ_{max}	Maximum value of the dimensionless stream function

1 Introduction

Natural convection in finite enclosures filled with fluid-saturated porous media is an essential transport mechanism encountered in a wide range of engineering and scientific applications. These applications include geothermal engineering, thermal insulation systems, packed bed chemical reactors, porous heat exchangers, oil separation from sand by steam, underground disposal of nuclear waste materials, food storage, and electronic device cooling. Due to these applications, several investigations have been made on convective heat transfer in porous media and a detailed literature on this topic can be found in the books of Vafai (2005), Ingham and Pop (2005) and Nield and Bejan (2006).

Investigations on natural convection heat transfer in porous enclosures of different shapes are abundant in the literature. Based on the requirement of an application, the enclosure of appropriate shape will be chosen to estimate the flow field, temperature distribution, and heat transfer, and in turn these data can be used in design process. Among the different enclosures, natural convection in a vertical porous annulus has been widely studied and well-documented in the literature, owing to its importance in building insulation, porous heat exchangers, and many others applications. Using the perturbation and finite difference methods, Havstad and Burns (1982) analyzed convective flow and heat transfer rates in a vertical porous annulus, and reported the heat transfer correlations. Hickox and Gartling (1985) performed the finite element investigation of natural convection in a vertical annular enclosure filled with porous media for a wide range of radius and aspect ratios. Also, using parallel flow approximations for the case of tall annular enclosure, the closed form solutions are obtained for the Nusselt number. Later, Prasad (1986) numerically investigated natural convection in a vertical porous annulus with constant heat flux conditions at the inner wall for a wide range of physical and geometrical parameters. Natural convection in a vertical porous annulus has also been experimentally investigated by many researchers. Reda (1983) experimentally investigated natural convection in a vertical porous annulus formed by constant-heat-flux inner cylinder and constant-temperature outer cylinder. Further, the numerical solutions are obtained using the finite element method and the comparisons between the experimental and numerical results are in good agreement. Using several fluid–solid combinations, Prasad et al. (1985, 1986) performed experiments on free convection in a vertical porous annulus for three aspect ratios for a fixed radius ratio of 5.338. Also, a theoretical model has been developed based on the Darcy's law, and an iterative scheme to estimate the effective thermal conductivity of the porous medium from the overall heat transfer has been presented.

Hasnaoui et al. (1995) reported the combined analytical and numerical study of natural convection in a vertical annular porous layer with the inner wall maintained at a constant heat flux and insulated outer wall. The finite difference scheme has been used to obtain the numerical solution, while the parallel flow assumption is used for analytical solution. Natural convection heat transfer in a vertical cylindrical annulus filled with fluid-saturated porous medium has been numerically investigated by Marpu (1995) using the Brinkman-extended Darcy–Forchheimer model. Among the models used in the literature to study the convective flow and heat transfer in porous media, the Brinkman-extended Darcy model with the convective terms is extensively used in modeling the flow and heat transfer in finite porous enclosures. Using the Brinkman-extended Darcy model, Shivakumara et al. (2002) numerically investigated natural convection in a vertical porous annulus. Kiwan and Al-Zahrani (2008) examined the effect of three porous inserts on natural convection in a vertical open-ended annulus. Along the inner cylindrical wall of the annulus, the porous inserts are attached in the form of porous rings and the flow equations are solved by considering two different set of equations for fluid and porous domains. Numerical investigation of natural convection heat transfer in a vertical porous annulus with internal heat generation has been performed by Reddy and Narasimhan (2010). Sankar et al. (2011a, b) examined the effects of size and location of a discrete heater on the natural convective heat transfer in a vertical porous annulus. They found that the size and location of heat source have profound influence on fluid flow and heat transfer rates. Later, Sankar et al. (2013) investigated the combined effect of discrete heating and internal heat generation on natural convection in vertical porous annulus. More recently, Sankar et al. (2014) reported the influence of two discrete heat sources on natural convective heat transfer in a vertical porous annulus for different aspect and radius ratios.

In many industrial applications, the enhancement of heat transfer performance is essential from the energy saving perspective. Through the numerical and experimental investigations, it has been established that the heat transfer rates are relatively higher for nonuniform thermal boundary conditions compared to uniform thermal conditions. Therefore, in recent years, rigorous attention has been paid to natural convection fluid flow and heat transfer characteristics in porous and nonporous enclosures with nonuniform temperature distributions on the active walls of the enclosure. In particular, among other nonuniform temperature distributions, substantial attention has been devoted to understand the influence of sinusoidally varying thermal boundary conditions on natural convection in porous and nonporous enclosures. Bilgen and Yedder (2007) considered a rectangular enclosure with the left wall that is heated and cooled by a sinusoidal temperature profile and other walls are insulated. The sinusoidal temperature profile forms two different types of heating and they found that the heat transfer rates are higher for lower half heating of inner wall. Deng and Chang (2008) analyzed the influence of two sinusoidal temperature profiles on natural convection in a rectangular enclosure and presented the results for wide range of amplitude ratio, phase deviation, aspect ratio, and Rayleigh number. The combined effects of magnetic field and sinusoidal thermal boundary conditions on mixed convection in a lid-driven cavity have been investigated by Sivasankaran et al. (2011).

Natural convection in a porous cavity with sinusoidal thermal boundary conditions has also received considerable attention in recent years. Saeid and Mohamad (2005) performed a numerical study on natural convection in a square cavity with left wall heated by a sinusoidal temperature profile; right wall cooled at a constant temperature, insulated horizontal walls, and found that the average heat transfer rate increases with amplitude. Natural convection under the influence of a heat source with sinusoidal temperature variation placed at the bottom wall of a square enclosure filled with fluid-saturated porous media is investigated by Saeid (2005). Using Darcy model, Saeid (2006) conducted a numerical study in a square porous enclosure with sinusoidal temperature variation with time and found that the Nusselt number becomes negative for high amplitude and frequency. Varol et al. (2008) investigated steady natural convection in a rectangular porous enclosure with a sinusoidal temperature variation at the bottom wall, insulated top and side walls. They showed that the heat transfer increases with amplitude, but decreases with aspect ratio and found multicellular flow pattern for all parameters of the problem. Using Darcy–Forchheimer–Brinkman model, Wang et al. (2010) carried out a numerical study of three-dimensional natural convections in an inclined porous cubic enclosure with oscillating sinusoidal thermal boundary conditions with time. They found that the flow and heat transfer rate are significantly altered by two inclined angles of the enclosure and suggested the appropriate oscillating frequencies and inclination angles to augment the convection heat transfer.

Selamat et al. (2012) applied finite difference technique to investigate natural convection in an inclined porous square cavity with a sinusoidally heated side wall and found the optimum inclination angle and wave number for maximum heat transfer rate. Using the Brinkman–Forchheimer-extended Darcy model, the influence of sinusoidal temperature profiles on mixed convection in a lid-driven porous square enclosure has been investigated by Sivasankaran and Pan (2012). Khansila and Witayangkurn (2012) studied natural convection in a rectangular porous enclosure whose left wall is heated and cooled by sinusoidal temperature and other walls are insulated. Sivasankaran and Bhuvaneswari (2013) performed natural convection in a porous square enclosure with sinusoidal temperature distribution on both side walls and found that the two sinusoidal temperature distributions on the side walls enhance the heat transfer rate compared to a single sinusoidal temperature profile. A similar work has been carried out Zahmatkesh (2014) using the Darcy model and also discussed the entropy generation phenomena in the enclosure. Recently, using the local thermal nonequilibrium model, Wu et al. (2015, 2016) reported a numerical study of natural convection heat transfer in a rectangular cavity filled with a heat-generating porous medium and the side walls of cavity are subjected to one or two sinusoidal temperature profiles. The above studies on natural convection heat transfer in porous enclosure indicate that the sinusoidal thermal boundary conditions can significantly augment the heat transfer rate. In a similar way, natural convection heat transfer of nanofluids has also been investigated in the literature with sinusoidally varying temperature profiles on the side walls of porous and nonporous enclosures. Ben-Cheikh et al. (2013) made a numerical study of natural convection of nanofluids in a square enclosure with sinusoidal temperature distribution on the bottom wall and the remaining walls are maintained at lower temperature. Using Buongiorno's model, Sheremet and Pop (2014) analyzed natural convection in a porous square enclosure filled with nanofluid and the side walls are heated sinusoidally. More recently, Sheremet and Pop (2015) performed a numerical investigation of natural convection of nanofluid in a wavy porous enclosure with sinusoidal thermal conditions on both side walls.

The above-detailed literature review reveals that most of the research efforts to understand the influence of sinusoidal thermal boundary condition on natural convection are mainly devoted to square or rectangular or wavy porous enclosures. However, natural convection in a vertical porous annulus with sinusoidally varying temperature profiles at the side walls aptly describes the physical configuration of several applications. Hence, the knowledge of convective flow and heat transfer rate in a vertical porous annular enclosure is a topic of fundamental importance in these applications. Also, it is difficult to predict the nature of natural convection flows as they depend strongly on the geometric characteristics of the enclosure. Unfortunately, the authors could not find any studies on natural convection in a porous annulus with sinusoidal temperature profiles at both the walls. Therefore, the main objective of the present study is to numerically investigate the natural convection flow in a porous annulus with side walls having sinusoidal temperature distributions and the horizontal top and bottom walls to be adiabatic.

2 Mathematical Formulation

2.1 Governing Equations

We consider a two-dimensional, vertical concentric annular enclosure filled with homogeneous, isotropic and fluid-saturated porous medium as shown schematically in Fig. 1. The annulus is formed by an inner cylinder of radius r_i and an outer cylinder of radius r_o. The cylindrical coordinate system (r, z), the corresponding velocity components (u, w), and the thermal conditions are also indicated in Fig. 1. Here, D and H respectively denote the width and height of the annular enclosure. Further, the top and bottom boundaries of the annulus are assumed to be closed and insulated. However, the inner and outer walls of the annulus are subjected to spatially varying sinusoidal temperature distribution of different amplitudes and phase deviations as given below:

$$\theta_1(z) = \theta_0 + A_l \sin\left(\frac{2\pi z}{D}\right) \quad \text{temperature distribution on inner wall}$$

$$\theta_2(z) = \theta_0 + A_r \sin\left(\frac{2\pi z}{D} + \phi\right) \quad \text{temperature distribution on outer wall}$$

Also, we assume that the flow to be axisymmetric, the fluid is Newtonian with negligible viscous dissipation and gravity acts in the negative z-direction. Further, the porous medium is assumed to be rigid, and is in local thermodynamic equilibrium with the fluid. In porous medium, the Brinkman-extended Darcy model is

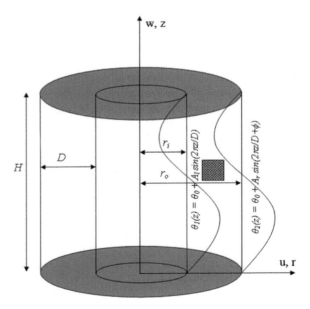

Fig. 1 Schematic diagram of the porous annulus and coordinate system

assumed to hold and the Forchheimer quadratic drag term of the momentum equation is neglected. Further, it is assumed that all thermophysical properties of fluid and solid matrix are constant except the effect of density variations in the buoyancy term, which is treated according to the Boussinesq approximation. Hence, the heat capacity ratio σ is taken to be unity.

Based on the above assumptions, the governing conservation equations describing the fluid flow and heat transfer phenomenon in the porous annulus are as follows:

$$\frac{\partial u}{\partial r} + \frac{\partial w}{\partial z} + \frac{u}{r} = 0, \tag{1}$$

$$\frac{1}{\varphi}\frac{\partial u}{\partial t^*} + \frac{1}{\varphi^2}\left[u\frac{\partial u}{\partial r} + w\frac{\partial u}{\partial z}\right] = -\frac{1}{\rho_0}\frac{\partial p}{\partial r} + \frac{v_e}{\varphi}\left[\nabla_1^2 u - \frac{u}{r^2}\right] - \frac{v_f}{K}u, \tag{2}$$

$$\frac{1}{\varphi}\frac{\partial w}{\partial t^*} + \frac{1}{\varphi^2}\left[u\frac{\partial w}{\partial r} + w\frac{\partial w}{\partial z}\right] = -\frac{1}{\rho_0}\frac{\partial p}{\partial z} + \frac{v_e}{\varphi}\nabla_1^2 w - g\beta(\theta - \theta_0) - \frac{v_f}{K}w, \tag{3}$$

$$\sigma\frac{\partial \theta}{\partial t^*} + u\frac{\partial \theta}{\partial r} + w\frac{\partial \theta}{\partial z} = \kappa\nabla_1^2\theta, \tag{4}$$

where $\nabla_1^2 = \frac{\partial^2}{\partial r^2} + \frac{1}{r}\frac{\partial}{\partial r} + \frac{\partial^2}{\partial z^2}$.

In the above equations, u and w are velocity components in r and z directions respectively p is the pressure, v_f is fluid kinematic viscosity, v_e is effective kinematic viscosity, ρ is fluid density, θ is temperature, and κ is the thermal diffusivity. Due to the azimuthal symmetry, the flow depends spatially on two cylindrical coordinates (r, z), and as a result, the pressure is eliminated via cross-differentiation. Hence, by introducing the Stokes stream function, the Navier–Stokes equations can be written in terms of stream function (ψ)—vorticity (ζ) formulation. Therefore, the dimensionless vorticity-stream function equations are as follows:

$$\sigma\frac{\partial T}{\partial t} + U\frac{\partial T}{\partial R} + W\frac{\partial T}{\partial Z} = \nabla^2 T, \tag{5}$$

$$C_1\left\{\frac{1}{\varphi}\frac{\partial \zeta}{\partial t} + \frac{1}{\varphi^2}\left[U\frac{\partial \zeta}{\partial R} + W\frac{\partial \zeta}{\partial Z} - \frac{U\zeta}{R}\right]\right\} = \frac{C_2 Pr\Lambda}{\varphi}\left[\nabla^2\zeta - \frac{\zeta}{R^2}\right] \\ - \frac{Pr}{Da}\zeta - PrRa\frac{\partial T}{\partial R}, \tag{6}$$

$$\zeta = \frac{1}{R}\left[\frac{\partial^2\psi}{\partial R^2} - \frac{1}{R}\frac{\partial \psi}{\partial R} + \frac{\partial^2\psi}{\partial Z^2}\right], \tag{7}$$

$$U = \frac{1}{R}\frac{\partial \psi}{\partial Z}, \quad W = -\frac{1}{R}\frac{\partial \psi}{\partial R}, \tag{8}$$

where $\nabla^2 = \frac{\partial^2}{\partial R^2} + \frac{1}{R}\frac{\partial}{\partial R} + \frac{\partial^2}{\partial Z^2}$.

In the present study, the values of fluid kinematic viscosity (v_f) and effective kinematic viscosity (v_e) are assumed to be equal ($v_f = v_e = v$). This approximation provides good agreement with the experimental data available in the literature.

The dimensionless variables used in the above equations are as follows:

$$(R, Z) = (r, z)/D, (U, W) = (u, w)D/\kappa, t = t^*\kappa/D^2, T = (\theta - \theta_0)/\Delta\theta, P = pk/\mu\kappa,$$
$$\zeta = \zeta^* D^2/\kappa, \psi = \psi*/r_i\kappa, \text{ where } D = r_o - r_i, \Delta\theta = A_l \text{ (amplitude)}. \tag{9}$$

The constants C_1 and C_2 in Eq. (6) can be set equal to 0 or 1 to obtain the Darcy or Darcy–Brinkman models, respectively.

2.2 Initial and Boundary Conditions

The dimensional form of initial and boundary conditions is given as follows:

$$t^* = 0: \quad u = w = 0, \theta = \theta_0; \quad r_i \leq r \leq r_o, 0 \leq z \leq H$$

$$t^* > 0: \quad u = w = 0, \theta_1(z) = \theta_0 + A_l \sin\left(\frac{2\pi z}{D}\right); \quad r = r_i$$

$$u = w = 0, \theta_2(z) = \theta_0 + A_r \sin\left(\frac{2\pi z}{D} + \phi\right); \quad r = r_o$$

$$u = w = 0, \frac{\partial \theta}{\partial z} = 0; \quad z = 0 \text{ and } H,$$

where A_l and A_r are amplitude of sinusoidal temperatures of inner and outer walls respectively. Using the nondimensional variables defined in Eq. (9), the dimensionless initial and boundary conditions are as follows:

$$t = 0: \quad U = W = T = 0, \psi = 0; \quad \frac{1}{\lambda - 1} \leq R \leq \frac{\lambda}{\lambda - 1} \text{ and } 0 \leq Z \leq Ar$$

$$t > 0: \quad \psi = \frac{\partial \psi}{\partial R} = 0, T_1(Z) = \sin(2\pi Z); \quad R = \frac{1}{\lambda - 1} \text{ and } 0 \leq Z \leq Ar$$

$$\psi = \frac{\partial \psi}{\partial R} = 0, T_2(Z) = \varepsilon \sin(2\pi Z + \phi); \quad R = \frac{\lambda}{\lambda - 1} \text{ and } 0 \leq Z \leq Ar$$

$$\psi = \frac{\partial \psi}{\partial Z} = 0, \frac{\partial T}{\partial Z} = 0; \quad Z = 0 \text{ and } Z = Ar.$$

2.3 Dimensionless Parameters and Heat Transfer Rate

The nondimensional parameters appearing in the present problem are as follows:

$Ra = \frac{g\beta\Delta\theta D^3}{v_f \kappa}$, the Rayleigh number, $Pr = \frac{v_f}{\kappa}$, the Prandtl number,

$Da = \frac{K}{D^2}$, the Darcy number, $\Lambda = \frac{v_f}{v_e}$, viscosity ratio, $Ar = \frac{H}{D}$, the aspect ratio,

$\varepsilon = \frac{A_r}{A_l}$, the amplitude ratio, ϕ = phase deviation, and $\lambda = \frac{r_o}{r_i}$, the radius ratio.

The total heat transfer rate across the enclosure can be estimated from the average Nusselt number and is equal to the sum of the averaged Nusselt number along the heating halves of both inner and outer walls. Hence, the average Nusselt number is defined as

$$\overline{Nu} = \int_{heating\,half} Nu_l dZ + \int_{heating\,half} Nu_r dZ,$$

where Nu_l and Nu_r are the local Nusselt numbers along the inner and outer walls and is defined as

$$Nu_l = \left(-\frac{\partial T}{\partial R}\right)_{R=\frac{1}{\lambda-1}} \quad \text{and} \quad Nu_r = \left(-\frac{\partial T}{\partial R}\right)_{R=\frac{\lambda}{\lambda-1}}.$$

3 Solution Procedure and Validation

3.1 Numerical Method

The dimensionless form of the vorticity transport and energy equations are solved by an implicit finite difference method based on the Alternating Direction Implicit (ADI) method. The time and spatial derivatives are approximated respectively by first-order and second-order finite difference representations. The nonlinear convection terms in the temperature and vorticity equations are also approximated by the second-order finite difference representations. However, the Successive Line Over Relaxation (SLOR) method is employed to solve the stream function equation by properly choosing the relaxation factor as 1.78. The SLOR method converges in less iteration than the usual point iteration methods and immediately transmits the boundary condition information to interior domain. The resulting finite difference equations can be arranged in tri-diagonal matrix form and can be solved by the

Thomas algorithm. For the solution of vorticity equation, the values of vorticity at the boundary are obtained by expanding the stream function in Taylor series. Using central difference approximation, the velocity components are evaluated from Eq. (8). Finally, the Simpson's rule is performed for the numerical integration of the average Nusselt number. A uniform grid in R and Z directions is used in the calculation domain and all numerical results are checked for the grid independency. After testing with several grid sizes, the grid independence has been achieved with the grid sizes of 252×126 for $Ar = 0.5$, 126×126 for $Ar = 1$ and 126×252 for $Ar = 2$. The unsteady terms are retained for the purpose of numerical calculations, and the large-time converged solutions will be taken as the steady state values. An in-house FORTRAN code has been developed for the present problem and several tests have been done for the validation of numerical code.

3.2 Code Validation

The FORTRAN code employed in this investigation has been successfully used in our recent works to study natural convection in porous and nonporous annulus. However, to validate the code used in this problem, natural convection in a differentially heated porous and nonporous enclosures of the earlier investigations have been solved. First, the streamline and isotherms for a square ($\lambda = 1$) nonporous enclosure with sinusoidally varying temperature profiles are obtained from the present code and are compared with that of Deng and Chang (2008). Figure 2a reveals the excellent agreement between the present results and that of Deng and Chang (2008) in a nonporous square enclosure. Next, in Fig. 2b, we compared the present streamlines and isotherms with that of a uniformly heated and cooled porous annulus investigated by Shivakumara et al. (2002) for $Ra = 10^6$, $Da = 10^{-2}$, $\Lambda = 3$, $\lambda = 2$, $Ar = 1$, and $Pr = 7.0$. From Fig. 2b, it has been observed that the present results are in good agreement with the results of Shivakumara et al. (2002). Finally, we set $C_1 = C_2 = 0$ in Eq. (6) and performed the simulations for the Darcy model in a porous annulus whose inner and outer walls are respectively maintained at uniform heat flux and uniform temperature. Table 1 shows the comparison of average Nusselt numbers between the present study and that of Prasad (1986) for different Darcy–Rayleigh numbers and radius ratios. From the table, it can be observed that both results are agreed within the 0.5% maximum relative difference. From Fig. 2 and Table 1, the correspondence between the present results and literature data is widely satisfactory. Through these validation tests, the accuracy of the present numerical computation is assured.

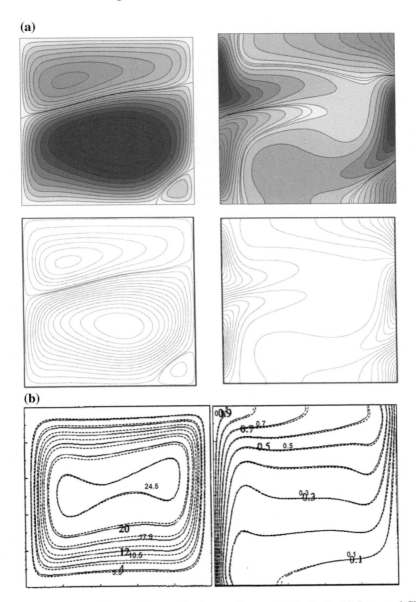

Fig. 2 a Comparison of streamlines and isotherms of present study (top) with Deng and Chang (2008) (bottom) for $Ra = 10^5$, $\varepsilon = 1$, $\phi = \pi/2$, $Ar = 1$, $\lambda = 1$ and $Da = \infty$, **b** comparison of streamlines and isotherms between the present results and that of Shivakumara et al. (2002) for $Ra = 10^6$, $Da = 10^{-2}$, $\Lambda = 3$, $\lambda = 2$, $Ar = 1$ and $Pr = 7.0$. Discontinuous lines correspond to present results and continuous lines correspond to Shivakumara et al. (2002)

Table 1 Comparison of average Nusselt number with the results of Prasad (1986) in a uniformly heated porous annulus at $Ar = 1$ (Darcy model results)

Radius ratio (λ)	Darcy–Rayleigh number (Ra_D)	Prasad (1986)	Present study	Relative difference (%)
2	10^3	6.4934	6.4815	0.18
	10^4	16.0498	16.0271	0.14
3	10^3	7.1659	7.1804	0.20
	10^4	17.2691	17.2226	0.27
5	10^3	8.0036	8.0262	0.28
	10^4	18.8055	18.8631	0.31
10	10^3	9.3975	9.4452	0.51
	10^4	20.7498	20.8325	0.40

4 Results and Discussion

The numerical simulations are performed to understand the influence of the amplitude ratio and phase deviation of two sinusoidally temperature profiles at the side walls on the flow fields, temperature distributions and heat transfer rate in the cavity for wide range of physical and geometrical parameters. In the present investigation, we have a total of nine nondimensional parameters. However, the influence of only five parameters are investigated and the remaining four parameters are kept at a fixed value. In particular, we choose the values of the Prandtl number (Pr) to be 0.7, radius ratio (λ) to 2.0, viscosity ratio (Λ) to 1.0 and the heat capacity ratio (σ) to 1.0 throughout the study. The simulations are carried out for the Rayleigh number (Ra) ranging from 10^3 to 10^6, the Darcy number (Da) ranging from 10^{-5} to 10^{-1}, amplitude ratio (ε) from 0 to 1, phase deviation (ϕ) from 0 to π and aspect ratio (Ar) from 0.5 to 2. The effects of the Rayleigh and Darcy numbers, the amplitude ratio, the phase deviation, and the aspect ratio are analyzed and the numerical results are presented in the form of streamline and isotherm contour plots, the local and average Nusselt numbers along the side walls.

4.1 Effect of Darcy Number

In this section, the combined influence of Darcy number, phase deviation, and amplitude ratio on the flow pattern and temperature fields is analyzed in the porous annulus. To understand these effects, streamlines and isotherms are plotted in Fig. 3 for three different Darcy numbers by fixing the values of other parameters. The influence of sinusoidal thermal boundary condition is vividly revealed on the flow and temperature fields for all Darcy numbers. As the phase deviation is fixed at 0, the sinusoidal temperature variation at the outer wall differs from the inner wall in terms of amplitude ratio. At lower Darcy number considered in this study

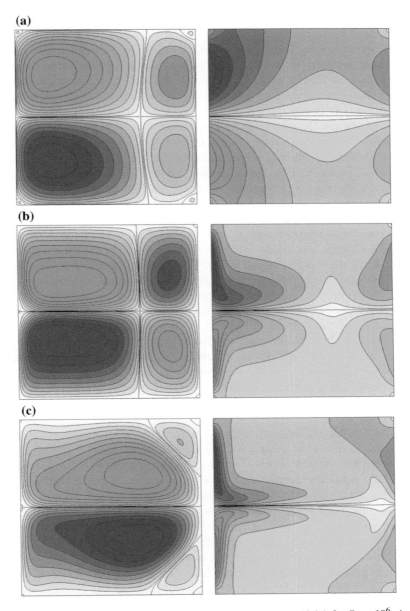

Fig. 3 Effect of Darcy number on streamlines (left) and isotherms (right) for $Ra = 10^6$, $Ar = 1$, $\phi = 0$ and $\varepsilon = 0.25$. **a** $Da = 10^{-6}$, ($\psi_{min} = -0.003$, $\psi_{max} = 0.003$), **b** $Da = 10^{-3}$, ($\psi_{min} = -11.23$, $\psi_{max} = 11.23$), **c** $Da = 10^{-1}$, ($\psi_{min} = -35.40$, $\psi_{max} = 35.36$)

($Da = 10^{-6}$), the streamlines exhibit a weak symmetrical flow with four eddies, of which two eddies are circulating clockwise and the remaining two eddies are circulating anticlockwise. The two eddies near the inner wall are larger in size and the eddies near the outer wall are smaller in size. The clockwise eddies formed by warmer fluids are located diagonally along the top left and bottom right corners of the enclosure, while anticlockwise eddies due to cooler fluids are placed diagonally from bottom left to top right of the enclosure. The flow circulation rate in the annulus is measured by the maximum and minimum value of the stream function. It indicates that even at high Rayleigh number, the flow movement is very weak and the flow is unable to penetrate deeper into the porous medium due to high resistance experienced by the smaller Darcy number value. The isotherms reveal that the movement of hot fluid is confined to bottom portion of the annulus and top region is occupied by cooler fluids. Also, the isotherms does not show much variation, which indicates that the heat transfer is mainly controlled by the conduction-dominated mechanism due to the porous drag. As the value of Da is increased to $Da = 10^{-3}$, the viscous forces, which are dominant over the porous resistance, increase the magnitude of the velocity. For $Da = 10^{-3}$, the flow intensity increases and the streamlines exhibit a strong flow pattern and isotherms reveal perceptible variation. As the Darcy number is increased to the extreme value considered in this study ($Da = 10^{-1}$), we noticed that the strength of convective flow becomes stronger compared to the cases shown in Figs. 3a, b. Also, as the Darcy number increases, the size of the eddies near the outer wall has been significantly reduced and the flow penetrates deeper into the porous medium. The presence of relatively stronger gradients can be found from the isotherms. In general, as the permeability of the porous medium is strongly influenced by the Darcy number, an overview of Fig. 3 reveals that the Darcy number significantly affects the flow and thermal distribution in the annulus.

To examine the influence of amplitude ratio on the local Nusselt number measured along the inner and outer walls, the local Nu profiles are illustrated in Fig. 4 by choosing five different values of ε and fixing other parameters. Due to sinusoidal temperature boundary condition at both walls, the local Nu curves reveal the undulating nature. A careful observation of the local Nusselt number variation along the inner wall reveals that the heat transfer rate at the inner wall does not show appreciable change by varying the amplitude ratio of the outer wall. On contrast, the local Nusselt number along the outer wall varies significantly with the amplitude ratio of sinusoidal temperature profile at the outer wall. In particular, the magnitude of local Nu increases with amplitude ratio and the maximum value of local Nu occurs for $\varepsilon = 1$. From this figure, it can be found that the amplitude ratio of the sinusoidal temperature profile affects the heat transfer at inner wall; but does not change the heat transfer along the inner wall. Figure 5 illustrates the influence of phase deviation (ϕ) on the local Nusselt number profiles estimated along the inner and outer walls. The local Nu profiles along the inner wall are modified in a similar fashion by the phase deviation of thermal boundary condition at the outer wall. For each value of ϕ, the value of Nu at the inner wall increases and decreases alternatively in the interval of 0.2 units. That is, the value of Nu increases from

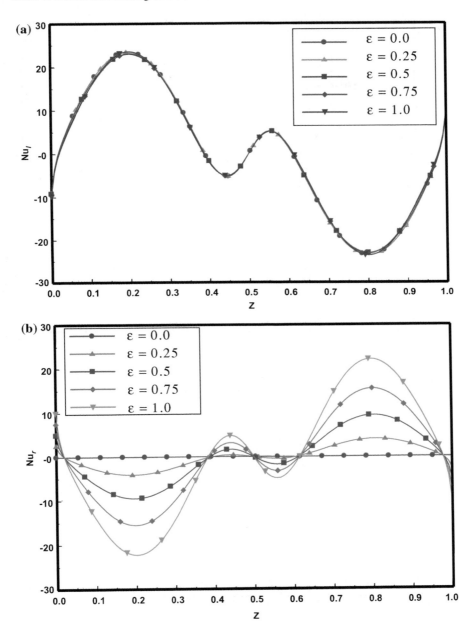

Fig. 4 Effect of amplitude ratio on the local Nusselt number on the inner (**a**) and outer (**b**) walls for $Da = 10^{-1}$, $Ar = 1$, $Ra = 10^6$, $\phi = 0$, $\lambda = 2$

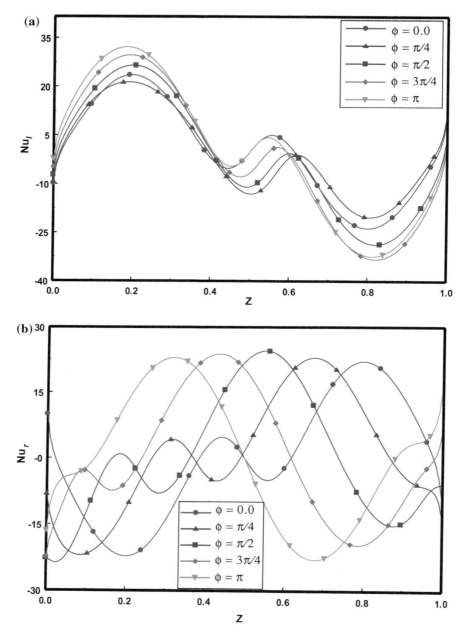

Fig. 5 Effect of phase deviation on the local Nusselt number on the inner (**a**) and outer (**b**) walls for $Da = 10^{-1}$, $Ar = 1$, $Ra = 10^6$, $\varepsilon = 1$, $\lambda = 2$

Z = 0 to Z = 0.2, but decreases from Z = 0.2 to Z = 0.4 and then increases from Z = 0.4 to Z = 0.6 and so on. For the case of outer wall, the phase deviation strongly influences the local Nusselt number profiles and for each value of ϕ, the behavior of Nu profile is completely different. On comparing the variation of Nu profiles on the inner and outer walls, it can be found that the heat transfer rate on the inner wall does not vary much with value of ϕ, whereas the heat transfer on outer wall greatly depends on the phase deviation. Also, it can be observed that the heat transfer at the inner wall does not change significantly with amplitude ratio; however, it changes considerably with phase deviation.

The combined influence of Darcy number with amplitude ratio and phase deviation on the average heat transfer rate for various values of ε and ϕ is depicted in Fig. 6. An overview of the figure reveals that the heat transfer rate increases with an increase in the Darcy number for all amplitude ratios and phase deviations. However, the variation of $\overline{Nu}$ up to $Da = 10^{-5}$ is very low, but a steep increase in average Nusselt number can be observed for $10^{-4} < Da < 10^{-2}$. The average Nusselt number has not changed appreciably with the amplitude ratio and this is consistent with the variation of local Nusselt number with ε. For sinusoidal temperature boundary condition at the side walls, the average Nusselt number is defined as the sum of $\overline{Nu}$ estimated at the inner and outer walls, the modest variation of average Nusselt number is due to the contribution of $\overline{Nu}$ from outer wall. However, lower heat transfer rate is achieved for $\varepsilon = 1$, while the higher heat transfer can be expected for $\varepsilon = 0$–0.5.

The variation of average Nusselt number with the Darcy number for various phase deviations is also measured and is shown in Fig. 6. In general, the heat transfer rate can be enhanced by increasing the values of Da and ϕ. As noted earlier for the case of amplitude ratio, the average Nusselt number is almost invariant with ϕ for $Da \leq 10^{-5}$ increases steadily $10^{-4} < Da < 10^{-2}$. Further, it can be observed that the maximum heat transfer occurs for $\phi = \pi$ for all the Darcy numbers considered in this study.

4.2 Effect of Amplitude Ratio (ε)

In order to analyze the influence of amplitude ratio (ε) of the sinusoidal temperature boundary condition on the flow and thermal fields, the streamlines and isotherms are displayed in Fig. 7 for four different values of ε and fixing other parameters. In general, it is observed that the flow and thermal fields are having horizontal symmetric for all the four amplitude ratios. For the case of zero amplitude ratio, which represents a uniformly cooled outer wall, a symmetric bicellular flow exists in which hot fluid lying above the cold fluid, and the isotherms also reveal a similar structure. As the amplitude ratio is increased ($\varepsilon > 0$), the sinusoidal temperature at the outer wall also produces the flow movement and hence the number of eddies in the porous annulus has been increased to four; two eddies near the inner wall and

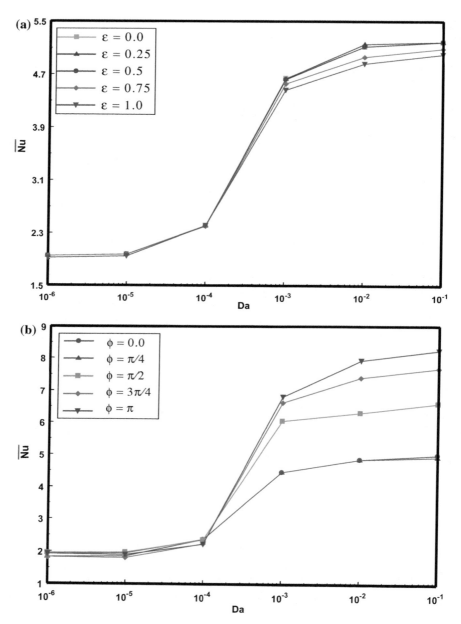

Fig. 6 Variation of average Nusselt number with Darcy number for different amplitude ratios (top) and phase deviation (bottom) at $Ar = 1$, $\lambda = 2$, $Ra = 10^6$. For the variation of ε (top), $\phi = 0$ and for the variation of ϕ (bottom) $\varepsilon = 1$

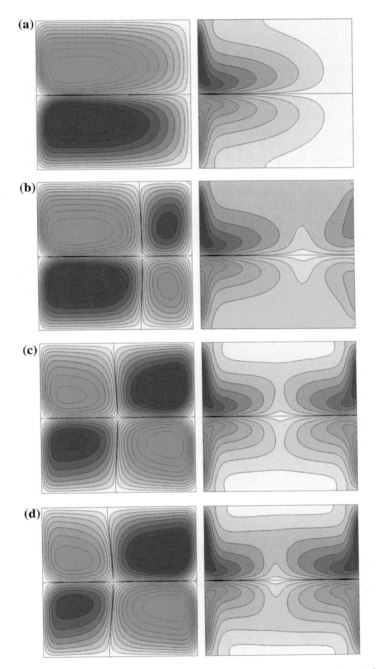

Fig. 7 Effect of amplitude ratio on streamlines (left) and isotherms (right) for $Ra = 10^6$, $Ar = 1$, $\phi = 0$ and $Da = 10^{-3}$. **a** $\varepsilon = 0$, ($\psi_{min} = -11.27$, $\psi_{max} = 11.27$), **b** $\varepsilon = 0.25$, ($\psi_{min} = -11.23$, $\psi_{max} = 11.23$), **c** $\varepsilon = 0.75$, ($\psi_{min} = -15.71$, $\psi_{max} = 15.71$), **d** $\varepsilon = 1$, ($\psi_{min} = -17.91$, $\psi_{max} = 17.91$)

another two eddies near the outer wall. The two larger and smaller eddies each near the inner and outer walls are due to the sinusoidal temperature profiles respectively at the inner and outer walls. It is interesting to observe that the eddies generated by the sinusoidal temperature at outer wall dominate over the eddies near the inner wall as the amplitude ratio increases. However, for $\varepsilon = 1$, the inner and outer walls have similar sinusoidal temperature profiles, and as a result, four equally sized eddies are observed in the annulus.

The variation of average Nusselt number has been calculated for different modified Rayleigh numbers and amplitude ratios by fixing the values of the aspect and radius ratios, Darcy number, phase deviation and their effect has been illustrated in Fig. 8. An overview of the variation of average Nusselt number reveals that the heat transfer rates are increasing with the Rayleigh number and do not change appreciably with the amplitude ratio of the outer sinusoidal temperature profile. The variation of average Nusselt number with the amplitude ratio reveals that the amplitude ratio may not be a useful parameter for the enhancement of heat transfer.

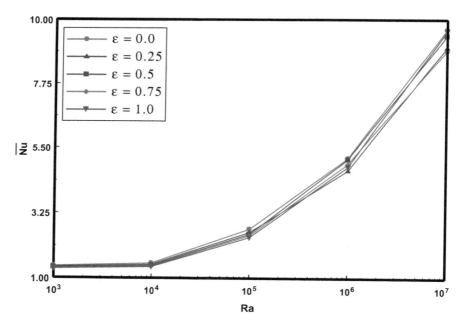

Fig. 8 Variation of the average Nusselt number for different Rayleigh numbers and amplitude ratios at $Ar = 1$, $Da = 10^{-2}$, $\phi = 0$, $\lambda = 2$

4.3 Effect of Phase Deviation (ϕ)

As regards to the influence of phase deviation (ϕ) on the flow and thermal distribution in the annulus, the streamlines and isotherms are displayed in Fig. 9 for three different values of ϕ and fixing the values of Ra, Da, ε and A. In the absence of phase deviation ($\phi = 0$), a four-eddy flow structure is observed in the annulus with hot fluid eddies and cold fluid flow circulations are located diagonally opposite to each other. Also, the streamlines and isotherms reveal the horizontal symmetry for the case of $\phi = 0$. As the value of phase deviation is increased to $\phi = \pi/2$, the symmetric structure of streamlines and isotherms has been destroyed and the diagonally located colder flow circulations are merged, and are sandwiched between the hotter eddies. However, the symmetric structure of streamlines and isotherms is retained back when the phase deviation is further increased to $\phi = \pi$. Also, the hot and cold fluid circulations are placed at top and bottom portions of the annulus respectively and two small eddies on top and bottom corner of the right wall are observed. In general, the number of eddies has been reduced as the phase deviation increases from $\phi = 0$ to $\phi = \pi$.

The important quantity of practical interest in heat transfer applications is the measurement of average heat transfer rate across the enclosure, given by the average Nusselt number. Figure 10 depicts the variation of average Nusselt number with Rayleigh numbers to illustrate the effects of phase deviation by fixing other parameters, such as the aspect ratio, Darcy number, Rayleigh number, and amplitude ratio. In general, the average Nusselt number is found to be increasing with Rayleigh number at all phase deviations. The average Nusselt number does not vary significantly for $\phi = 0$ to $\phi = \pi/2$, however, considerable variation can be found for $\phi > \pi/2$. An overview of the figure confirms that the heat transfer rate could be enhanced with higher values of phase deviation.

4.4 Effect of Aspect Ratio

To examine the influence of aspect ratio on the flow and thermal fields, numerical simulations are illustrated in Fig. 11 for three different aspect ratios. The impact of aspect ratio on the flow and thermal fields is very interesting for the case of sinusoidal temperature boundary conditions. For shallow aspect ratio ($Ar = 0.5$), a strong clockwise unicellular flow with a small secondary vortex at the bottom corner exists and isotherms show stratified structure with a formation of thin boundary layer along the side walls.

For the square annulus ($Ar = 1.0$), a symmetric bicellular flow structure is observed with hot and cold fluids respectively which occupies the top and bottom regions of the annulus. As the aspect ratio is increased to $Ar = 2$, the flow field consists of four eddies with hot and cold fluids located alternatively and similar thermal structure is observed from the isotherms. An overview of the contour plots

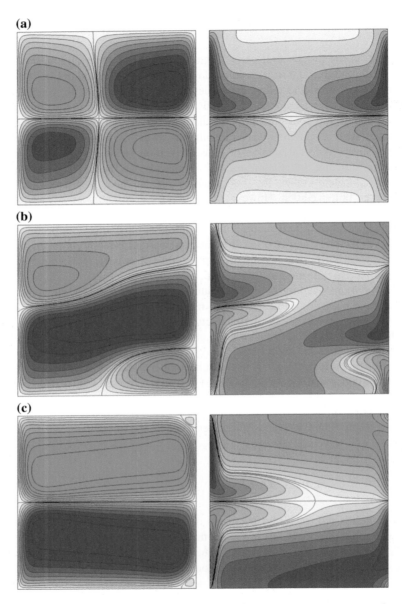

Fig. 9 Effect of phase deviation on streamlines (left) and isotherms (right) for $Ra = 10^6$, $Ar = 1$, $Da = 10^{-3}$, $\varepsilon = 1$. **a** $\phi = 0$, ($\psi_{min} = -17.91$, $\psi_{max} = 17.91$), **b** $\phi = \pi/2$, ($\psi_{min} = -17.61$, $\psi_{max} = 12.04$), **c** $\phi = \pi$, ($\psi_{min} = -12.57$, $\psi_{max} = 12.57$)

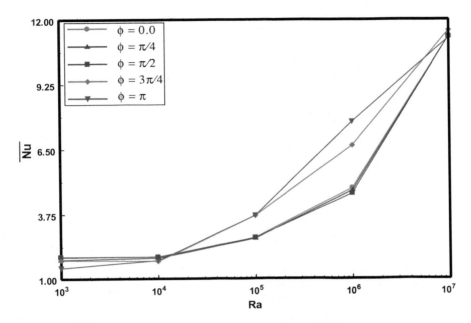

Fig. 10 Variation of the average Nusselt number for different Rayleigh numbers and phase deviations at $Ar = 1$, $Da = 10^{-2}$, $\varepsilon = 1$, $\lambda = 2$

shows that the flow and thermal fields are having horizontal symmetry for all aspect ratios and the number of eddies increases with aspect ratio for the sinusoidal temperature condition at the side walls. Also, as observed from the extreme value of stream function, the flow circulation rate increases as aspect ratio increases. Interestingly, for the case of tall enclosure ($Ar = 2$), Shivakumara et al. (2002) observed a single main recirculation in their study of natural convection in a uniformly heated vertical porous annulus. This indicates that the multicellular flow structure is caused by the sinusoidal temperature boundary conditions. The variation of overall heat transfer rate with aspect ratio is measured in terms of average Nusselt number for different Darcy and Rayleigh numbers and is depicted in Fig. 12 for fixed values of other relevant parameters. It is observed that, for all aspect ratios, the heat transfer rate increases with Darcy and Rayleigh numbers. Since an increase in Darcy number causes a reduction of porous drag, it is expected an enhancement in the heat transfer rate. However, it is observed that higher heat transfer is achieved for taller porous enclosure rather than a smaller enclosure and this prediction is consistent with the results of Deng and Chang (2008).

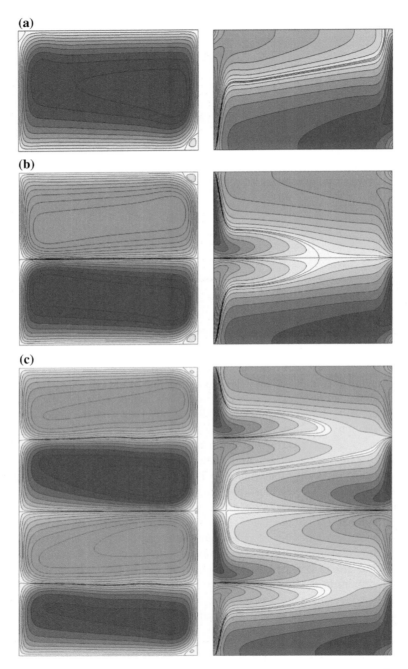

Fig. 11 Effect of aspect ratio on streamlines (left) and isotherms (right) for $Ra = 10^6$, $\phi = \pi$, $Da = 10^{-3}$, $\varepsilon = 1$. **a** $Ar = 0.5$, ($\psi_{min} = -13.25$, $\psi_{max} = 0.25$), **b** $Ar = 1$, ($\psi_{min} = -12.57$, $\psi_{max} = 12.57$), **c** $Ar = 2$, ($\psi_{min} = -18.21$, $\psi_{max} = 18.21$)

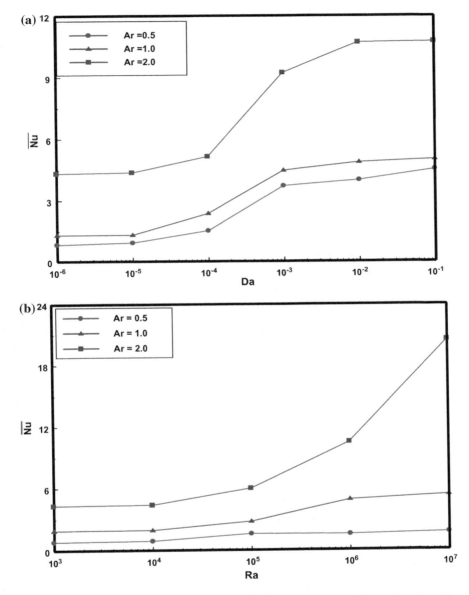

Fig. 12 Variation of the average Nusselt number with aspect ratios. **a** average Nusselt number versus Darcy numbers and for $\varepsilon = 1$, $Ra = 10^6$, $\phi = \pi$, $\lambda = 2$, **b** average Nusselt number versus Rayleigh number for $\varepsilon = 1$, $Da = 10^{-2}$, $\phi = \pi$, $\lambda = 2$

5 Heat Transfer Correlations

From the numerical simulations obtained over wide range of parameters, it is proposed to derive correlations for the average Nusselt number in the enclosure. For this, two correlations of the form $\overline{Nu} = CDa^m \varepsilon^n$ and $\overline{Nu} = CDa^m \phi^n$, where C is a coefficient and m and n are exponents, are derived for the average Nusselt number. The least square regressions are performed to obtain the values of C, m, n. The derived correlations are as follows:

$$\overline{Nu} = \begin{cases} 7.2673 Da^{0.1030} \varepsilon^{-0.0183} & \text{for } 10^{-6} \leq Da \leq 10^{-1}, \phi = 0 \\ 9.8641 Da^{0.0286} \phi^{0.4746} & \text{for } 10^{-6} \leq Da \leq 10^{-1}, \varepsilon = 1. \end{cases}$$

It needs to be mentioned that the above correlation is valid for $Ra = 10^6$, $Ar = 1$ and $= 2$. Similarly, another correlation for the average Nusselt number as a function of Ra, ε and ϕ is derived in the form $\overline{Nu} = CRa^m \varepsilon^n$ and $\overline{Nu} = CRa^m \phi^n$ and the correlations are as follows:

$$\overline{Nu} = \begin{cases} 0.1234 Ra^{0.2641} \varepsilon^{-0.0176} & \text{for } 10^3 \leq Ra \leq 10^6, \phi = 0 \\ 0.2440 Ra^{0.2411} \phi^{0.2977} & \text{for } 10^3 \leq Ra \leq 10^6, \varepsilon = 1. \end{cases}$$

The above correlation is valid for $Da = 10^{-2}$, $Ar = 1$ and $= 2$.

6 Conclusions

In the present investigation, natural convection in a vertical porous annular enclosure with two sinusoidal temperature profiles of different amplitudes and phase deviation on the inner and outer walls has been numerically studied. The governing equations are modeled using the Brinkman-extended Darcy model, which includes viscous and convective effects. The influence of porosity, amplitude ratio, phase deviation and aspect ratio on the streamlines, isotherms, local and average Nusselt numbers has been presented for a wide range of physical and geometrical parameters. The present results for nonuniform heating thermal condition show some interesting differences in the flow pattern and temperature distributions as compared to uniform heating. It is observed that the flow pattern and thermal fields have undergone drastic change with amplitude ratio, however, the heat transfer rate has not changed appreciably. As regards to the influence of phase deviation, the simulation results reveal that it is possible to enhance the heat transfer rate with a proper choice of phase deviation. Also, the total heat transfer rate increases with Rayleigh number, but decreases with Darcy number. The sinusoidal temperature boundary condition causes the multicellular flow in the porous annulus. Further, the number of eddies in the porous annulus increases as the aspect ratio of the enclosure increases and the heat transfer rate also increases with aspect ratio.

Acknowledgements The authors M. Sankar and S. Kiran are respectively grateful to the Managements and R&D Centers of Presidency University, Bangalore, Sapthagiri College of Engineering, Bangalore and to VTU, Belgaum, India for the support and encouragement. Y. D. was supported by the National Research Foundation of Korea (NRF) grant funded by the Korea government (MSIP) (No. NRF-2016R1A2B4011009).

References

Ben-Cheikh N, Chamkha AJ, Ben-Beya B, Lili T (2013) Natural convection of water-based nanofluids in a square enclosure with non-uniform heating of the bottom wall. J Mod Phys 4:147–159

Bilgen E, Yedder RB (2007) Natural convection in enclosure with heating and cooling by sinusoidal temperature profiles on one side. Int J Heat Mass Transf 50:139–150. https://doi.org/10.1016/j.ijheatmasstransfer.2006.06.027

Deng Q-H, Chang J-J (2008) Natural convection in a rectangular enclosure with sinusoidal temperature distributions on both side walls. Numer Heat Transf Part A Appl 54:507–524. https://doi.org/10.1080/01457630802186080

Hasnaoui M, Vasseur P, Bilgen E, Robillard L (1995) Analytical and numerical study of natural convection heat transfer in a vertical porous annulus. Chem Eng Commun 131:141–159

Havstad MA, Burns PJ (1982) Convective heat transfer in vertical cylindrical annuli filled with a porous medium. Int J Heat Mass Transf 25:1755–1766

Hickox CE, Gartling DK (1985) A numerical study of natural convection in a vertical annular porous layer. Int J Heat Mass Transf 28:720–723

Ingham DB, Pop I (2005) Transport phenomena in porous media. Elsevier, Oxford

Khansila P, Witayangkurn S (2012) Visualization of natural convection in enclosure filled with porous medium by sinusoidally temperature on the one side. Appl Math Sci 6(97):4801–4812

Kiwan S, Al-Zahrani MS (2008) Effect of porous inserts on natural convection heat transfer between two concentric vertical cylinders. Numer Heat Transf 53:870–889

Marpu DR (1995) Forchheimer and brinkman extended darcy flow model on natural convection in a vertical cylindrical porous annulus. Acta Mech 109:41–48

Nield DA, Bejan A (2006) Convection in porous media. Springer, New York

Prasad V, Kulacki FA, Keyhani M (1985) Natural convection in porous media. J Fluid Mech 150:89–119

Prasad V, Kulacki FA, Kulkarni AV (1986) Free convection in a vertical porous annulus with constant heat flux on the inner wall—experimental results. Int J Heat Mass Transf 29:713–723

Prasad V (1986) Numerical study of natural convection in a vertical, porous annulus with constant heat flux on the inner wall. Int J Heat Mass Transf 29:841–853

Reda DC (1983) Natural convection experiments in a liquid saturated porous medium bounded by vertical coaxial cylinders. ASME J Heat Transf 105:795–802

Reddy BVK, Narasimhan A (2010) Heat generation effects in natural convection inside a porous annulus. Int Commun Heat Mass Transf 37:607–610

Saeid NH, Mohamad AA (2005) Natural convection in a porous cavity with spatial sidewall temperature variation. Int J Numer Meth Heat Fluid Flow 15:555–566

Saeid NH (2005) Natural convection in porous cavity with sinusoidal bottom wall temperature variation. Int Commun Heat Mass Transf 32:454–463

Saeid NH (2006) Natural convection in a square porous cavity with an oscillating wall temperature. Arab J Sci Eng 31:35–46

Sankar M, Park Y, Lopez JM, Do Y (2011a) Numerical study of natural convection in a vertical porous annulus with discrete heating. Int J Heat Mass Transf 54:1493–1505. https://doi.org/10.1016/j.ijheatmasstransfer.2010.11.043

Sankar M, Venkatachalappa M, Do Y (2011b) Effect of magnetic field on the buoyancy and thermocapillary driven convection of an electrically conducting fluid in an annular enclosure. Int J Heat Fluid Flow 32:402–412. https://doi.org/10.1016/j.ijheatfluidflow.2010.12.001

Sankar M, Park J, Kim D, Do Y (2013) Numerical study of natural convection in a vertical porous annulus with internal heat source: effect of discrete heating. Numer Heat Transf Part A Appl 63:687–712

Sankar M, Jang B, Do Y (2014) Numerical study of non-darcy natural convection from two discrete heat sources in a vertical annulus. J Porous Media 17:373–390

Selamat MS, Roslanand I, Hashim R (2012) Natural convection in an inclined porous cavity with spatial sidewall temperature variations. J Appl Math 939620:10

Sheremet MA, Pop I (2014) Natural convection in a square porous cavity with sinusoidal temperature distributions on both side walls filled with a nanofluid: Buongiorno's mathematical model. Transp Porous Media 105:411–429. https://doi.org/10.1007/s11242-014-0375-7

Sheremet MA, Pop I (2015) Natural convection in a wavy porous cavity with sinusoidal temperature distributions on both side walls filled with a nanofluid: Buongiorno's mathematical model. J Heat Transf 137 (Trans ASME 072601-8/)

Shivakumara IS, Prasanna BMR, Rudraiah N, Venkatachalappa M (2002) Numerical study of natural convection in a vertical cylindrical annulus using a non-Darcy equation. J Porous Media 5:87–102

Sivasankaran S, Malleswaran A, Lee J, Sundar P (2011) Hydro-magnetic combined convection in a lid-driven cavity with sinusoidal boundary conditions on both sidewalls. Int J Heat Mass Transf 54:512–525

Sivasankaran S, Pan KL (2012) Numerical simulation on mixed convection in a porous lid-driven cavity with nonuniform heating on both side walls. Numer Heat Transf Part A Appl Int J Comput Methodol 61:101–112

Sivasankaran S, Bhuvaneswari M (2013) Natural convection in a porous cavity with sinusoidal heating on both sidewalls. Numer Heat Transf Part A Appl Int J Comput Methodol 63:14–30

Vafai K (2005) Handbook of porous media. Taylor & Francis, New York

Varol Y, Oztop HF, Pop I (2008) Numerical analysis of natural convection for a porous rectangular enclosure with sinusoidally varying temperature profile on the bottom wall. Int Commun Heat Mass Transf 35:56–64

Wang QW, Yang J, Zeng M, Wang G (2010) Three-dimensional numerical study of natural convection in an inclined porous cavity with time sinusoidal oscillating boundary conditions. Int J Heat Fluid Flow 31:70–82

Wu F, Zhou W, Ma X (2015) Natural convection in a porous rectangular enclosure with sinusoidal temperature distributions on both side walls using a thermal non-equilibrium model. Int J Heat Mass Transf 85:756–771

Wu F, Wang G, Zhou W (2016) Numerical study of natural convection in a porous cavity with sinusoidal thermal boundary condition. Chem Eng Technol 39:767–774

Zahmatkesh I (2014) Natural convection and entropy generation in a porous enclosure with sinusoidal temperature variation on the side walls. J Fluid Flow Heat Mass Transf 1:23–29

Natural Convection in an Inclined Parallelogrammic Porous Enclosure

Bongsoo Jang, R. D. Jagadeesha, B. M. R. Prasanna and M. Sankar

Nomenclature

A	Aspect ratio
L	Width of the enclosure
H	Height of the enclosure
g	Gravitational acceleration
k	Thermal conductivity
K	Permeability
Nu	Local Nusselt number
$\overline{Nu}$	Average Nusselt number
p	Pressure
Ra	Darcy–Rayleigh number
t	Time
T	Temperature
T_r	Reference temperature
(x, y)	Cartesian coordinates
(X, Y)	Transformed coordinates
(u, v)	Velocity components in the x- and y-directions

B. Jang
Department of Mathematical Sciences, Ulsan National Institute of Science and Technology (UNIST), Ulsan, Republic of Korea

R. D. Jagadeesha
Department of Mathematics, Government Science College, Hassan 573201, India

B. M. R. Prasanna
Department of Mathematics, Siddaganga Institute of Technology, Tumkur, India

M. Sankar (✉)
Department of Mathematics, School of Engineering, Presidency University, Bangalore 560089, India
e-mail: manisankarir@yahoo.com

© Springer Nature Singapore Pte Ltd. 2018
N. Narayanan et al. (eds.), *Flow and Transport in Subsurface Environment*,
Springer Transactions in Civil and Environmental Engineering,
https://doi.org/10.1007/978-981-10-8773-8_9

Greek Letters

α	Inclined angle of the cavity
α_m	Thermal diffusivity of the porous medium
β	Coefficient of thermal expansion
υ	Kinematic viscosity
ϕ	Parallelogrammic angle
μ	Dynamic viscosity
ρ	Fluid density
σ	Heat capacity ratio
(ξ, η)	Dimensionless transformed coordinates
τ	Dimensionless time
θ	Dimensionless temperature
$\psi*$	Stream function
ψ	Dimensionless stream function

Subscripts

c	Cold wall
h	Hot wall
r	Reference value
max	Maximum value

1 Introduction

Natural convection heat transfer in the presence of porous media has wide range of applications in the field of engineering and geophysics that includes, among other important applications, high-performance building insulations, cooling of electronic devices, solar power collectors, food processing and storage, and metallurgy. A detailed review on most of the studies in porous media has been summarized exceptionally in the recent monographs by Nield and Bejan (2013), Vafai (2005), and Ingham and Pop (2005). In particular, the analysis of natural convection in finite porous enclosures has been received considerable attention. This interest stems from the fact that the convective flows are very frequently encountered in various industrial and engineering applications such as cooling of electronic equipments, nuclear reactors, thermal insulation, air conditioning and ventilation, crystal growth, and semiconductor production. Numerous studies in this field have been published on various finite enclosures (Saeid 2006; Sivasankaran et al. 2011; Ghorab 2015).

The convective flow and corresponding heat transfer characteristics in finite enclosures are highly sensitive to the geometry of enclosure. By virtue of different applications, studies on convection heat transfer in finite nonrectangular geometries have increased in recent years. The rate of heat transfer through a nonrectangular enclosure may effectively be controlled by any of its geometrical parameters. Among the finite nonrectangular enclosures, a parallelogrammic-shaped enclosure is a special type of enclosure that has several important practical applications such as thermal insulation devices, solar collection panels, civil construction, and double-glazed windows, to mention a few. Consequently, several theoretical and experimental studies have been investigated on different aspects of this enclosure filled with clear fluids and fluid-saturated porous media. Hyun and Choi (1990) numerically investigated transient convective heat transfer in a differentially heated parallelogrammic enclosure. They investigated the influences of tilt angle of the sidewall, aspect ratio and Rayleigh number on the flow structure, thermal fields, and heat transfer rate. Later, natural convection in a parallelogrammic enclosure with inclined hot and cold walls and insulated horizontal walls has been studied by Aldridge and Yao (2001) for high Rayleigh and Prandtl numbers. Baïri et al. (2010, 2013) have performed a detailed numerical investigation of transient natural convection in a parallelogrammic enclosure for high Rayleigh numbers with different sets of thermal boundary conditions. Further, the numerical results are accompanied and validated with experimental results. During their analysis, it has been found that the parallelogrammic enclosure, depending on the inclination angle, can be considered either as "insulating" or "conductive" enclosure. In addition, the results are compiled in the form of heat transfer correlation in the parallelogrammic enclosure with isothermal, isoflux, and discrete heat source boundary conditions. Recently, Baïri et al. (2014) made a detailed review on natural convection in finite enclosures with a special emphasis to parallelogrammic enclosure.

Using Darcy model, Baytas and Pop (1999) performed numerical simulations of free convection in an oblique enclosure filled with fluid-saturated porous medium. Double-diffusive natural convection in a parallelogrammic enclosure has been numerically investigated by Costa (2005) for various aspect ratios, inclination angles, buoyancy ratios, and Rayleigh numbers. A general analysis on natural convection in a parallelogrammic porous enclosure with isothermal horizontal walls and insulated inclined sidewalls has been investigated by Han and Hyun (2008). In recent times, natural convection in parallelogrammic porous enclosures using nanofluid also received a great deal of attention (Nayak et al. 2015; Ghalambaz et al. 2015).

In addition to parallelogrammic enclosure, several investigations are concerned with the analysis of natural convection in other nonrectangular enclosures, such as trapezoidal and triangular enclosures. Basak et al. (2013) performed numerical investigation in a porous trapezoidal enclosure to study the influences of nonuniform heating, entropy generation, and thermal efficiency on natural convection within the trapezoidal enclosure. Natural convection heat transfer and entropy generation effects in a right-angled inclined triangular porous enclosure have been investigated by Bhardwaj and Dalal (2013).

Numerous studies have been performed to explore natural convection in inclined enclosures due to its variety of applications in practical problems. It was found that the inclination angle of the enclosure plays a vital role on the flow pattern and heat transfer performances and it can also be considered as an effective parameter to control the heat transfer processes in the enclosure. Using Darcy model, Moya and Ramos (1987) numerically examined natural convection in an inclined fluid-saturated rectangular porous material for different tilt angles and aspect ratios. They found three different convective modes and multiple solutions in their parameter regimes and also presented the local and global Nusselt numbers. Hslao et al. (1994) analyzed the behavior of natural convection from a discrete heat source in an inclined square porous enclosure. The influence of inclination on natural convection in a square enclosure has been reported by Rasoul and Prinos (1997) for wide range of Prandtl numbers. Later, Ece and Büyük (2007) analyzed the effect of magnetic field on natural convection in an inclined square enclosure. Some of the recent studies on inclined porous square enclosures reveal the effect of heat generating porous medium (Tian et al. 2014) and discrete heating (Elsherbiny and Ismail 2015) on natural convection in the enclosure.

Apart from inclined square enclosure, a great deal of attention has also been focused on natural convection in inclined enclosures of other cross sections such as trapezoidal and triangle shapes. Varol et al. (2008b) analyzed natural convection in an inclined trapezoidal porous enclosure, using Darcy model. Later, Varol et al. (2008a) performed a numerical study to explore the effects of inclination angle on natural convection in an inclined triangular enclosure filled with porous medium. Recently, using LBM technique, Mejri et al. (2016) investigated natural convection of water in an inclined triangular enclosure.

A careful review of the existing literature reveals that the standard problem of natural convection in a parallelogrammic enclosure with differentially heated inclined walls and insulated horizontal walls has been extensively studied with different conditions. These investigations have been completely devoted to non-inclined parallelogrammic enclosures. However, for an inclined parallelogrammic enclosure, the flow structure and the heat transfer performances in the enclosure are significantly altered due to the buoyancy forces present in both horizontal and transverse directions. Moreover, an inclined parallelogrammic enclosure amply describes the physical configuration of numerous technological applications that involves a thermal diode wall. Also, in an inclined parallelogrammic enclosure, the flow and thermal fields are influenced by the effect of two tilt angles, one from the inclination of enclosure and other is due to inclined side walls. Recognizing this important fact, the objective of the present study is to comprehend the combined effect of tilt angles on the flow structure and heat transfer performances. Hence, the behavior of natural convection heat transfer with different combinations of the geometrical and physical parameters of the problem remains to be addressed adequately and this constitutes the motivation for this study. To the best of authors' knowledge, the findings of this investigation are new and unique which have not been reported in the literature. In this paper, we numerically investigate the natural convection inside an inclined parallelogrammic enclosure to explore the influence of two inclined angles on flow pattern and heat transfer performance. A stable implicit finite difference technique has been employed to solve the governing equations, and an in-house code has been developed and successfully validated with many benchmark solutions before performing the simulations.

2 Mathematical Formulation

2.1 Governing Equations

Consider the two-dimensional inclined parallelogrammic enclosure filled with a fluid-saturated porous medium, as shown in Fig. 1a. The height and width of the parallelogrammic enclosure are H and L, respectively, and the enclosure is tilted at an angle α from the horizontal. Let ϕ be the inclination angle made by the vertical walls with y-axis. The left and right vertical inclined walls of the enclosure are, respectively, kept at uniform, but different temperatures T_h and T_c, where $T_h > T_c$, whereas the top and bottom walls are kept adiabatic.

It is assumed that the fluid is Newtonian and has negligible viscous dissipation, and the gravity is chosen in the negative y-direction. The flow is assumed to be

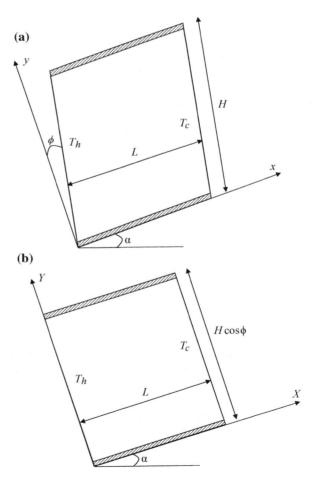

Fig. 1 a Physical configuration and coordinate system. b Transformed computational domain

laminar, and the thermophysical properties are taken as constant, except for the density variations in the buoyancy term of the momentum equations, where the Boussinesq approximation is invoked. It is also assumed that the local thermal equilibrium model is valid; accordingly, the temperature of the fluid is equal to the temperature of the solid matrix in the porous medium. Further, the porous medium is considered to be homogeneous and isotropic. In the porous medium, the Darcy's law has been adopted to form the governing equations of the problem and hence the inertial viscous drag terms of the momentum equations are neglected. By employing the afore-stated assumptions, the unsteady-state equations governing the conservation of mass, momentum in Darcy regime, and energy can be written as (Nield and Bejan 2013; Baytas and Pop 1999)

$$u = -\frac{K}{\mu}\left[\frac{\partial p}{\partial x} + \rho g \sin \alpha\right], \quad (1)$$

$$v = -\frac{K}{\mu}\left[\frac{\partial p}{\partial y} + \rho g \cos \alpha\right], \quad (2)$$

$$\sigma\frac{\partial T}{\partial t} + u\frac{\partial T}{\partial x} + v\frac{\partial T}{\partial y} = \alpha_m\left(\frac{\partial^2 T}{\partial x^2} + \frac{\partial^2 T}{\partial y^2}\right), \quad (3)$$

where σ is the heat capacity ratio and defined as $\sigma = \frac{\varepsilon(\rho c)_f + (1-\varepsilon)(\rho c)_p}{(\rho c)_f}$ (Nield and Bejan 2013). Here, $(\rho c)_f$ and $(\rho c)_p$ are, respectively, the heat capacities of the fluid and saturated porous medium.

In the present study, the flow depends only on two spatial coordinates (x, y). Hence, by introducing the Stokes stream function in the usual way to eliminate the pressure terms from Eqs. (1) and (2), the modified governing equations in terms of stream function are

$$\frac{\partial^2 \psi^*}{\partial x^2} + \frac{\partial^2 \psi^*}{\partial y^2} = \frac{gK\beta}{\nu}\left(\frac{\partial T}{\partial y}\sin \alpha - \frac{\partial T}{\partial x}\cos \alpha\right), \quad (4)$$

$$\sigma\frac{\partial T}{\partial t} + \frac{\partial \psi^*}{\partial y}\frac{\partial T}{\partial x} - \frac{\partial \psi^*}{\partial x}\frac{\partial T}{\partial y} = \alpha_m\left(\frac{\partial^2 T}{\partial x^2} + \frac{\partial^2 T}{\partial y^2}\right), \quad (5)$$

where the stream function ψ^* is defined as $u = \frac{\partial \psi^*}{\partial y}$ and $v = -\frac{\partial \psi^*}{\partial x}$.

Since the inclined vertical side walls are not aligned with the y-axis, a rectangular grid mesh that fits all four boundaries of the enclosure cannot be generated (see Fig. 1a). Hence, an axis transformation, as suggested by Baytas and Pop (1999), has been used to transform the grids in the nonrectangular physical domain into the orthogonal grids in the rectangular computational domain Fig. 1b. For this, we use the following transformation:

$$X = x - y\tan\phi, \quad Y = y \tag{6}$$

Using the above transformation (9), the governing Eqs. (4) and (5) are transformed in the computational domain as

$$\left(\frac{\partial^2\psi^*}{\partial X^2} - 2\sin\phi\cos\phi\frac{\partial^2\psi^*}{\partial X\partial Y} + \cos^2\phi\frac{\partial^2\psi^*}{\partial Y^2}\right)$$
$$= \frac{gK\beta}{\nu}\cos^2\phi\left[\left(\frac{\partial T}{\partial Y} - \tan\phi\frac{\partial T}{\partial X}\right)\sin\alpha - \frac{\partial T}{\partial X}\cos\alpha\right] \tag{7}$$

$$\sigma\frac{\partial T}{\partial t} + \frac{\partial\psi^*}{\partial Y}\frac{\partial T}{\partial X} - \frac{\partial\psi^*}{\partial X}\frac{\partial T}{\partial Y} = \frac{\alpha_m}{\cos^2\phi}\left(\frac{\partial^2 T}{\partial X^2} - 2\sin\phi\cos\phi\frac{\partial^2 T}{\partial X\partial Y} + \cos^2\phi\frac{\partial^2 T}{\partial Y^2}\right). \tag{8}$$

Further, the following dimensionless variables are introduced to non-dimensionalize Eqs. (7) and (8):

$$\xi = \frac{X}{a}, \quad \eta = \frac{Y}{L\cos\phi}, \quad \tau = \frac{t\alpha_m}{\sigma a L\cos\phi}, \quad \psi = \frac{\psi^*}{\alpha_m}, \quad \theta = \frac{(T-T_r)}{T_h - T_c},$$

where $T_r = (T_h + T_c)/2$ is the reference temperature.
Using the above nondimensional variables, Eqs. (7) and (8) can be written as

$$\frac{\partial^2\psi}{\partial\xi^2} - 2\frac{\sin\phi}{A}\frac{\partial^2\psi}{\partial\xi\partial\eta} + \frac{1}{A^2}\frac{\partial^2\psi}{\partial\eta^2} = Ra\cos\phi\left[\frac{\sin\alpha}{A}\frac{\partial\theta}{\partial\eta} - \cos(\phi-\alpha)\frac{\partial\theta}{\partial\xi}\right] \tag{9}$$

$$\frac{\partial\theta}{\partial\tau} + \frac{\partial\psi}{\partial\eta}\frac{\partial\theta}{\partial\xi} - \frac{\partial\psi}{\partial\xi}\frac{\partial\theta}{\partial\eta} = \frac{A}{\cos\phi}\left(\frac{\partial^2\theta}{\partial\xi^2} - 2\frac{\sin\phi}{A}\frac{\partial^2\theta}{\partial\xi\partial\eta} + \frac{1}{A^2}\frac{\partial^2\theta}{\partial\eta^2}\right) \tag{10}$$

In the above equations, A is the aspect ratio of the enclosure and Ra is the Darcy–Rayleigh number which are defined as

$$A = \frac{L}{a}, \quad Ra = \frac{gK\beta\Delta Ta}{\nu\alpha_m}. \tag{11}$$

2.2 Boundary Conditions and Heat Transfer Rate

The dimensionless initial and boundary conditions of the present study are

$$\tau = 0: \quad \psi = 0, \; \theta = 0; \qquad 0 \leq \xi \leq 1, \; 0 \leq \eta \leq 1$$
$$\tau > 0: \quad \psi = 0, \; \theta = +0.5; \qquad \xi = 0, \; 0 \leq \eta \leq 1$$
$$\psi = 0, \; \theta = -0.5; \qquad \xi = 1, \; 0 \leq \eta \leq 1$$
$$\begin{cases} \psi = 0, \\ \frac{\partial \theta}{\partial \eta} - A \sin\phi \frac{\partial \theta}{\partial \xi} = 0 \end{cases} \begin{cases} \eta = 0 \text{ and } 1, \\ 0 \leq \xi \leq 1. \end{cases}$$

The quantified measure of heat transfer rate is the local (Nu) and average ($\overline{Nu}$) Nusselt numbers, which are defined as

$$Nu = -\frac{1}{\cos\phi}\left(\frac{\sin\phi}{A}\frac{\partial \theta}{\partial \eta} - \frac{\partial \theta}{\partial \xi}\right), \quad \overline{Nu} = \int_0^1 Nu \, d\eta.$$

3 Solution Methodology

The governing model equations, namely stream function and energy equations, are coupled, nonlinear partial differential equations with mixed partial derivatives. Hence, these equations along with the initial and boundary conditions are numerically solved using an implicit finite difference technique. In particular, the energy equation is solved by alternating direction implicit (ADI) method and the stream function equation has been solved by successive line over relaxation (SLOR) method. A special care has been taken while discretizing the mixed order partial derivatives in these methods. The time derivative is approximated using forward difference, whereas the convection and diffusion terms are discretized using a second-order central difference. The resulting finite difference equations can be arranged in the form of tridiagonal matrix, which can efficiently solved by Thomas algorithm. The steady-state results are obtained as an asymptotic limit to the transient solutions when the converge criterion $\left|\frac{\chi_{i,j}^{n+1} - \chi_{i,j}^n}{\chi_{i,j}^{n+1}}\right| < \varepsilon$ is satisfied. Here, χ is either ψ or θ, ε is the prespecified constant set at 10^{-6}.

3.1 Grid Independency

In the present study, a uniform grid has been used in the $\xi - \eta$ plane of the computational domain, and a proper grid independence study has been carried out to decide an appropriate grid size. For this, the grid sizes are varied from a coarse grid size of 51×51 to a finer grid size of 201×201, and observed the variations of average Nusselt number and maximum stream function values as the sensitivity measures. After a thorough investigation on the variation of average Nusselt number and maximum stream function for several grid sizes, we have chosen the grids that give a good compromise between solution accuracy and CPU time. Based on the grid independence tests, all simulations are performed with a 201 × 101 grids for $A = 0.5$, 101 × 101 grids for $A = 1$, and 101 × 201 grids for $A = 2$. For brevity, the details of the grid independence tests are not presented in the paper. Further, to perform the simulations, we have developed an in-house FORTRAN code for the present model and have been successfully validated against the various benchmark solutions available in the literature before obtaining the simulations.

3.2 Validation

First, we have obtained the streamlines and isotherms of a differentially heated non-inclined parallelogrammic enclosure ($\alpha = 0$). In a similar enclosure, Baytas and Pop (1999) have carried out a numerical solution of natural convection based on the Darcy's law. To assess the accuracy of the present numerical algorithm, computations are carried out under the same conditions of Baytas and Pop (1999) and the comparisons are depicted in Fig. 2. Figure 2 exhibits an excellent agreement between the present streamlines and isotherms and those of Baytas and Pop (1999) in a non-inclined parallelogrammic enclosure. Next, a comparison is presented with natural convection in an inclined rectangular porous enclosure ($\phi = 0$). For this, the average Nusselt numbers are determined for two different aspect ratios and various inclination angles. Table 1 provides comparison of average Nusselt numbers between present results and Moya et al. (1987) in an inclined rectangular porous enclosure. A careful observation of Table 1 reveals that the agreement is found to be good, since the difference between our results and Moya et al. (1987) is less than 5% for all the compared cases. Finally, we compared the average Nusselt numbers from our simulation with Saeid and Pop (2004) and Baytas (2000) in a square porous enclosure and the comparisons are given in Table 2. It can be seen that the agreement between our study and the existing studies is in good agreement. Therefore, from Fig. 2 and Tables 1 and 2, the agreement between the present results and benchmark solutions is quite acceptable.

4 Results and Discussion

The results of numerical simulations are presented in this section to understand the influence of the physical and geometrical parameters on the flow pattern, thermal fields, and heat transfer rates. Due to the presence two inclination angles, one from the tilting of enclosure and another is due to the sloping walls, the flow structure and heat transfer rate in the inclined parallelogrammic porous enclosure are significantly different from those of a non-inclined parallelogrammic porous enclosure. The present study involves one physical parameter, namely Rayleigh number (Ra), and three geometrical parameters, such as aspect ratio (A), inclined angle of the

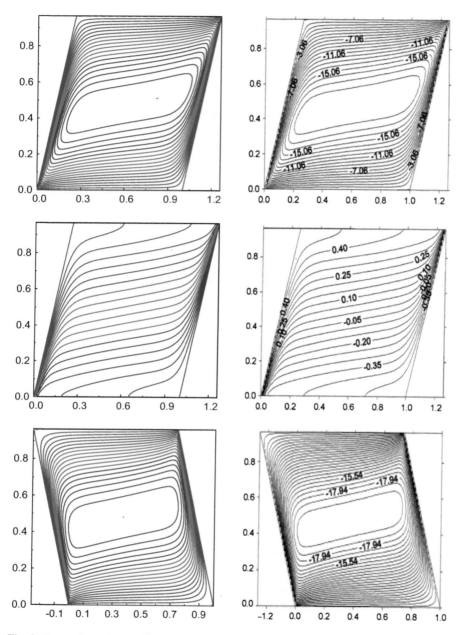

Fig. 2 Comparison of streamlines and isotherms between the present study (left) and Baytas and Pop (1999) (right) for $Ra = 10^3$, $\phi = 15$ (I and II Row), $\phi = -15$ (III and IV Row), $A = 1$, and $\alpha = 0$

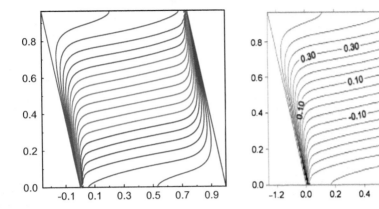

Fig. 2 (continued)

Table 1 Comparison of present average Nusselt numbers with that of Moya et al. (1987) in an inclined rectangular porous enclosure for $Ra = 100$ and $\phi = 0$

A	α	Moya et al. (1987)	Present study	Difference in (%)
1	0	2.517	2.503	0.5
	45	3.564	3.486	2.2
	90	2.873	2.813	2.1
2	0	2.46	2.41	2.0
	5	2.53	2.45	3.2
	10	2.55	2.49	2.4
	15	2.39	2.27	5.0

Table 2 Comparison of present average Nusselt number with that of Saeid and Pop (2004) and Baytas (2000) in a square porous enclosure for $A = 1$, $\alpha = 0$ and $\phi = 0$

Ra	Saeid and Pop (2004)	Baytas (2000)	Present study
100	3.002	3.160	3.102
1000	13.726	14.060	13.914
10,000	43.953	48.330	42.316

enclosure (α), and inclined angle of the sloping walls (ϕ). To get complete and thorough understanding of natural convection in the chosen geometry, we performed the detailed and systematic simulations for wide range of the parameters considered in our study. In particular, the Darcy–Rayleigh number (Ra) and aspect ratio (A) are, respectively, varied in the range $10 \leq Ra \leq 5 \times 10^3$ and $0.5 \leq A \leq 2$.

To investigate the true effects of the inclination angles, the enclosure tilt angle (α) is varied in the range of $-60° \leq \alpha \leq 60°$, while the tilt angle of the inclined walls (ϕ) is examined in the range $-60° \leq \phi \leq 60°$. The influence of all parameters on the flow pattern and temperature distributions inside the inclined parallelogrammic porous enclosure are comprehensively discussed through the streamline and isotherm contours. Further, the variation of heat transfer rate with respect to the parameters of the problem is explained in terms of average Nusselt numbers.

4.1 Effect of Darcy–Rayleigh Number (Ra)

In this section, the results of numerical simulations are presented to illustrate the effect of Darcy–Rayleigh number on the flow pattern and temperature distribution. The streamlines and isotherms are illustrated in Fig. 3 for three different Darcy–Rayleigh numbers and fixing other parameters. For a small value of Ra ($Ra = 10^2$), as illustrated in Fig. 3, the heat transfer inside the parallelogrammic enclosure is dominated by conduction mechanism as evidenced by the less distorted isotherms. The streamlines reveal a simple circulating flow pattern of regularly spaced streamlines with the main vortex at the center of parallelogrammic enclosure. However, as the Darcy–Rayleigh number is increased, the intensity of fluid motion increases indicating a more vigorous convection. Also, the streamlines in Fig. 3 reveal that the flow circulation rate measured by the extremum value of stream function increases in many order of magnitudes as the value of Ra becomes larger. At higher values of Ra, the streamlines and isotherms, respectively, reveal a sharp velocity and temperature gradients, indicating the formation of velocity and thermal boundary layers. Further, the streamlines indicate that the stagnant flow appears in the core of the enclosure and isotherms tend to follow the same pattern. This phenomenon can be attributed to the inclination of the enclosure, which causes the fluid flow mainly along the sloping side walls. On comparing the streamlines and isotherms of the present study with those of a nonrectangular enclosure ($\alpha = 0$), as discussed by Baytas and Pop (1999), it has been found that the flow and thermal behavior have undergone drastic changes. When $\alpha = 0$, the flow is diagonal and is from bottom to top, however, for $\alpha \neq 0$, diagonal flow is from top left corner to bottom right corner.

4.2 Effect of Tilt Angle of the Sloping Wall (ϕ)

In this section, we discuss the influence of tilt angle (ϕ) of sloping walls on the flow pattern, temperature distribution, and heat transfer characteristics. As regards to effect of ϕ on the flow and thermal fields, the streamlines and isotherms are exhibited in Fig. 4 for three different inclination angles (ϕ) by fixing the values of $A = 1$, $\alpha = 30$ and $Ra = 3 \times 10^3$. A careful examination of the streamlines and isotherms reveals an important fact that the tilt angle of the enclosure (ϕ) is dominant in significantly altering the flow pattern and thermal fields compared to the tilt angle of the sloping wall (α). In other words, the influence of tilt angle, α,

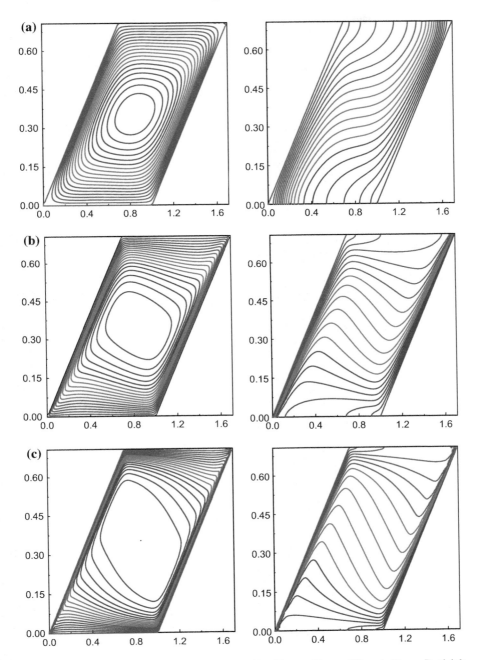

Fig. 3 Streamlines and isotherms for $A = 1$, $\phi = 45$, $\alpha = 45$ at different Darcy–Rayleigh numbers. **a** $Ra = 10^2$, $|\psi_{\max}| = 5.1$, **b** $Ra = 10^3$, $|\psi_{\max}| = 25.6$, **c** $Ra = 5 \times 10^3$, $|\psi_{\max}| = 60.8$

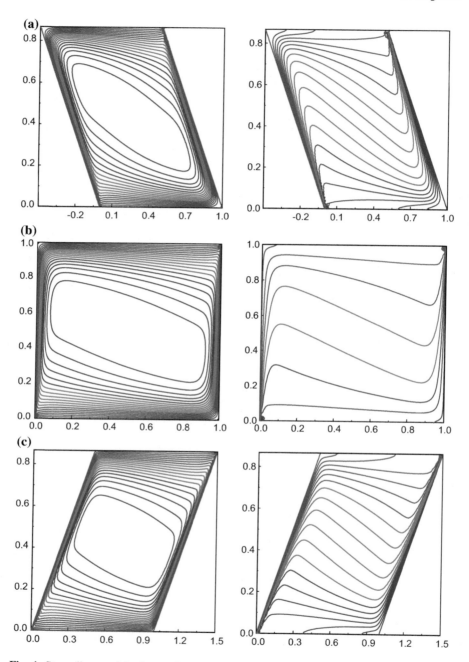

Fig. 4 Streamlines and isotherms for $A = 1$, $\alpha = 30$, $Ra = 3 \times 10^3$ at different tilt angles of sloping walls. **a** $\phi = -30$, $|\psi_{max}| = 37.5$, **b** $\phi = 0$, $|\psi_{max}| = 42.2$, **c** $\phi = 30$, $|\psi_{max}| = 43.6$

can be observed for all inclination of the sloping walls. Also, it can be seen that the value of extremum stream function is lower when both the tilt angles are in opposite directions.

On contrast, the flow circulation rate is higher for the case of positive tilt angles. For $\phi = 0$, due to the absence of sloping wall, the flow pattern and thermal distributions are akin to an inclined square enclosure. These predictions are in good agreement with the results of inclined porous square enclosures (Moya et al. 1987; Rasoul et al. 1997). The isotherms reveal strong thermal stratification, and formation of velocity and thermal boundary layers can be noticed along the sloping walls.

To examine the effect of tilt angle of the sloping walls on the heat transfer rate, the average Nusselt numbers are displayed in Fig. 5 for two different enclosure tilt

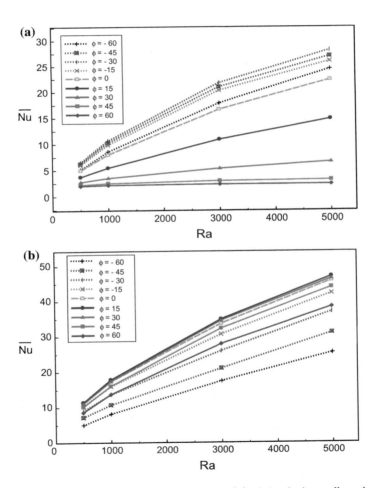

Fig. 5 Effect of Darcy–Rayleigh number and tilt angle (ϕ) of the sloping wall on the average Nusselt number for A =1. **a** $\alpha = -30$, **b** $\alpha = 30$

angles. The average Nusselt numbers are calculated in the range of $-60° \leq \phi \leq 60°$ for a positive and a negative enclosure tilting angles. The average Nusselt number corresponding to positive tilt angles (ϕ) of the sloping wall is plotted as continuous curves, while the curves of $\overline{Nu}$ corresponding to negative values of ϕ are plotted as discontinuous curves. An overview of the graph reveals that the heat transfer rate increases with Darcy–Rayleigh number for all the inclination angles ϕ and α. The influence of two tilt angles on the average Nusselt number is highly complex. In other words, the tilt angle, ϕ, at which a minimum or a maximum heat transfer occurs strongly depends on the enclosure tilt angle, α. In general, for any tilt angle (ϕ) of the sloping wall, the heat transfer rates are found to be higher when the parallelogrammic enclosure is tilted in the positive direction rather than in negative direction.

Also, when the enclosure is tilted in clockwise (negative) direction, it is interesting to note that the heat transfer rates are higher for negative tilt angles of the sloping wall. On contrast, as the enclosure is tilted in counterclockwise (positive) direction, the heat transfer rates depend on the tilt angles of the sloping wall. For positive inclination of the enclosure, higher heat transfer rates are achieved for $\phi = 15$, while minimum heat transfer rates are found for $\phi = -60$. A careful review of Fig. 5 reveals that the heat transfer rates are higher for a parallelogrammic enclosure compared to square enclosure. The heat transfer rate can be effectively modified using the tilt angle of the sloping walls and it can be considered as an independent parameter to control the heat transfer rate.

4.3 Effect of Enclosure Tilt Angle (α)

The influence of tilt angle of the enclosure on the flow pattern, thermal field, and heat transfer rates is presented in Fig. 6a–e. To understand the true effect of α, we have varied the tilt angle of the enclosure in the range $-60° \leq \alpha \leq 60°$, by keeping other parameters fixed. The effect of α on the flow and thermal fields can be apparently noticeable from the streamlines and isotherms. As the enclosure is tilted in clockwise (negative) direction to the maximum value considered in the present study ($\alpha = -60°$), the flow circulation rate in the enclosure is extremely low with two weak eddies in the middle of enclosure. The isotherms do not show any variation and are parallel to the sloping walls, indicating conduction is the major mode of heat transfer at the extreme value of α. For $\alpha = -60°$, the hot wall approaches to top position, and as a result the flow movement has been decelerated. As the inclination of the enclosure increases ($\alpha > -60°$), two eddies in the middle of enclosure merge to a unicellular structure and the flow circulation rate measured by the extreme stream function value increases. Also, the orientation of main vortex completely depends on the tilt angle of parallelogrammic enclosure.

It is interesting to observe that these flow structures are entirely different from the non-inclined parallelogrammic enclosure. For negative tilt angles of the

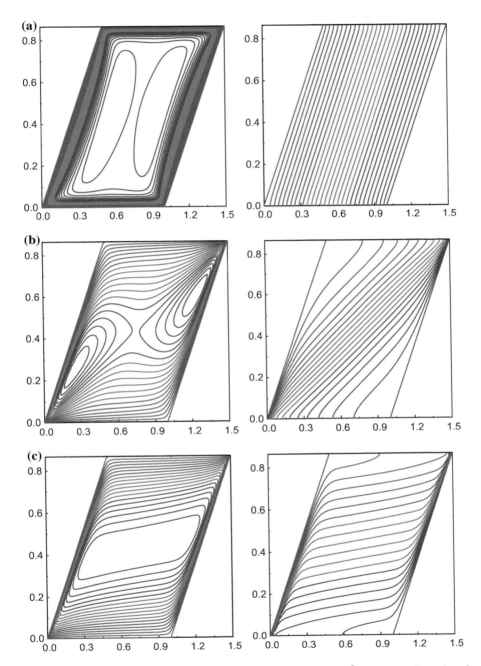

Fig. 6 Streamlines and isotherms for $A = 1$, $\phi = 30$ and $Ra = 3 \times 10^3$ at different tilt angles of the enclosure. **a** $\alpha = -60$, $|\psi_{max}| = 0.6$, **b** $\alpha = -30$, $|\psi_{max}| = 7.1$, **c** $\alpha = 0$, $|\psi_{max}| = 30.3$, **d** $\alpha = 30$, $|\psi_{max}| = 43.6$, **e** $\alpha = 60$, $|\psi_{max}| = 48.5$

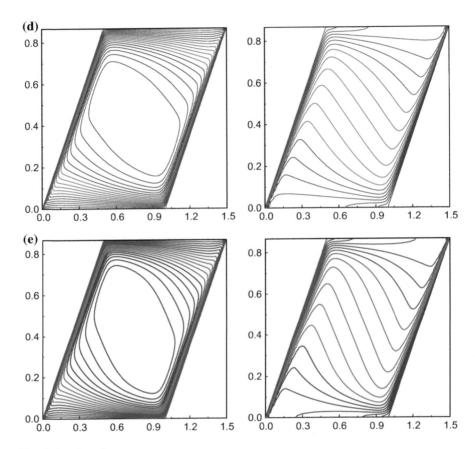

Fig. 6 (continued)

enclosure, the gravitationally stable condition of top heating exists, where the hot wall is facing downward and cold wall is facing upward. In such situation, convection heat transfer exists in the upper zone, whereas conduction heat transfer prevails in the lower part of the enclosure. However, as the tilt angle (α) is increased, the isotherm contour reveals that the stratification of the temperature intensified by the stronger flow motion. Further, the influence of tilt angle on the temperature pattern is significant and is manifested through the isotherms. The variation of streamline and isotherm patterns with the tilt angles of parallelogrammic enclosure are in good agreement with the streamline and isotherm patterns in tilted enclosures. An overview of the figure reveals that the geometrical parameter, namely the tilt angle of the enclosure, plays a key role in controlling the flow and thermal patterns. In general, reduced flow intensity and temperature stratification exist for negative tilt angles of the enclosure, whereas higher flow intensity and well-established, thermally stratified temperature pattern exist for positive tilt angles.

To know the influence of tilt angle on the heat transfer rate, the variation of average Nusselt number for various tilt angles (α) is illustrated in Fig. 7, which is helpful for proper design of thermal diodes. The effect of α on the average Nusselt number is presented in Fig. 7 for different values of Ra at two different tilt angles of the sloping walls, namely $\phi = -45°$ and $45°$. As expected, due to enhanced convective strength at higher values of Ra, heat transfer rate increases with Darcy–Rayleigh number at all tilt angles of the enclosure and sloping walls. When the sloping walls are tilted to $-45°$, maximum heat transfer is obtained for $\alpha = 0$. Also,

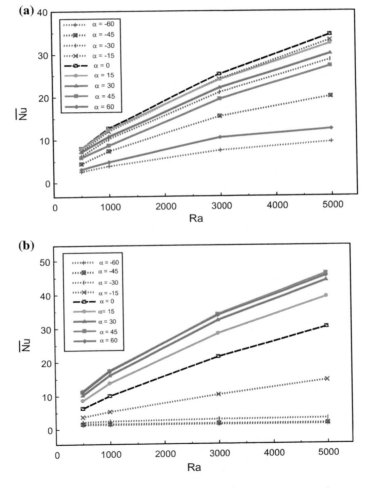

Fig. 7 Effect of Darcy–Rayleigh number and tilt angle of the enclosure on the average Nusselt number for A = 1. **a** $\phi = -45$, **b** $\phi = 45$

as the tilt angle (ϕ) of the enclosure is increased in positive or negative directions, it results in the reduction of heat transfer. In particular, heat transfer declined rapidly for negative tilt angles of the enclosure and this can be attributed to the existence of stagnation flow at these tilt angles. On the other hand, for $\phi = +45°$, the heat transfer rate increases with positive values of α, but decreases with negative values of α. For $\alpha = -45°$ and $-60°$, the heat transfer rate does not show any variation with Ra and α. This can be anticipated at these tilt angles, since the location of hot wall changes to top position and this gives rise to the reduced convection currents in the enclosure. In general, a careful examination of the figure reveals that maximum heat transfer is achieved when the sloping walls are inclined in positive direction.

When the engineers fabricate a new design of thermal diodes, they greatly rely on the available theoretical results, which can greatly enhance their insight while building the new design. As mentioned earlier, the main objective of the present study is to understand the combined influence of two tilt angles on the heat transfer rates and is illustrated in Fig. 8. For this, the variation of average Nusselt number for different values of tilt angles ϕ and α is depicted in Fig. 8 for the fixed values of $Ra = 10^3$ and $A = 1$. In general, the heat transfer rate increases monotonically with enclosure tilt angle for all values of ϕ except at $\phi = -30°$ and $0°$, at which the value of $\overline{Nu}$ increases steadily up to $\alpha = 0°$ and $30°$, respectively, and then decreases thereafter. At each tilt angle of the enclosure and sloping walls in the range $-60° \leq \alpha \leq 60°$ and $-60° \leq \phi \leq 60°$, there exists a minimum and maximum average heat transfer rate. In other words, the minimum and maximum values of $\overline{Nu}$ at $\alpha = -60°$ occur for $\phi = 30°$ and $-60°$, respectively, whereas the minimum and maximum values of $\overline{Nu}$ at $\alpha = +60°$ occur for $\phi = -60°$ and $30°$, respectively.

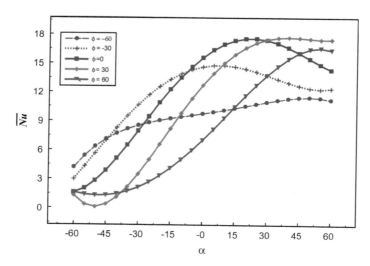

Fig. 8 Effect of inclination angles of the wall (ϕ) and the enclosure (α) on the average Nusselt number for $Ra = 10^3$ and $A = 1$

Hence, the average Nusselt number is a complex function of the two tilt angles: one from the enclosure and another from the sloping walls. However, from the entire range of tilt angles considered in the present analysis ($-60° \leq \alpha \leq 60°$ and $-60° \leq \phi \leq 60°$), it can be found that the minimum heat transfer rate occurs at $\alpha = \phi = -60°$, whereas the maximum heat transfer rates take place at $\alpha = 60°$ and $\phi = 30°$. A careful observation of the results in Fig. 8 reveals that it is possible to control the heat transfer rate by the making appropriate choices on the tilt angles.

4.4 Effect of Enclosure Aspect Ratio (A)

In this section, we discuss the effects of aspect ratio and Darcy–Rayleigh number on the flow pattern, thermal distribution, and heat transfer rate. In Fig. 9, we examine the influence of aspect ratio on the flow and thermal fields for three different parallelogrammic enclosures, namely shallow, square, and tall enclosures. For a shallow enclosure ($A = 0.5$), the flow intensity is very low as given by the extreme value of stream function ($|\psi_{max}| = 1.8$) and the corresponding undistorted isotherms reveal the conduction mode of heat transfer. Also, the appearance of two vortices can be seen in the core of the enclosure, one near the hot wall and another near the cold wall. As the aspect ratio is increased to $A = 1$, the vortex near cold wall moves toward the top wall, while the hot wall vortex moves to bottom adiabatic wall. Interestingly, as the aspect ratio is increased to $A = 2$, two vortices in the streamlines merge to produce a strong flow and the strength of the convective flow ($|\psi_{max}| = 21.7$) increases. Therefore, it can be concluded that the appearance of the secondary vortex is due to the inclination of enclosure. Also, the streamline and isotherm patterns retain the symmetric structure for all aspect ratios considered in our analysis. To investigate the effect of aspect ratio, Fig. 10 demonstrates the influence of aspect ratio 0.5, 1.0, and 2.0 on the average Nusselt numbers. Figure 10 exemplifies the variation of heat transfer rate in the enclosure due to a change in aspect ratio by fixing other parameters of the problem. An overview of the figure reveals that the heat transfer rates are higher for a low aspect ratio (shallow enclosure) compared to square and tall enclosures. The results reveal that the heat transfer enhancement for the case of $A = 0.5$ has been always higher, and the enhancement for the case of $A = 1.0$ is higher than that for $A = 2.0$. The variation of heat transfer rate with aspect ratio is in sharp contrast with the variation of flow circulation rate in different aspect ratios. It has been found that the flow circulation rate for shallow enclosure is lower compared to square and tall enclosures. This indicates that the higher circulation rate does not need to induce or enhance the heat transfer rate.

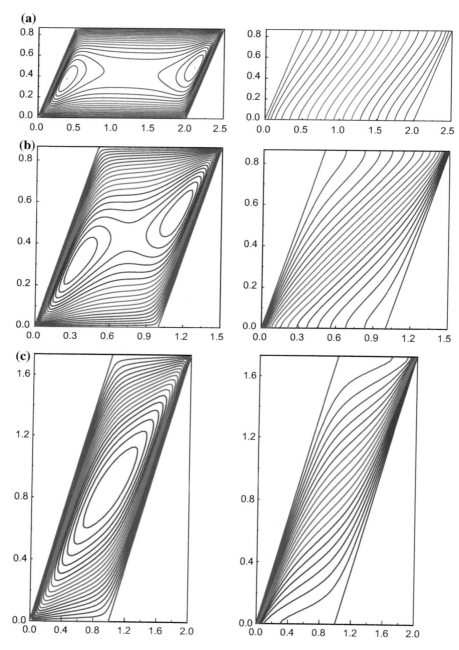

Fig. 9 Streamlines and isotherms for $Ra = 10^3$, $\phi = 30$, $\alpha = -30$ at different aspect ratios. **a** $A = 0.5$, $|\psi_{max}| = 1.8$, **b** $A = 1$, $|\psi_{max}| = 4.4$, **c** $A = 2$, $|\psi_{max}| = 21.7$

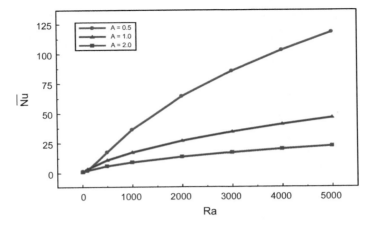

Fig. 10 Effect of Darcy–Rayleigh number and aspect ratio on the average Nusselt number for $\phi = 30$, $\alpha = 30$

5 Conclusions

The present numerical investigation aims to explore the combined influences of tilt angles on natural convection heat transfer in an inclined parallelogrammic cavity. The present study is first of its kind, and a similar work has not been reported in the literature as per the author's knowledge. After a meticulous and systematic numerical simulations covering wide range of physical and geometrical parameters, the following important conclusions have been drawn from our analysis:

1. The numerical results reveal that the flow circulation and heat transfer rates in an inclined parallelogrammic enclosure are relatively higher compared to a non-inclined parallelogrammic enclosure.
2. The inclination angle, α, of the enclosure has strong influence on the flow pattern, thermal distribution, and average heat transfer rate. Also, for a positive inclination of the cavity, the heat transfer rate is found to be higher, whereas relatively lower heat transfer rates are found for negative inclination.
3. When compared the influences of two tilt angles on the flow and thermal fields, we found that the tilt angle of the enclosure has profound influence on the flow pattern and temperature distributions compared to the tilt angle of the sloping wall.
4. When the two tilt angles are of opposite sign, the flow circulation and heat transfer rate have been decreased. However, the positive inclinations of the enclosure and sloping walls augment the flow intensity and heat transfer rates.
5. The enclosure tilt angle has distinct influence on the heat transfer performance when it is combined with the tilt angle of the sloping walls. For a fixed value of the tilt angle of sloping wall, we found that the maximum and minimum heat transfer rates greatly depend on the enclosure tilt angle.

6. As the aspect ratio of the enclosure is varied, interesting changes in flow pattern and temperature distribution are found compared to non-inclined parallelogrammic enclosure. As far as the heat transfer rates are concerned, the lower aspect ratio ($A = 0.5$) is useful to achieve a higher heat transfer rates.
7. For $Ra = 10^3$ and $A = 1$, the numerical simulations of present analysis have revealed that the minimum heat transfer rate occurs at $\alpha = 30°$ and $\phi = -50°$. However, the maximum heat transfer rates takes place at $\alpha = 40°$ and $\phi = 30°$.
8. Depending upon the need of applications, the two tilt angles can be utilized as an effective control parameter to either enhance or decrease the heat transfer in an inclined parallelogrammic enclosure.

Acknowledgements This research was supported by Basic Science Research Program through the National Research Foundation of Korea (NRF) funded by the Ministry of Education (NRF-2014R1A1A2A16051147). The authors MS, RDJ, and BMRP are, respectively, grateful to the Presidency University, Bangalore, Government Science College, Hassan and SIT, Tumkur and to VTU, Belgaum, India for the support and encouragement.

References

Aldridge KD, Yao H (2001) Flow features of natural convection in a parallelogrammic enclosure. Int Comm Heat Mass Transfer 28(7):923–931

Baïri A, Garcia de Maria JM (2013) Nu–Ra–Fo correlations for transient free convection in 2D convective diode cavities with discrete heat sources. Int J of Heat Mass Transfer 57:623–628

Baïri A, Garcia de Maria JM, Laraqi N (2010) Transient natural convection in parallelogrammic enclosures with isothermal hot wall. Experimental and numerical study applied to on-board electronics. Appl Therm Eng 30:1115–1125

Baïri A, Zarco-Pernia E, García de María J-M (2014) A review on natural convection in enclosures for engineering applications. The particular case of the parallelogrammic diode cavity. Appl Thermal Eng 63:304–322

Basak T, Anandalakshmi R, Roy S, Pop I (2013) Role of entropy generation on thermal management due to thermal convection in porous trapezoidal enclosures with isothermal and non-isothermal heating of wall. Int J Heat Mass Transfer 67:810–828

Baytas AC (2000) Entropy generation for natural convection in an inclined porous cavity. Int J Heat Mass Transfer 43:2089–2099

Baytas AC, Pop I (1999) Free convection in oblique enclosures filled with a porous medium. Int J Heat Mass Transfer 42:1047–1057

Bhardwaj S, Dalal A (2013) Analysis of natural convection heat transfer and entropy generation inside porous right-angled triangular enclosure. Int J Heat Mass Transfer 65:500–513

Costa VAF, Oliveira MSA, Sousa ACM (2005) Laminar natural convection in a vertical stack of parallelogrammic partial enclosures with variable geometry. Int J Heat Mass Transfer 48:779–792

Ece MC, Büyük E (2007) Natural convection flow under a magnetic field in an inclined square enclosure differentially heated on adjacent walls. Meccanica 42:435–449

Elsherbiny SM, Ismail OI (2015) Heat transfer in inclined air rectangular cavities with two localized heat sources. Alexandria Eng J 54(4):917–927

Ghalambaz M, Sheremet MA, Pop I (2015) Free convection in a parallelogrammic porous cavity filled with a nanofluid using Tiwari and Das' nanofluid model. PLoS ONE 10(5):e0126486

Ghorab MG (2015) Modeling mixing convection analysis of discrete partially filled porous channel for optimum design. Alexandria Eng J 54(4):853–869

Han HS, Hyun JM (2008) Buoyant convection in a parallelogrammic enclosure filled with a porous medium—general analysis and numerical simulations. Int J Heat Mass Transfer 51:2980–2989

Hslao SW, Chen CK, Cheng P (1994) A numerical solution for natural convection in an inclined porous cavity with a discrete heat source on one wall. Int J Heat Mass Transfer 37(15):2193–2201

Hyun JM, Choi BS (1990) Transient natural convection in a parallelogram-shaped enclosure. Int J Heat Fluid Flow 11(2):129–134

Ingham DB, Pop I (2005) Transport phenomena in porous media. Elsevier, Oxford, USA

Mejri I, Mahmoudi A, Abbassi MA, Omri A (2016) LBM simulation of natural convection in an inclined triangular cavity filled with water. Alexandria Eng J. https://doi.org/10.1016/j.aej.2016.03.020

Moya SL, Ramos E (1987) Numerical study of natural convection in a tilted rectangular porous material. Int J Heat Mass Transfer 30(4):741–756

Nayak RK, Bhattacharyya S, Pop I (2015) Numerical study on mixed convection and entropy generation of Cu–water nanofluid in a differentially heated skewed enclosure. Int J Heat Mass Transfer 85:620–634

Nield DA, Bejan A (2013) Convection in porous media. Springer, New York

Rasoul J, Prinos P (1997) Natural convection in an inclined enclosure. Int J Numer Methods Heat Fluid Flow 7(5):438–478

Saeid NH (2006) Natural convection from two thermal sources in a vertical porous layer. ASME J Heat Transfer 128:104–109

Saeid NH, Pop I (2004) Transient free convection in a square cavity filled with a porous medium. Int J Heat Mass Transfer 47:1917–1924

Sivasankaran S, Do Y, Sankar M (2011) Effect of discrete heating on natural convection in a rectangular porous enclosure. Transp Porous Media 86:261–281

Tian L, Ye C, Xue SH, Wang G (2014) Numerical investigation of unsteady natural convection in an inclined square enclosure with heat-generating porous medium. Heat Transfer Eng 35(6–8):620–629

Vafai K (2005) Handbook of porous media. Taylor & Francis, New York

Varol Y, Oztop HF, Pop I (2008a) Influence of inclination angle on buoyancy-driven convection in triangular enclosure filled with a fluid-saturated porous medium. Heat Mass Transfer 44:617–624

Varol Y, Oztop HF, Pop I (2008b) Numerical analysis of natural convection in an inclined trapezoidal enclosure filled with a porous medium. Int. J Thermal Sci 47:1316–1331

Natural Convection of Cold Water Near Its Density Maximum in a Porous Wavy Cavity

S. Sivasankaran

Nomenclature

A	Amplitude
g	Gravitational acceleration (m s^{-2})
L	Enclosure length (m)
K	Permeability of the porous medium (m^2)
Nu_{loc}	Local Nusselt number
$\overline{Nu}_h$	Average Nusselt number
P	Pressure (Pa)
Ra_D	Darcy–Rayleigh number
T	Temperature (K)
u, v	Velocity components in x- and y-directions (m s^{-1})
x, y	Cartesian coordinates (m)
X, Y	Dimensionless Cartesian coordinates

Greek symbols

α	Thermal diffusivity (m^2 s^{-1})
β	Volumetric coefficient of thermal expansion (K^{-1})
λ	Number of undulations
μ	Dynamic viscosity (m^2 s^{-1})
v	Kinematic viscosity, (kg m^{-1} s^{-1})
ψ	Stream function (m^2 s^{-1})
Ψ	Dimensionless stream function
Θ	Dimensionless temperature

S. Sivasankaran (✉)
Department of Mathematics, King Abdulaziz University, Jeddah 21589, Saudi Arabia
e-mail: sd.siva@yahoo.com; sdsiva@gmail.com

© Springer Nature Singapore Pte Ltd. 2018
N. Narayanan et al. (eds.), *Flow and Transport in Subsurface Environment*,
Springer Transactions in Civil and Environmental Engineering,
https://doi.org/10.1007/978-981-10-8773-8_10

Subscript

c Cold
h Hot
m At density maximum

1 Introduction

Convective flow and heat transfer through a porous medium have been of great interest to researchers due to the wide range of applications in chemical and biomedical engineering, biological sciences, geothermal engineering, and cooling technology. The modeling of convective flow through a porous medium have been clearly demonstrated in the books by Kaviany (1995), Vafai (2005), and Nield and Bejan (2013). Natural convection in enclosures has been analyzed over four decades under different geometrical and other parameters (Murthy et al. 1997; Ratish Kumar 2000; Ratish Kumar and Shalini 2003; Das et al. 2003; Misirlioglu et al. 2005; Bhuvaneswari et al. 2011; Sankar et al. 2011). The non-rectangular geometry is more realistic for the designing of buildings and containers in some situation. The study on corrugated cavities can improve the understanding on circulation of cardiovascular system, design of solar energy collectors, and geothermal power plants. Murthy et al. (1997) investigated natural convective flow and heat transfer of a porous wavy cavity using the Darcy model. They found that the waviness of the wall reduces the heat transfer into the system. Ratish Kumar (2000) studied convection process in a porous cavity with vertical wavy wall subject to heat flux and observed that the flow and convection process are sensitive to the amplitude and number of undulations. Using the Darcy–Forchheimer extended model on convection in enclosure, Ratish Kumar and Shalini (2003) found that rough wavy surface causes secondary circulation zones in the region adjacent to the wavy wall, which leads to a decrease in the heat transfer due to convection. Das et al. (2003) numerically investigated natural convection in a wavy cavity with cold top and hot bottom wavy walls. They observed that the heat transfer rate increases with the rise of wall waviness at high aspect ratio.

Misirlioglu et al. (2005) studied natural convection in a vertically wavy porous cavity heated and cooled from the side walls using the Darcy model. Misirlioglu et al. (2006) also considered natural convection in an inclined wavy porous cavity and they found that the heat transfer rate is highly dependent on wall waviness and Rayleigh number. Sultana and Hyder (2007) investigated natural convection in a convex wavy cavity filled with a porous medium with differentially heated vertical wavy walls. They found that the effect of walls waviness on heat transfer is less significant as compared to Darcy and Rayleigh numbers. Chen et al. (2007) numerically studied convective flow in a vertically in-phase wavy cavity using the Darcy–Brinkman–Forchheimer extended model. They found that a recirculation zone appears in both the top and bottom regions at low values of the Darcy–Rayleigh number. Khanafer et al. (2009) studied natural convection in a porous cavity with a hot wavy side wall and a cold straight side wall with different models of convective flow through a porous medium. They reported that the amplitude and number of undulations affect the heat transfer inside the cavity. Natural convection in a concave porous cavity with wavy top and bottom walls was studied by Mansour et al. (2011). They observed that the total average Nusselt number decreases when the modified conductivity ratio of fluid to solid material increases. Natural convection in a porous cavity with wavy sidewalls was investigated by Sojoudi et al. (2014). They found that the enhancement of the average heat transfer rate is higher on increasing amplitude than that of a number of undulations of the wavy walls.

Water is the most common compound on Earth's surface. In nature, water exists in the liquid, solid, and gaseous states. It is in dynamic equilibrium between the liquid and gas states at 0 °C and 1 atm of pressure. At room temperature (approximately 25 °C), it is a tasteless, odorless, and colorless liquid. It is commonly referred to as the universal solvent because many substances dissolve into water. It is well known, in general, that volume of a liquid increases and hence density decreases with increase in temperature. But as far as water is concerned, when a mass of water is heated from 273 K (0 °C) to 277 K (4 °C), volume decreases (hence density increases) and with further rise in temperature (that is above 277 K), volume increases and density decreases, see Fig. 1. This is due to the temperature effects on the density of water. This is called density anomalous behavior. In most of the analysis pertaining to the convection of water in enclosures, a linear temperature density relationship is taken. But, in practice, this will never happen as the density of water increases on increasing temperature up to 4 °C and then decreases, that is, density of water varies with temperature in a nonlinear fashion, attaining its maximum at 4 °C, more precisely at 3.98 °C. In this region, a linear temperature–density relationship does not work effectively. So a nonlinear

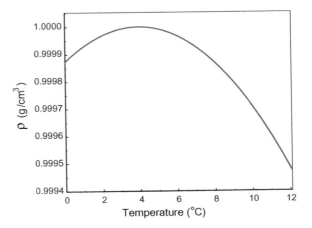

Fig. 1 Density of water

temperature–density relationship is taken to attack the problems effectively. The transformation from ice to water is accompanied by the breaking of some of the hydrogen bonds, leading to a dramatic increase in density. Note that the density at 0 °C is less than that at 3.98 °C. Because of this relationship, ice floats.

Convective heat transfer of cold water near its density maximum region occurs commonly in our environment and in many engineering and technical process. Convective flow and heat transfer of water around its density maximum are complicated. The effect of density inversion on steady natural convection of water between horizontal concentric cylinders was experimentally investigated by Seki et al. (1975). Their result clearly showed that counter secondary eddy appeared in the lower part of the cylinder due to the effect of density inversion at 4 °C. Vasseur and Robillard (1980) investigated the inversion of flow patterns caused by the maximum density of water around 4 °C enclosed in rectangular cavities. They concluded that the convective heat transfer is greatly influenced by the presence of density maximum in the convective fluid. Robillard and Vasseur (1981) studied maximum density effect and supercooling of transient laminar natural convection heat transfer of water in a rectangular cavity. They found out that convection in the absence of maximum density effect is enhanced by an increase in the physical parameters and Biot number. Natural convection of a mass of water near 4 °C confined within a closed cavity was investigated by Robillard and Vasseur (1982). They compared the experimental results with the numerical results for the case of a horizontal circular pipe. An experimental investigation was made by Inaba and Fukuda (1984) to study the effect of density inversion of water on steady natural convective flow patterns and heat transfer in an inclined rectangular cavity. From their results, it was clear that two counter eddies due to the density inversion disturbed the convective heat transfer from the hot wall to cold wall and the influence of two counter eddies was strongly dependent on the inclination angle.

Lankford and Bejan (1986) studied experimentally the natural convection of water near 4 °C in a vertical cavity. The natural convection of cold water near the density maximum in an enclosure was numerically analyzed by Lin and Nansteel (1987). Nansteel et al. (1987) numerically investigated natural convection of cold water in the vicinity of its density maximum in a rectangular enclosure in the limit of small Rayleigh number. They observed that the strength of the counter-rotating flow decreases with decreasing aspect ratio. Ho and Lin (1988, 1990) analyzed the steady laminar natural convection of cold water within a (horizontal/vertical) annulus. They found that the occurrence of density inversion was mainly dependent on the density inversion parameter. An experimental and theoretical investigation of transient natural convection of water near its maximum density in a rectangular cavity was reported by Braga and Viskanta (1992). They demonstrated that the density inversion of water has a great influence on the natural convection in the cavity. McDonough and Faghri (1994) studied both experimentally and numerically the transient and steady-state natural convection of water around its density maximum in a rectangular enclosure with the vertical end walls. They obtained that the effect of the density inversion on the convective flow pattern was dominant. Tong and Koster (1994) numerically investigated the density inversion effect on transient natural convection in a rectangular enclosure filled with water. The effect of the aspect ratio on natural convection of water subject to density inversion had been investigated by Tong (1999). Kandaswamy and Kumar (1999) numerically studied the effect of a magnetic field on the buoyancy-driven flow of water in a square cavity at hot wall temperature from 4 to 12 °C.

Ishikawa et al. (2000) numerically investigated the natural convection with density inversion in a two-dimensional square cavity with variable fluid properties. They chose the cavity size from 1 to 10 cm for different hot wall temperature. Sundaravadivelu and Kandaswamy (2000) numerically investigated combined temperature and species gradients induced buoyancy-driven natural convection flow of cold water at temperatures in the neighborhood of the maximum density. Ho and Tu (2001) numerically and experimentally investigated the natural convection of water near its maximum density in a differentially heated rectangular enclosure. They found that the periodic traveling wave motion of the maximum density contour of water in the temperature visualization experiment. Benhadji et al. (2002) numerically studied the natural convection heat transfer of cold water near 4 °C in a rectangular cavity filled with a porous medium subject to Dirichlet–Neumann thermal boundary conditions. Their results clearly showed the influence of the inversion parameter on the flow pattern and the Nusselt number. Saeid and Pop (2004a, b) examined the natural convection of cold water at temperatures around the temperature of maximum density in a two-dimensional porous cavity with the isothermal discrete heater on the right wall. Three-dimensional natural convection of water with variable physical properties was investigated by Moraga and Vega (2004). Hossain and Rees (2005) studied unsteady laminar natural convection flow of water subject to density inversion in a rectangular cavity with internal heat generation. They found that the flow and temperature fields depend very strongly on the internal heat generation parameter and mean temperature of solid walls.

The effect of aspect ratio, baffle length, and baffle position on convection of cold water near its density extremum in a rectangular cavity is numerically investigated by Sivasankaran and Kandaswamy (2007). They found that multiple fluid vortices exist inside the enclosure due to the density inversion effect, and the size of those vortices strongly depends on the hot wall temperature. Double-diffusive convection of anomalous density of water in the enclosure is investigated by Sivasankaran et al. (2008). Sivasankaran and Ho (2008a, b) studied the effect of temperature-dependent fluid properties on natural convection of water around density maximum region in the rectangular enclosure. They found that the average Nusselt number for considering variable viscosities is higher than the value of average Nusselt number for considering both variable viscosity and thermal conductivity. They also investigated the similar problem in the presence of magnetic field (Sivasankaran and Ho 2008a, b), and in the presence of thermocapillary effect (Sivasankaran and Ho 2010). Varol et al. (2010) investigated numerically the natural convection heat transfer of cold water near 4 °C in a thick-bottom-walled cavity filled with a porous medium. The obtained results show that multiple circulation cells are formed in the cavity and the local Nusselt numbers at the bottom wall and solid–fluid interface are highly affected by formed cells. Sivasankaran et al. (2011) investigated the influence of temperature-dependent properties of the water near its density maximum on buoyancy and thermocapillary-induced magneto-convection in an open cavity.

There is no study on convection flow and heat transfer of cold water near its temperature of maximum density in wavy porous enclosures in the literature. Therefore, this study numerically investigates the effect of density maximum of water and waviness of the wall of the cavity on free convection in a wavy cavity filled with cold-water-saturated Darcy porous medium.

2 Mathematical Modeling

2.1 *Governing Equations*

Consider a two-dimensional enclosure of height and width L filled with cold-water-saturated porous medium as shown in the Fig. 2. The top and bottom walls of the enclosure are insulated. The left wall is maintained with higher temperature (θ_h), while the right wall is cooled with lower temperature (θ_c) with $\theta_h > \theta_c$. The gravity acts in the vertically downward direction. The velocity components, u and v are taken in the x- and y-directions, respectively. The fluid in the enclosure is incompressible and Newtonian. The fluid properties are constant except the density variation and the Boussinesq approximation is valid. Water is the most

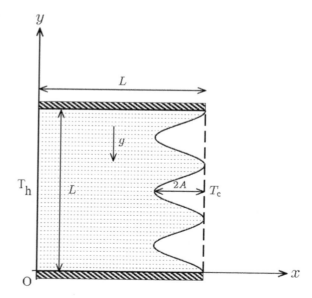

Fig. 2 Schematic diagram of the physical system

common fluid which has a density maximum, we use a quadratic equation for density–temperature relation as

$$\rho - \rho_m = -\beta \rho_m (\theta - \theta_m)^2, \tag{1}$$

where β is a constant and its value is equal to 8.0×10^{-6} (°C)$^{-2}$ and $\rho_m = 1.0$ g/cm^3 when $\theta_m = 3.98$ °C. The above relation is valid within 4% in the range of 0 °C $\leq \theta_m \leq 8$ °C.

The porous medium is assumed to be homogeneous, isotropic, and in thermal equilibrium with the fluid. Furthermore, we assume that the flow is steady and the viscous dissipation is negligible. The Darcy model is adopted to explain the fluid flow through the porous medium. By the law of conservation for mass, momentum, and energy, the governing equations are

$$\frac{\partial u}{\partial x} + \frac{\partial v}{\partial y} = 0, \tag{2}$$

$$u = -\frac{K}{\mu}\frac{\partial P}{\partial x}, \tag{3}$$

$$v = -\frac{K}{\mu}\frac{\partial P}{\partial y} + \frac{K\beta g}{\upsilon}(\theta - \theta_m)^2, \tag{4}$$

$$u\frac{\partial \theta}{\partial x} + v\frac{\partial \theta}{\partial y} = \alpha\left(\frac{\partial^2 \theta}{\partial x^2} + \frac{\partial^2 \theta}{\partial y^2}\right). \tag{5}$$

The governing equations are then expressed in terms of stream functions, which are defined as $u = \partial \psi/\partial y$ and $v = -\partial \psi/\partial x$. Then, the momentum equations are reduced as follows after eliminating the pressure term:

$$\frac{\partial^2 \psi}{\partial x^2} + \frac{\partial^2 \psi}{\partial y^2} = -2\frac{K\beta g}{\upsilon}(\theta - \theta_m)\frac{\partial \theta}{\partial x}. \tag{6}$$

The following dimensionless variables are introduced

$$X = \frac{x}{L}, Y = \frac{y}{L}, \Psi = \frac{\psi}{\alpha}, T = \frac{\theta - \theta_c}{\theta_h - \theta_c}, Ra_D = \frac{K\beta g(T_h - T_c)^2 L}{\alpha \upsilon}, T_m = \frac{\theta_m - \theta_c}{\theta_h - \theta_c} \tag{7}$$

where Ra_D is the modified Darcy–Rayleigh number based on the enclosure width, L and T_m is the density inversion parameter.

Using Eq. (7), the governing equations are non-dimensionalized and dimensionless governing equations are

$$\frac{\partial^2 \Psi}{\partial X^2} + \frac{\partial^2 \Psi}{\partial Y^2} = -2\mathrm{Ra_D}(T - T_\mathrm{m})\frac{\partial T}{\partial X}, \tag{8}$$

$$\frac{\partial^2 T}{\partial X^2} + \frac{\partial^2 T}{\partial Y^2} = \frac{\partial \Psi}{\partial Y}\frac{\partial T}{\partial X} - \frac{\partial \Psi}{\partial X}\frac{\partial T}{\partial Y}. \tag{9}$$

2.2 Boundary Conditions

The boundary conditions are

$$\begin{array}{lll} x = 0, & 0 \le y \le L; & u = v = 0, \theta = \theta_h \\ x = L - AL\left(1 - \cos\left(\frac{2\pi\lambda y}{L}\right)\right), & 0 \le y \le L; & u = v = 0, \theta = \theta_c \\ y = 0 \text{ and } L, & 0 \le x \le L; & u = v = 0, \frac{\partial \theta}{\partial y} = 0. \end{array} \tag{10}$$

The boundary conditions are non-dimensionalized using Eq. (7) and they are

$$\begin{array}{lll} X = 0, & 0 \le Y \le 1; & \Psi = 0, T = 1 \\ X = 1 - A(1 - \cos(2\pi\lambda Y)), & 0 \le Y \le 1; & \Psi = 0, T = 0 \\ Y = 0 \text{ and } 1, & 0 \le X \le 1; & \Psi = 0, \frac{\partial T}{\partial Y} = 0. \end{array} \tag{11}$$

The heat transfer rate in the enclosure is measured by the Nusselt number. It shows the ratio of convection to conduction heat transfer. The local Nusselt number along the left and right walls is calculated as

$$Nu_l = -\frac{\partial T}{\partial X}, \quad Nu_r = -\frac{\partial T}{\partial N}, \tag{12}$$

where N is the normal plane of the sidewall considered. The average Nusselt number on the left sidewall is defined as

$$\overline{Nu} = \int_0^1 Nu_l \mathrm{d}Y. \tag{13}$$

3 Solution Procedure

The finite volume method is used to discretize the governing equations (8) and (9) subject to boundary conditions (11). The discretized equations are solved using the Gauss–Seidel iterative method. In order to carry out the computations using uniform grids in the X- and Y-directions, a grid test is performed in the range of

41 × 41 to 281 × 281 grids for $Ra_D = 10^3$. It is found that the grid size of 201 × 201 is sufficient to perform the computation as good as the finer mesh sizes. The converged solution is obtained by the condition $\sum_{i,j} \left| \xi_{i,j}^{n+1} - \xi_{i,j}^n \right| < 10^{-6}$, where ξ is either Ψ or T and n denotes the iteration number. The trapezoidal rule is used to calculate the average Nusselt number.

The validation of the computer code is very important in the study of numerical simulation. An in-house code is written to find the solution of the considered system. The computer code is validated based on the previous literature using differentially heated square Darcy porous enclosure. It is seen from Table 1, the results predicted by current code agree well with the previous studies.

4 Results and Discussion

Numerical simulation on buoyancy-induced convective flow of cold water in a wavy enclosure filled with the porous medium is made for different combinations of parameters involved in the study. The present work is to investigate the interaction of wavy nature of wall and density maximum effect of cold water around 4 °C inside the wavy porous cavity. The values of the pertinent parameters involved in the investigation are carefully scrutinized. The density inversion parameter varies from 0 to 1, the amplitude of wavy wall from 0 to 0.2, the undulation of the wavy wall from 1 to 5, and the Rayleigh number from 0 to 10^3.

There exist two distinct horizontal water layers with different density gradients separated by the maximum density plane. The density inversion parameter helps us to locate the maximum density plane inside the enclosure. The density inversion parameter (T_m) lies between 0 and 1. Suppose $T_m = 0$, the cold wall is at temperature $\theta_m (\theta_m = \theta_c)$, so that density of water decreases monotonically with temperature everywhere in the enclosure. Suppose $T_m = 1$, the hot wall is at temperature $\theta_m (\theta_m = \theta_h)$, so that density of water increases monotonically with temperature everywhere in the enclosure. It is observed that the density maximum plane moves from the right (cold) wall to the left (hot) wall on increasing the density inversion parameter from 0 to 1. It is an interesting phenomenon and it affects the flow field as well as heat transfer inside the enclosure.

Table 1 Comparison of $\overline{Nu}$ for natural convection in a square porous cavity

References	$Ra_D = 10^2$	$Ra_D = 10^3$
Baytas and Pop (1999)	3.160	14.060
Saeid and Pop (2004a, b)	3.002	13.726
Misirlioglu et al. (2005)	3.050	13.150
Present study	3.101	13.280

4.1 Effect of Density Maximum

Figure 3a–g shows the flow pattern inside the wavy cavity for different values of density inversion parameters with $A = 0.1$ and $\lambda = 3$. The flow consists of a single eddy occupying the whole cavity for $T_m = 0$. Here, the cold wall is at a temperature θ_m, so that the fluid density decreases monotonically with temperature everywhere in the enclosure. When $T_m = 0$, the hot fluid along the hot wall goes up and cold fluid down along the cold wall ($\theta_m = \theta_c$), that is, a clockwise-rotating eddy exists; see Fig. 3a. The flow field slightly altered at the right bottom of the cavity on increasing the density inversion parameter to $T_m = 0.2$. When increasing the density inversion parameter $T_m = 0.4$, the density maximum plane moves inside the cavity, which causes the flow field and produces the dual cellular structure. A small counteracting eddy exists near the cold wall. When $T_m = 0.5$, the side wall temperatures are equally far from θ_m, which corresponds to the maximum density. Therefore, the density maximum plane exists in the middle of the cavity. There exist two counteracting cells, symmetrical about the vertical centerline. Both the isothermal walls generate an identical upward buoyancy force, which produces a two-cell flow structure inside the cavity. The fluid flows upward along the isothermal walls and down the middle of the cavity. This is clearly seen from the Fig. 3d. When increasing the density inversion parameter, $T_m = 0.6$, the clockwise hot cell along the hot wall decreases in its size and the cold cell size increases. The counter-rotating eddy occupies the majority of the cavity for $T_m = 0.6$. Further increasing the density inversion parameter to 0.8, the clockwise-rotating eddy disappears. The counter-rotating eddy occupies the whole cavity when $T_m = 1$; see Fig. 3g. It is observed that a single cell occupies the whole cavity when $T_m = 0$ or 1. Since the density maximum plane is at either cold wall or hot wall, it produces a clockwise- or counterclockwise-rotating cell due to the domination of the buoyancy force.

Isotherms for different values of the density inversion parameters with $A = 0.1$ and $\lambda = 3$ are displayed in Fig. 3a–g. The isotherms clearly show the effects of density inversion. It is obvious that the density maximum plane moves from right (cold) wall to left (hot) wall on increasing the density inversion parameter from 0 to 1. When $T_m = 0(1)$, the density maximum plane is at the isothermal right (left) wall. The location of the density maximum plane comes to mid-plane of the cavity when $T_m = 0.5$. The isotherms for all cases indicate that the heat transfer is dominated by the convection mode. The thermal boundary layer is formed near the right top and left bottom of the cavity for $T_m = 0$ and is formed near the right bottom and left top of the cavity for $T_m = 1.0$. When the density maximum plane comes to the middle of the cavity ($T_m = 0.5$), large temperature gradients around the density maximum plane in the upper region of the cavity are found. The vertical temperature stratification is observed for $T_m = 0$ and $T_m = 1.0$. The thermal boundary layers disappear at $T_m = 0.5$ and the strength of the convection is reduced.

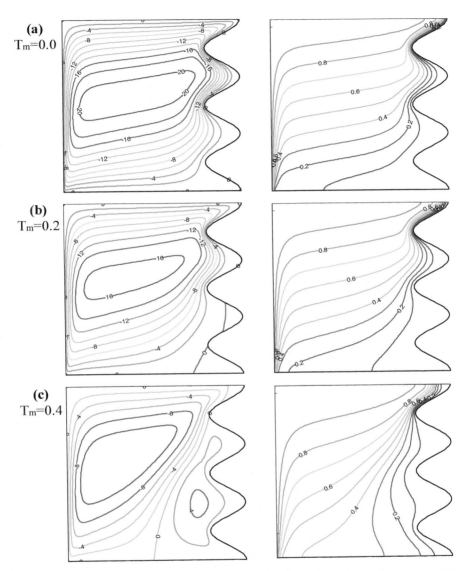

Fig. 3 Streamlines and isotherms for different density inversion parameter (T_m) with $A = 0.1$, $\lambda = 3$ and $Ra = 10^3$

(d) $T_m=0.5$

(e) $T_m=0.6$

(f) $T_m=0.8$

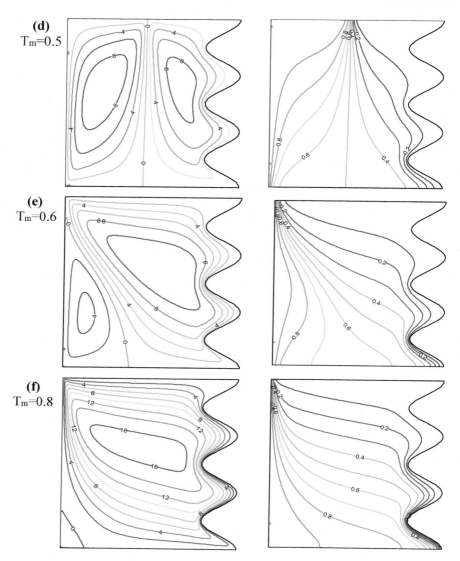

Fig. 3 (continued)

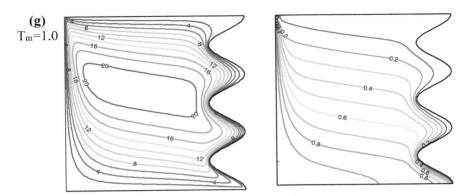

(g) $T_m = 1.0$

Fig. 3 (continued)

4.2 Effect of Wavy Wall

Figure 4 shows the effect of wall waviness on flow field and temperature distributions with $T_m = 0$ and $T_m = 0.5$. A single clockwise-rotating cell occupies the whole cavity for the case $T_m = 0$ when changing the shape of the wavy wall, that is, changing the values of amplitude and number of undulations. The shape of the eddy (flow field) is adjusted according to the waviness of the wall along the right wavy wall. The similar flow behavior is observed for the case $T_m = 1.0$. There is no secondary flow observed between crests of the wavy wall. However, the flow structure is drastically changed with waviness of the wall for the case $T_m = 0.5$. When $A = 0.05$, there exist two counteracting cells of the same size and strength, symmetrical about the vertical centerline. When changing the undulations for fixed values of amplitude, the flow field affects along the wavy wall only. The symmetrical dual-cell structure is altered much for high values of amplitude, ($A = 0.2$). The two cells are not of equal size and strength here. The counter-rotating cell affects much further increasing the number of undulations ($\lambda = 5$). Therefore, the dual-cell structure is more affected by the waviness of wall than the single-cell flow structure.

4.3 Heat Transfer Rate

The density anomalous behavior of water is the important phenomena, which affects not only the fluid flow but also the heat transfer. To study this behavior, the local and average Nusselt numbers are plotted for different values density inversion parameter. Figure 5a–b show the local Nusselt number along the hot and cold walls for different values of density inversion parameter. The local Nusselt number gets the minimum value at the bottom of the hot wall when $T_m \leq 0.5$, where the density

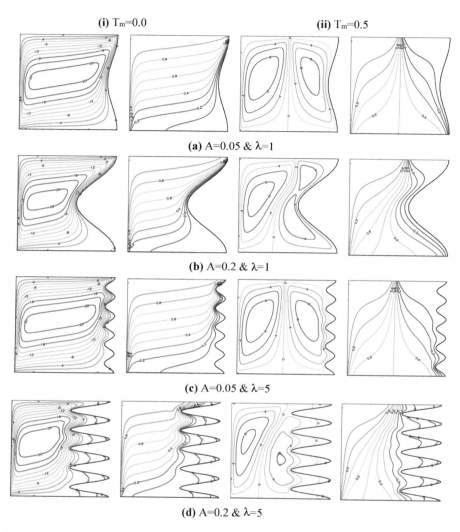

Fig. 4 Streamlines and isotherms for different A and ϕ with $T_m = 0.0$ and $T_m = 0.5$

maximum plane is at the cold wall to near to the cold wall of the cavity. The local Nusselt number decreases from the bottom to the top of the hot wall when $T_m \leq 0.5$. The opposite observation is found for $T_m > 0.5$, where the density maximum plane is at the hot wall or near to the hot wall. In this situation, the local Nusselt number increases with the hot wall height and reaches its maximum at the top of the hot wall. The local Nusselt number along the right wall is of wavy form due to the wavy nature of the Wall; see Fig. 5b. The higher heat transfer is observed at the crests of the wavy wall, while lower heat transfer is obtained at the troughs of the wavy wall. It is also found that the higher heat transfer is observed at the lower crests of the wavy wall for $T_m \leq 0.5$. The higher heat transfer is observed at the upper crests of the wavy wall for $T_m > 0.5$.

Figure 6 shows the average Nusselt number against density inversion parameter for different values of Darcy–Rayleigh number. The heat transfer rate behaves nonlinearly with density inversion parameter. The heat transfer rate reaches its minimum at $T_m = 0.5$ for all values of the Darcy–Rayleigh numbers. In this situation, flow consists of two counteracting cells. Since heat energy is transferred from one (hot) cell to another (cold) cell by conduction in dual-cell structure, the heat transfer rate is reduced. The exchange of heat energy takes place between these two cells only by conduction. That is, the dual-cell structure prohibits convection mode of heat transfer across the cavity. However, in the single-cell flow pattern, heat energy is directly transferred from hot region to cold region. Therefore, the average heat transfer rate is enhanced when $T_m = 0$ and 1. Figure 7 shows the effect of wall waviness on average heat transfer rate with $T_m = 0, 0.5,$ and 1 and $Ra_D = 10^3$. The average Nusselt number increases on increasing the values of amplitude and number of undulations when $T_m = 0$ and 1. But, this trend on average Nusselt number is not observed at $T_m = 0.5$ because of dual-cell structure due to density maximum effect. There is no consistent tendency on average heat transfer rate on increasing the values of amplitude and number of undulations. It can be seen that increasing the amplitude and number of undulations of the cold wall can increase the average heat transfer of the heated wall. Increasing the amplitude of the wavy wall gives more increment on the heat transfer than that of increasing the number of undulations.

The correlations for the average Nusselt number in the enclosure is derived from the numerical results obtained over a wide range of parameters. Since three different types of flow behavior are obtained from this study according to the values of density inversion parameter, the correlation equations are derived for $T_m = 0, 0.5,$ and 1.

For $T_m = 0$,

$$\overline{Nu} = -0.4070A^2\lambda^2 + 3.3380A^2\lambda - 0.00535A\lambda^2 + 0.4751A\lambda + 13.1691.$$

For $T_m = 0.5$,

$$\overline{Nu} = 2.8570A^2\lambda^2 + -16.6388A^2\lambda - 0.5999A\lambda^2 + 3.4972A\lambda + 25.3279A^2 \\ - 4.8930A + 0.02140\lambda - 0.1192\lambda + 3.4924.$$

For $T_m = 1$,

$$\overline{Nu} = -0.5059A^2\lambda^2 + 1.3904A^2\lambda - 0.0601A\lambda^2 + 1.1549A\lambda + 13.0611.$$

This is valid only for $Ra_D = 10^3$.

Fig. 5 Local Nusselt number for different density inversion parameters with $A = 0.1$, $\lambda = 3$, and $Ra = 10^3$

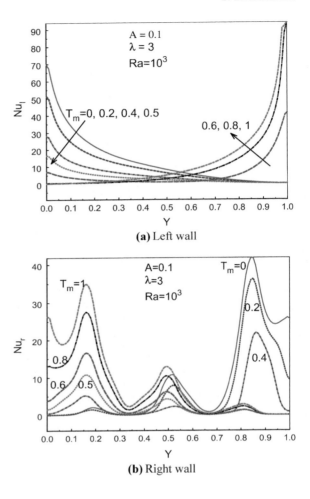

(a) Left wall

(b) Right wall

Fig. 6 Average Nusselt number for different density inversion parameters and Rayleigh number with $A = 0.1$ and $\lambda = 3$

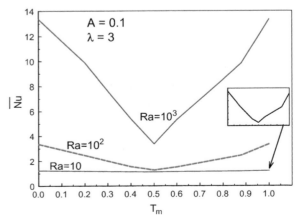

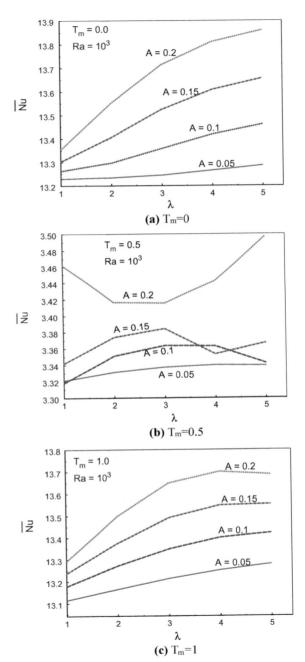

Fig. 7 Average Nusselt number for different amplitude and phase deviations with **a** $T_m = 0$, **b** $T_m = 0.5$, and **c** $T_m = 1$, for $Ra = 10^3$

5 Conclusions

The effects of density inversion and waviness of the wall on free convection of cold water near its density maximum region in a wavy porous cavity are numerically investigated. It is observed that the temperature of maximum density leaves strong effects on fluid flow and heat transfer due to the formation of bi-cellular structure. The formation of dual-cell structure and strength of each cell always depends on the density inversion parameter and waviness of the wall. The dual-cell structure prohibits convection mode of heat transfer across the cavity. The heat transfer rate behaves nonlinearly with density inversion parameter. The heat transfer rate increases on increasing the amplitude and number of undulations when the density maximum plane is at either left or right wall. However, there is no uniform trend on heat transfer rate on increasing the amplitude and number of undulations when the density maximum plane is at the middle of the cavity. The waviness of the wall affects the formation of dual-cell structure in this situation and heat transfer behaves nonlinearly. That is, the heat transfer rate increases or decrease according to the number of undulations and amplitude when the density maximum plane is at the middle of the cavity.

References

Baytas AC, Pop I (1999) Free convection in oblique enclosures filled with a porous medium. Int J Heat Mass Transf 42:1047–1057

Benhadji K, Robillard L, Vasseur P (2002) Convection in a Porous cavity saturated with water near 4 °C and subject to Dirichlet-Newmann thermal boundary conditions. Int Commun Heat Mass Transf 29(7):897–906

Bhuvaneswari M, Sivasankaran S, Kim YJ (2011) Effect of aspect ratio on natural convection in a porous enclosure with partially active thermal walls. Comput Math Appl 62:3844–3856

Braga SL, Viskanta R (1992) Transient natural convection of water near its density extremum in a rectangular cavity. Int J Heat Mass Transf 35(4):861–875

Chen XB, Yu P, Winoto SH, Low HT (2007) Free convection in a porous wavy cavity based on the Darcy-Brinkman-Forchheimer extended model. Numer Heat Transf, Part A 52(4):377–397

Das PK, Mahmud S, Tasnim SH, Islam AKMS (2003) Effect of surface waviness and aspect ratio on heat transfer inside a wavy enclosure. Int J Numer Methods Heat Fluid Flow 13(8): 1097–1122

Ho CJ, Lin YH (1988) Laminar natural convection of cold water enclosed in a horizontal annulus with mixed boundary conditions. Int J Heat Mass Transf 31:2113–2121

Ho CJ, Lin YH (1990) Natural convection of cold water in a vertical annulus with constant heat flux on the inner wall. ASME J Heat Transf 112:117–123

Ho CJ, Tu FJ (2001) Visualization and prediction of natural convection of water near its density maximum in a tall rectangular enclosure at high Rayleigh numbers. J Heat Transf 123:84–95

Hossain MA, Rees DAS (2005) Natural convection flow of water near its density maximum in a rectangular enclosure having isothermal walls with heat generation. Heat Mass Transf 41: 367–374

Inaba H, Fukuda T (1984) An experimental study of natural convection in an inclined rectangular cavity filled with water at its density extremum. J Heat Transf 106:109–117

Ishikawa M, Hirata T, Noda S (2000) Numerical simulation of natural convection with density inversion in a square cavity. Numer Heat Transf, Part A 37:395–406

Kandaswamy P, Kumar K (1999) Buoyancy-driven nonlinear convection in a square cavity in the presence of magnetic field. Acta Mech 136:29–39

Kaviany M (1995) Principles of heat transfer in porous media, 2nd edn. Springer-Verlag, New York

Khanafer K, Al-Azmi B, Marafie A, Pop I (2009) Non-Darcian effects on natural convection heat transfer in a wavy porous enclosure. Int J Heat Mass Transf 52(7–8):1887–1896

Lankford KE, Bejan A (1986) Natural convection in a vertical enclosure filled with water near 4 °C. ASME J Heat Transf 108:755–763

Lin DS, Nansteel MW (1987) Natural convection heat transfer in a square enclosure containing water near its density maximum. Int J Heat Mass Transf 30(11):2319–2329

Mansour MA, El-Aziz MMA, Mohamed RA, Ahmed SE (2011) Numerical simulation of natural convection in wavy porous cavities under the influence of thermal radiation using a thermal non-equilibrium model. Transp Porous Media 86:585–600

McDonough MW, Faghri A (1994) Experimental and numerical analyses of the natural convection of water through its density maximum in a rectangular enclosure. Int J Heat Mass Transf 37(5):783–801

Misirlioglu A, Baytas AC, Pop I (2005) Free convection in a wavy cavity filled with a porous medium. Int J Heat Mass Transf 48(9):1840–1850

Misirlioglu A, Baytas AC, Pop I (2006) Natural convection inside an inclined wavy enclosure filled with a porous medium. Transp Porous Media 64(2):229–246

Moraga NO, Vega SA (2004) Unsteady three-dimensional natural convection of water cooled inside a cubic enclosure. Numer Heat Transf, Part A 45:825–839

Murthy PVSN, Ratish Kumar BV, Singh P (1997) Natural convection heat transfer from a horizontal wavy surface in a porous enclosure. Numer Heat Transfer, Part A 31(2):207–221

Nansteel MW, Medjani K, Lin DS (1987) Natural convection of water near its density maximum in a rectangular enclosure: low Rayleigh number calculations. Phys Fluids 30(2):312–317

Nield DA, Bejan A (2013) Convection in porous media, 4th edn. Springer, New York

Ratish Kumar BV (2000) A study of free convection induced by a vertical wavy surface with heat flux in a porous enclosure. Numer Heat Transf, Part A 37(5):493–510

Ratish Kumar BV, Shalini (2003) Free convection in a non-Darcian wavy porous enclosure. Int J Eng Sci 41(16):1827–1848

Robillard L, Vasseur P (1981) Transient natural convection heat transfer of water with maximum density effect and supercooling. ASME J Heat Transf 103:528–534

Robillard L, Vasseur P (1982) Convection response of a mass of water near 4 °C to a constant cooling rate applied on its boundaries. J Fluid Mech 118:123–141

Saeid NH, Pop I (2004a) Maximum density effects on natural convection from a discrete heater in a cavity filled with a porous medium. Acta Mech 171:203–212

Saeid NH, Pop I (2004b) Transient free convection in a square cavity filled with a porous medium. Int J Heat Mass Transf 47:1917–1924

Sankar M, Bhuvaneswari M, Sivasankaran S, Do Y (2011) Buoyancy induced convection in a porous cavity with partially thermally active sidewalls. Int J Heat Mass Transf 54:5173–5282

Seki N, Fukusako S, Nakaoka M (1975) Experimental study on natural convection heat transfer with density inversion of water between two horizontal concentric cylinders. ASME J Heat Transf 75:556–561

Sivasankaran S, Ho CJ (2008a) Effect of temperature dependent properties on convection of water near its density maximum in enclosures. Numer Heat Transf, Part A 53:507–523

Sivasankaran S, Ho CJ (2008b) Effect of temperature dependent properties on MHD convection of water near its density maximum in a square cavity. Int J Thermal Sci 47:1184–1194

Sivasankaran S, Ho CJ (2010) Buoyancy and thermo-capillary induced convection of cold water in an open enclosure with variable fluid properties. Numer Heat Transf, Part A 58(6):457–474

Sivasankaran S, Kandaswamy P (2007) Double diffusive convection of water in a rectangular partitioned enclosure with conductive baffle on hot wall. Arabian J Sci Eng 32(1B):35–48

Sivasankaran S, Kandaswamy P, Ng CO (2008) Double diffusive convection of anomalous density fluids in a cavity. Transp Porous Media 71(2):133–145

Sivasankaran S, Bhuvaneswari M, Kim YJ, Ho CJ, Pan KL (2011) Magneto-convection of cold water near its density maximum in an open cavity with variable fluid properties. Int J Heat Fluid Flow 32:932–942

Sojoudi A, Saha SC, Khezerloo M, Gu YT (2014) Unsteady natural convection within a porous enclosure of sinusoidal corrugated side walls. Transp Porous Media 104:537–552

Sultana Z, Hyder MN (2007) Non-Darcy free convection inside a wavy enclosure. Int Commun Heat Mass Transf 34(2):136–146

Sundaravadivelu K, Kandaswamy P (2000) Double diffusive nonlinear convection in a square cavity. Fluid Dyn Res 27:291–303

Tong W (1999) Aspect ratio effect on natural convection in water near its density maximum temperature. Int J Heat Fluid Flow 20:624–633

Tong W, Koster JN (1994) Density inversion effect on transient natural convection in a rectangular enclosure. Int J Heat Mass Transf 37(6):927–938

Vafai K (2005) Handbook of porous media 2nd edn. CRC Press, Taylor & Francis

Varol Y, Oztop HF, Mobedi M, Pop I (2010) Visualization of heat flow using Bejan's heatline due to natural convection of water near 4 °C in thick walled porous cavity. Int J Heat Mass Transf 53:1691–1698

Vasseur P, Robillard L (1980) Transient natural convection heat transfer in a mass of water cooled through 4 °C. Int J Heat Mass Transf 23:1195–1205

Convective Mass and Heat Transfer of a Chemically Reacting Fluid in a Porous Medium with Cross Diffusion Effects and Convective Boundary

M. Bhuvaneswari and S. Sivasankaran

1 Introduction

Free convection flows resulting from combined mass and heat transfers in a porous medium have been examined extensively owing to a rich spectrum of applications in engineering and geophysical transport processes. A good number of publications are established in the literature, (Nield and Bejan 2013; Vafai 2005; Kaviany 1995). Free convection boundary layer flow in a porous medium is explored under several conditions (Ferdows et al. 2009; Pal and Mondal 2012; Bhuvaneswari et al. 2012; Khader and Megahed 2014; Najafabadi and Gorla 2014). In the existence of chemical reaction, the mass and heat transfer problems are of importance in many chemical engineering procedures (Chambre and Young 1958; Dass et al. 1994; Muthucumaraswamy 2003; Prasad et al. 2003; Bhuvaneswari et al. 2009). Chambre and Young (1958) premeditated chemical reaction on convective flow from a horizontal plate. Hayat et al. (2015) studied Dufour and Soret effects in convective flow over a plate with heat source/sink, chemical reaction and porous medium.

It is well acknowledged that the heat transfer characteristics depend on the thermal boundary conditions imposed on the system. In general, there are four common heating procedures requiring the wall-to-ambient heat diffusion, viz. wall temperature (Dirichlet condition), surface heat flux (Neumann condition), conjugate conditions and Newtonian heating. The consequence of Newtonian heating is analysed under different conditions on free convection boundary layer flow (Merkin 1994; Lesnic et al. 1999, 2000, 2004; Chaudhary and Jain 2007; Rajesh 2012). Mahanta and Shaw (2015) investigated the MHD Casson fluid flow over a linearly

---

M. Bhuvaneswari · S. Sivasankaran (✉)
Department of Mathematics, King Abdulaziz University, Jeddah 21589, Saudi Arabia
e-mail: sd.siva@yahoo.com; sdsiva@gmail.com

M. Bhuvaneswari
e-mail: msubhuvana@yahoo.com

© Springer Nature Singapore Pte Ltd. 2018
N. Narayanan et al. (eds.), *Flow and Transport in Subsurface Environment*,
Springer Transactions in Civil and Environmental Engineering,
https://doi.org/10.1007/978-981-10-8773-8_11

stretching porous sheet with convective boundary. Das et al. (2015) examined the mass and heat transfers of nanofluid over a heated stretching sheet in a porous medium with convective boundary. They detected that a rise in the surface convection, thermal radiation, Brownian motion and thermophoretic parameters lead to a rise in the thermal boundary layer thickness. Uddin et al. (2015) investigated the thermal radiation and Newtonian heating effects on mass and heat transfers of nanofluids in a porous medium.

Diffusion of heat affected by concentration gradients, called Dufour effect, and diffusion of substance affected by thermal gradients, called Soret effect, acquire very important while concentration and temperature gradients are prominent. Mass and heat transfers with thermo-diffusion consequence are a content of rigorous investigation due to extensive presentations (Bhuvaneswari et al. 2011; Khidir and Sibanda 2014). These include oil reservoirs, isotope separation, multicomponent melts, geophysics and in combination among gases. Beg et al. (2009) inspected the magneto-convection mass and heat transfers from a stretching plate to a porous medium with Dufour and Soret effects. They figured out that the concentration increases on increasing the Soret parameter and it decreases on increasing the Dufour parameter. Tsai and Huang (2009) premeditated the effects of Dufour and Soret on mass and heat transfers of Hiemenz flow.

Nawaz et al. (2012) examined the cross diffusion effects on MHD flow with Joule heating, viscous dissipation, and chemical reactions. Raman et al. (2012) premeditated the cross diffusion properties on magneto-convection with chemical reaction and thermophoresis over a porous stretching surface. They observed that liquid velocity decreases due to the uniform porosity at the wall surface in the occurrence of thermophoresis element confession. Pal and Mondal (2013) explored the impact of cross diffusion effect on MHD mass and heat transfers. They concluded that temperature increases on decreasing the Soret parameter (or increasing Dufour parameter).

As discussed above, the convective flow, mass and heat transfers over a stretching plate have been extensively examined. However, the convective flow of chemically reacting fluid over a stretching surface in a porous medium in the existence of Dufour and Soret effect with convective boundary condition is not investigated. So, the present study deals this extensively under different combination of parameters taken in the study.

2 Mathematical Modelling

The unsteady 2D laminar incompressible convective boundary layer flow, mass and heat transfers of a viscous fluid over a stretching plate in a porous medium are considered. Figure 1 describes the coordinate system and physical model of the problem. The x-axis alongside the surface and y-axis vertical to the plate are considered. Then, u and v are acquired as the velocity components along x and y directions, respectively. The fluid is of constant properties and the density variation is

Fig. 1 Physical configuration

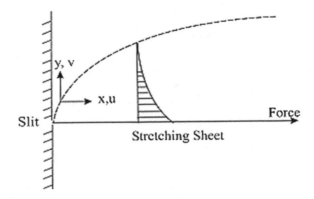

negligible except in the buoyancy term. The Boussinesq approximation is valid for the body force acting on the system, which is the buoyancy force resulting from the density variation with concentration and temperature. The porous medium is assumed to be homogeneous, isotropic and in thermal equilibrium with the fluid. Thus, the Darcy model is implemented to describe the flow through the porous medium. The porous medium is generating heat at a uniform rate throughout the system. In addition, first-order homogeneous chemical reaction, Soret and Dufour effects are considered.

2.1 *Governing Equations*

Based on the boundary layer assumptions (1968), the equations are given by

$$u_x + v_y = 0 \tag{1}$$

$$u_t + uu_x + vu_y = vu_{yy} - \left(\frac{v}{\overline{K}}\right)u + g\beta^*(C - C_\infty) + g\beta(T - T_\infty) \tag{2}$$

$$T_t + uT_x + vT_y = \alpha T_{yy} + Q(T - T_\infty) + \frac{D_e k}{c_s c_p} C_{yy} \tag{3}$$

$$C_t + uC_x + vC_y = DC_{yy} - \Gamma_0(C - C_\infty) + \frac{D_e k}{T_m} T_{yy} \tag{4}$$

2.2 *Boundary Conditions*

The boundary conditions play an important role which dictate the particular solutions to be obtained from the governing equations in all boundary layer problems. The boundary conditions for the above-said problem are as follows:

$$u = U_w, \quad v = V_w, \quad C \to C_w, \quad -k_T T_y = h_f(T_f - T) \quad \text{at } y = 0,$$
$$u \to 0, \quad v \to 0, \quad C \to C_\infty, \quad T \to T_\infty, \quad \text{as } y \to \infty, \tag{5}$$

where C is the concentration of the fluid, T is the temperature of the fluid, k is the thermal conductivity and v $(= \mu/\rho)$ kinematic viscosity of fluid, respectively.

3 Similarity Solution

The symmetry of systems, the property of remaining unchanged when certain transformations are performed, has important consequences such as the conservation of physical quantities. This gives the advantage to simplify when solving the physical model. The similarity solutions are solutions which depend on certain groupings of the independent variables, rather than on each variable separately. The use of the word similar was explained in Evans (1968). For most flows, the shape of velocity profile varies gradually as x-changes. This means that the manner of increase or decrease of velocity in the x-direction is the same for all values of y. Since the velocity profiles for these flows have a similar shape, such boundary layer may be described as similar and the corresponding solutions to the governing equations as similarity solution.

The similarity variable and other non-dimensional variables are introduced as follows.

$$\eta = y\sqrt{\frac{a}{v(1-ct)}}, \quad u = \frac{ax}{(1-ct)} f'(\eta), \quad v = -\sqrt{\frac{av}{(1-ct)}} f(\eta),$$
$$\theta(\eta) = \frac{T - T_\infty}{T_f - T_\infty}, \quad \phi(\eta) = \frac{C - C_\infty}{C_w - C_\infty}, \tag{6}$$

where η is the similarity variable. Substituting Eq. (6) into Eqs. (1–5), we attain the ensuing coupled ODEs with boundary conditions:

$$f''' - A\left(f' + \frac{1}{2}\eta f''\right) - f'^2 + ff'' - Kf' + Gr_T\theta + Gr_c\phi = 0 \tag{7}$$

$$\theta'' - Pr\, A\left(\theta + \frac{1}{2}\eta\theta'\right) + Pr(f\theta' - \theta f') + Pr\, D_f\phi'' + Pr\, S\theta = 0 \tag{8}$$

$$\phi'' - Sc(\phi f' - f\phi') - Sc\, A\left(\phi + \frac{1}{2}\eta\phi'\right) - Sc\, Cr\phi + Sr\, Sc\, \theta'' = 0 \tag{9}$$

$$f = f_w, \quad f' = 1, \quad \phi = 1, \quad \theta' = -Bi(1-\theta), \quad \text{at } \eta = 0,$$
$$f' = 0, \quad \phi = 0, \quad \theta = 0, \quad \text{as } \eta \to \infty, \tag{10}$$

where

$$Bi = \frac{h_f}{k}\sqrt{\frac{v}{a}} \text{ is the Biot number,}$$

$$Cr = \frac{\Gamma_0 x}{U_w} \text{ is the chemical reaction parameter,}$$

$$K = \frac{v}{c\tilde{K}} \text{ is the permiability parameter,}$$

$$Df = \frac{D_m k(C_w - C_\infty)}{c_s c_p (T_w - T_\infty) v} \text{ is the Dufour parameter,}$$

$$Gr_T = \frac{g\beta(T_\omega - T_\infty)x^3}{v^2} \text{ is the thermal Grashof number,}$$

$$Gr_c = \frac{g\beta^*(C_\omega - C_\infty)x^3}{v^2} \text{ is the solutal Grashof number,}$$

$$Pr = \frac{v}{\alpha} \text{ is the Prandtl number,}$$

$$S = \frac{Qx}{U_w} \text{ is the heat generation/absorbtion parameter,}$$

$$Re_x = \frac{U_w x}{v} \text{ is the local Reynolds number,}$$

$$Sr = \frac{D_m k(T_w - T_\infty)}{T_m \alpha (C_w - C_\infty)} \text{ is the Soret parameter,}$$

$$Sc = \frac{v}{D} \text{ is the Schmidt number.}$$

The coupled Eqs. (7–9) cannot get the closed loop solution. Therefore, the system of equations is explicated numerically by shooting method along with Runge–Kutta fourth-order technique.

4 Physical Parameters

From the process of numerical calculation, the physical quantities describing the local skin friction coefficient, Sherwood number and Nusselt number are wall shear stress, rates of mass and heat transfers. They are given by the expressions

$$C_f = \frac{2\tau_\omega}{\rho U_\infty^2}, \quad Nu = \frac{xq_\omega}{k(T_f - T_\infty)}, \quad Sh = \frac{xq_m}{D(C_\omega - C_\infty)}, \quad (11)$$

where

$$\tau_\omega = \mu u_y\big|_{y=0}, \quad q_\omega = -kT_y\big|_{y=0}, \quad q_m = -DC_y\big|_{y=0} \qquad (12)$$

Finally, we get the non-dimensional local skin friction (C_f), Sherwood number (Sh) and Nusselt number (Nu) as follows:

$$\frac{\sqrt{Re_x}C_f}{2} = -f''(0)$$
$$\frac{Sh}{\sqrt{Re_x}} = -\phi'(0) \qquad (13)$$
$$\frac{Nu}{\sqrt{Re_x}} = -\theta'(0)$$

5 Results and Discussion

The consequences of cross diffusion and Newtonian heating on convective heat transfer and mixed convective flow over a stretching surface in a porous medium for various values of the pertinent parameter are elaborated in the study. The local skin friction coefficient, the rates of mass and heat transfers are deliberated in Table 1 for different values of K, A, Hg, Cr, Df, Sr, Bi and f_w with fixed values of $Gr_T = 1$, $Gr_c = 1$, $Pr = 0.7$, $Sc = 0.62$. The skin friction increases at the surface on rising the values of the parameters K, A, f_w. The skin friction decreases on rising the values of the parameters Hg, Df, Sr, Bi. The skin friction is high in the absence of chemical reaction. The rate of heat transfer drops on rising the values of K, Hg, S, Cr, Df parameters. The heat transfer rate grows on rising the values of the

Table 1 $f''(0)$, $-\phi'(0)$, $-\theta'(0)$ with fixed $Gc = 1$, $Gr = 2$, $Sc = 0.62$ and $Pr = 0.7$

K	A	Hg	Df	Sr	Cr	f_w	Bi	$f''(0)$	$-\theta'(0)$	$-\phi'(0)$
0	1	0.1	0.5	0.5	1	0.5	1	0.632456	0.424509	1.471390
1								1.050554	0.413908	1.445375
3								1.688202	0.399475	1.410778
5								2.184852	0.389868	1.387949
7								2.601578	0.382883	1.371305
1	0	0.1	0.5	0.5	1	0.5	1	0.599024	0.347485	1.306964
	1							1.050554	0.413908	1.445375
	3							1.701848	0.491935	1.727436

(continued)

Table 1 (continued)

K	A	Hg	Df	Sr	Cr	f_w	Bi	$f''(0)$	$-\theta'(0)$	$-\phi'(0)$
	5							2.166584	0.535188	1.986435
	10							3.010141	0.593537	2.532542
1	1	−0.5	0.5	0.5	1	0.5	1	1.129925	0.471649	1.421133
		−0.3						1.107140	0.454704	1.428358
		0.0						1.066382	0.425146	1.440742
		0.3						1.014470	0.388752	1.455587
		0.5						0.970682	0.359023	1.467363
1	1	0.1	0	0.5	1	0.5	1	1.227465	0.555485	1.382912
			0.5					1.050554	0.413908	1.445375
			1					0.868330	0.258531	1.514382
			1.5					0.677535	0.084997	1.592360
			2					0.474237	−0.11294	1.682823
1	1	0.1	0.5	0	1	0.5	1	1.051686	0.408141	1.492316
				0.5				1.050554	0.413908	1.445375
				1				1.049692	0.420254	1.394256
				1.5				1.049211	0.427327	1.337979
				2				1.049275	0.435340	1.275134
1	1	0.1	0.5	0.5	−0.5	0.5	1	1.051102	0.478390	0.939410
					−0.3			1.053464	0.467863	1.026314
					0			1.054665	0.453593	1.140540
					0.3			1.054306	0.440644	1.241672
					0.5			1.053545	0.432573	1.303789
1	1	0.1	0.5	0.5	1	−1	1	0.529808	0.365527	1.050848
						−0.5		0.664307	0.381677	1.167580
						0		0.835495	0.397710	1.298963
						0.5		1.050554	0.413908	1.445375
						1		1.315323	0.430550	1.606869
1	1	0.1	0.5	0.5	1	0.5	0.1	1.233131	0.066812	1.495919
							2	0.964346	0.583488	1.420466
							4	0.888864	0.734853	1.398128
							6	0.854445	0.804758	1.387780
							8	0.834728	0.845048	1.381807

parameters A, Sr, Bi and f_w. By rising the values of the parameters A, Hg, Df, f_w, the mass transfer rate enhances. The rate of mass transfer reduces on rising the values of the parameters K, Sr, Bi.

The consequence of permeability (K) parameter on the non-dimensional velocity, concentration and temperature for $A = 1.0$, $Hg = 0.1$, $Df = 0.5$, $Sr = 0.5$, $Cr = 1.0$, $f_w = 0.5$, $Bi = 1.0$ is shown in Fig. 2. It is noticed that the velocity distribution of the fluid drops on raising the permeability parameter. The temperature and concentration enhance on rising the values of permeability parameter. Figure 3a–c illustrates the profiles of velocity, concentration and temperature for diverse values of unsteady parameter. It is detected that the velocity, temperature and concentration diminish on raising the values of unsteady parameter. Figure 4a–b displays the distributions of velocity, and temperature profiles for several values of heat generation parameter. Figure 4a illustrated that the velocity increases on increasing the internal heat generation parameter (S). There is less effect observed in the case of heat absorption case than that of heat generation case. Figure 4b depicts that the heat generation parameter boosts up the temperature inside the boundary layer and results in thickening the thermal boundary layer. Figure 5a–b shows the effect of Dufour number on velocity and temperature distributions. The momentum and thermal boundary layer thicknesses increase on increasing the Dufour number. The Soret effect on the concentration is illustrated in Fig. 6. The concentration increases on rising the Soret number. There is no significant effect on velocity and temperature profiles on increasing the Cr parameter, which is clearly seen in Fig. 7a–b. It is also ascertained from Fig. 7c that the concentration profiles of the species decrease on increasing the chemical reaction parameter. Figure 8a–c illustrates the suction consequence on the profiles of velocity, concentration and temperature. The momentum, solutal and thermal boundary layer thickness decrease on enhancing the values of suction parameter. The consequence of Biot number on of velocity, concentration and temperature is exposed in Fig. 9a–c. The Biot number enhances the temperature, concentration and velocity.

6 Correlation

The correlation equations are very useful for thermal engineers. Based on the numerical calculations, the following correlation equations are derived.

$$f''(0) = (0.002K^3 - 0.038K^2 + 0.45K) + (0.002A^3 - 0.045A^2 + 0.49A) - 0.16Hg$$
$$- 0.37Df + 0.0006Sr - 0.005Cr + (0.091fw^2 + 0.39fw) + \frac{0.65}{Bi + 1.3} - 0.1$$
$$-\theta'(0) = -0.005K + (0.0004A^3 - 0.009A^2 + 0.073A) - 0.11Hg$$
$$+ (-0.035Df^2 - 0.26Df) + 0.014Sr - 0.042Cr + 0.032fw - \frac{1.39}{Bi + 1.4} + 1.1$$
$$-\phi'(0) = -0.014K + (-0.003A^2 + 0.15A) + 0.046Hg + 0.15Df - 0.11Sr$$
$$+ (-0.056Cr^2 + 0.36Cr) + (0.029fw^2 + 0.28fw) - \frac{0.15}{Bi + 1.5} + 0.89$$

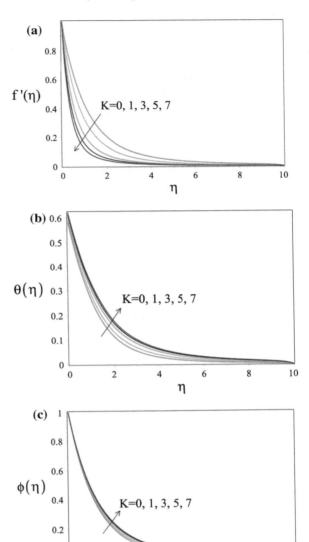

Fig. 2 Velocity (**a**) temperature (**b**) and concentration (**c**) profiles for different permeability parameters K with $A = 1.0$, $Hg = 0.1$, $Df = 0.5$, $Sr = 0.5$, $Cr = 1.0$, $f_w = 0.5$, $Bi = 1.0$

Fig. 3 Velocity (**a**) temperature (**b**) and concentration (**c**) profiles for different unsteady parameters A with $K = 1$, $Hg = 0.1$, $Df = 0.5$, $Sr = 0.5$, $Cr = 1.0$, $f_w = 0.5$, $Bi = 1.0$

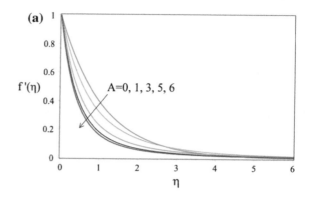
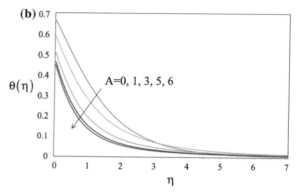
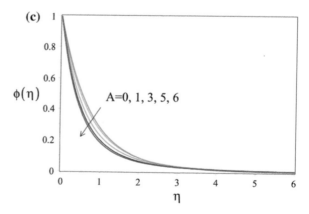

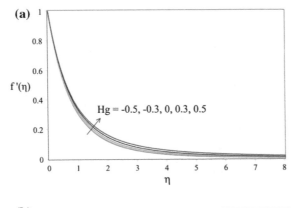

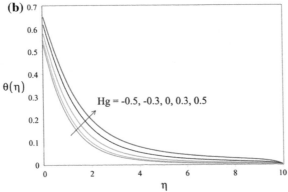

Fig. 4 Velocity (**a**) and temperature (**b**) profiles for different heat generation parameters Hg with $K = 1$, $A = 1.0$, $Df = 0.5$, $Sr = 0.5$, $Cr = 1.0$, $f_w = 0.5$, $Bi = 1.0$

Fig. 5 Velocity (**a**) and temperature (**b**) profiles for different Dufour parameters Df with $K = 1$, $A = 1.0$, $Hg = 0.1$, $Sr = 0.5$, $Cr = 1.0$, $f_w = 0.5$, $Bi = 1.0$

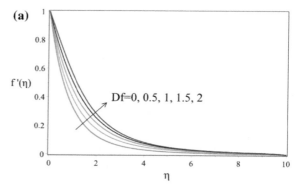

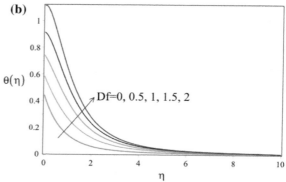

Fig. 6 Concentration profiles for different Soret parameters Sr with $K = 1$, $A = 1.0$, $Hg = 0.1$, $Df = 0.5$, $Cr = 1.0$, $f_w = 0.5$, $Bi = 1.0$

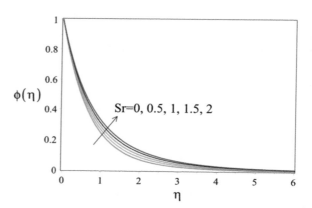

Fig. 7 Velocity (**a**), temperature (**b**) and concentration (**c**) profiles for different chemical reaction parameters Cr with $K = 1$, $A = 1.0$, $Hg = 0.1$, $Df = 0.5$, $Sr = 0.5$, $f_w = 0.5$, $Bi = 1.0$

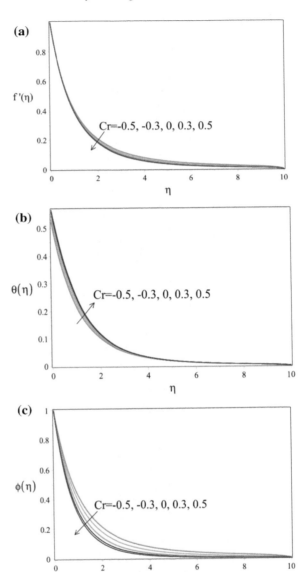

Fig. 8 Velocity (**a**), temperature (**b**) and concentration (**c**) profiles for different suction/injection parameters with $K = 1$, $A = 1$, $Hg = 0.1$, $Df = 0.5$, $Sr = 0.5$, $Cr = 1$, $Bi = 1$

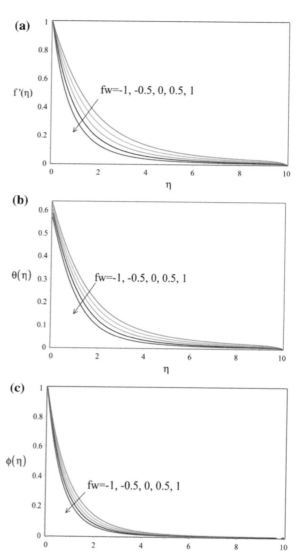

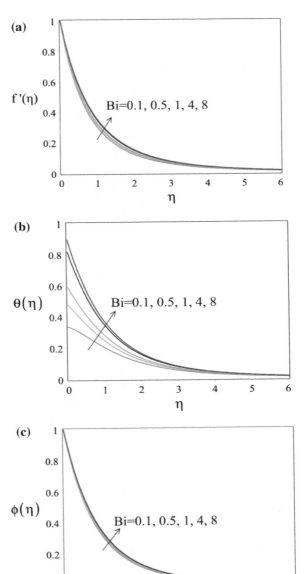

Fig. 9 Velocity (**a**), temperature (**b**) and concentration (**c**) profiles for different Biot number Bi with $K = 1$, $A = 1$, $Hg = 0.1$, $Df = 0.5$, $Sr = 0.5$, $Cr = 1$, $f_w = 0.5$

7 Conclusions

This chapter analyses the Soret and Dufour effects and Newtonian heating on convective flow, mass and heat transfers characteristics of viscous chemically reacting fluid over a stretching surface in a porous medium. The similarity transformation is employed to get the non-dimensional ordinary differential equations from the governing system of partial differential equations. Then, equations are numerically explicated by shooting method and Runge–Kutta fourth-order technique. The following implications are obtained from the results of this content. The chemical reaction enriches the rate of mass transfer and it diminishes the rate of heat transfer. The rate of heat transfer enhances on raising the values of Biot number, but, the mass transfer and skin friction decrease on increasing the Biot number. The physical quantities are boosted up with the suction of fluid. It is also remarked that the Dufour parameter enriches the mass transfer and the Soret effect enriches the heat transfer rate.

References

Beg OA, Bakier AY, Prasad VR (2009) Numerical study of free convection magnetohydrodynamic heat and mass transfer from a stretching surface to a saturated porous medium with Soret and Dufour effects. Comput Mater Sci 46:57–65

Bhuvaneswari M, Sivasankaran S, Ferdows M (2009) Lie group analysis of natural convection heat and mass transfer in an inclined surface with chemical reaction. Non-Linear Anal Hybrid Syst 3(4):536–542

Bhuvaneswari M, Sivasankaran S, Kim YJ (2011) Numerical study on double diffusive mixed convection in a two-sided lid driven cavity with Soret effect. Numer Heat Transf A 59:543–560

Bhuvaneswari M, Sivasankaran S, Kim YJ (2012) Lie group analysis of radiation natural convection flow over an inclined surface in a porous medium with internal heat generation. J Porous Media 15(12):1155–1164

Chambre PL, Young JD (1958) On the diffusion of a chemically reactive species in a laminar boundary layer flow. Phys Fluids 1:48–54

Chaudhary RC, Jain P (2007) An exact solution to the unsteady free-convection boundary-layer flow past an impulsively started vertical surface with Newtonian heating. J Eng Phys Thermophys 80:954–960

Das K, Duari PR, Kundu PK (2015) Numerical simulation of nanofluid flow with convective boundary condition. J Egypt Math Soc 23:435–439

Dass UN, Deka RK, Soundalgekar VM (1994) Effects of mass transfer on flow past an impulsively started infinite vertical plate with constant heat flux and chemical reaction. Forschung im Ingenieurwsen 60:284–287

Evans HL (1968) Laminar boundary layer theory. Addison-Wesley Publishing Company, Boton

Ferdows M, Kaino K, Sivasankaran S (2009) Free convection flow in an inclined porous surface. J Porous Media 12(10):997–1003

Hayat T, Muhammad T, Shehzad SA, Alsaedi A (2015) Soret and Dufour effects in three-dimensional flow over an exponentially stretching surface with porous medium, chemical reaction and heat source/sink. Int J Numer Meth Heat Fluid Flow 25(4):762–781

Kaviany M (1995) Principles of heat transfer in porous media, 2nd edn. Springer, New York, USA

Khader MM, Megahed AM (2014) Effect of viscous dissipation on the boundary layer flow and heat transfer past a permeable stretching surface embedded in a porous medium with a second-order slip using Chebyshev finite difference method. Transp Porous Media 105:487–501

Khidir AA, Sibanda P (2014) Effect of temperature-dependent viscosity on MHD mixed convective flow from an exponentially stretching surface in porous media with cross-diffusion. Spec Top Rev Porous Media—An Int J 5(2):157–170

Lesnic D, Ingham DB, Pop I (1999) Free convection boundary-layer flow along a vertical surface in a porous medium with Newtonian Heating. Int J Heat Mass Transf 42:2621–2627

Lesnic D, Ingham DB, Pop I (2000) Free convection from a horizontal surface in a porous medium with Newtonian heating. J Porous Media 3:227–235

Lesnic D, Ingham DB, Pop I, Storr C (2004) Free convection boundary-layer flow above a nearly horizontal surface in a porous medium with Newtonian heating. Heat Mass Transf 40:665–672

Mahanta G, Shaw S (2015) 3D Casson fluid flow past a porous linearly stretching sheet with convective boundary condition. Alexandria Eng J 54:653–659

Merkin JH (1994) Natural-convection boundary-layer flow on a vertical surface with Newtonian Heating. Int J Heat Fluid Flow 15:392–398

Muthucumaraswamy R (2003) Effects of chemical reaction on moving isothermal vertical plate with variable mass diffusion. Theoret Appl Mech 30(3):209–220

Najafabadi MM, Gorla RSR (2014) Mixed convection MHD heat and mass transfer over a nonlinear stretching vertical surface in a non-Darcian porous medium. J Porous Media 17(6):521–535

Nawaz M, Hayat T, Alsaedi A (2012) Dufour and Soret effects on MHD flow of viscous fluid between radially stretching sheets in porous medium. Appl Math Mech (English Edition) 33(11):1403–1418

Nield DA, Bejan A (2013) Convection in porous media, 4th edn. Springer, New York, USA

Pal D, Mondal H (2012) MHD non-Darcy mixed convective diffusion of species over a stretching sheet embedded in a porous medium with non-uniform heat source/sink, variable viscosity and Soret effect. Commun Nonlinear Sci Numer Simul 17:672–684

Pal D, Mondal H (2013) Influence of Soret and Dufour on MHD buoyancy-driven heat and mass transfer over a stretching sheet in porous media with temperature-dependent viscosity. Nucl Eng Des 256:350–357

Prasad KV, Abel S, Datti PS (2003) Diffusion of chemically reactive species of a non-Newtonian fluid immersed in a porous medium over a stretching sheet. Int J Non-Linear Mech 38:651–657

Rajesh V (2012) Effects of mass transfer on flow past an impulsively started infinite vertical plate with Newtonian heating and chemical reaction. J Eng Phys Thermophys 85(1):221–228

Raman NS, Prabhu KKS, Kandasamy R (2012) Soret and Dufour effects on MHD free convective heat and mass transfer with thermophoresis and chemical reaction over a porous stretching surface. J Appl Mech Tech Phys 53(6):871–879

Tsai R, Huang JS (2009) Heat and mass transfer for Soret and Dufour's effects on Hiemenz flow through porous medium onto a stretching surface. Int J Heat Mass Transf 52:2399–2406

Uddin MJ, Beg OA, Khan WA, Ismail AI (2015) Effect of Newtonian heating and thermal radiation on heat and mass transfer of nanofluids over a stretching sheet in porous media. Heat Transf Asian Res 44(8):681–695

Vafai K (2005) Handbook of porous media, 2nd edn. CRC Press, Taylor and Francis, USA

Local Non-similar Solution of Induced Magnetic Boundary Layer Flow with Radiative Heat Flux

M. Ferdows and Sakawat Hossain

1 Introduction

The study of the flow and heat transfer problem of a viscous incompressible fluid on a plane surface moving in an ambient fluid is important in several engineering processes such as materials manufactured by extrusion processes, heat-treated materials traveling between a feed roll and a wind-up roll or on a conveyer belt. The foundations for the appreciation of the behavior of viscous, incompressible, electrically conducting fluid in boundary layer were first given by Greenspan and Carrier (1959) in the presence of a symmetrically oriented semi-infinite flat plate. They found the solution using Fourier transformation together with asymptotic analysis and claim that the velocity gradient at the plate approaches zero due to increase in the applied magnetic field intensity. Free convection heat transfer with radiation effect over the isothermal stretching sheet and a flat sheet near the stagnation point has been investigated by Ghaly and Elbarbary (2002). Hossain and Takhar (2009) studied the radiation effects on mixed convection along a vertical plate with uniform surface temperature. Cortell (2007) studied the effects of viscous dissipation and radiation on the thermal boundary layer over a nonlinearly stretching sheet. Magnetohydrodynamic-free convection flow of gas past a semi-infinite vertical plate with variable thermophysical properties has been solved by Aboeldahab and Gendy (2002) for high temperature difference numerically using shooting method. The steady laminar incompressible flow of a viscous electrically conducting fluid with constant properties past a semi-infinite flat plate with aligned magnetic field, without heat transfer, has been studied by Glauert (1961) and Gribben (1965).

M. Ferdows (✉) · S. Hossain
Research Group of Fluid Flow Modeling and Simulation, Department of Applied Mathematics, University of Dhaka, Dhaka 1000, Bangladesh
e-mail: ferdows@du.ac.bd

The boundary layer flow due to a stretching vertical surface in a quiescent viscous and incompressible fluid when the buoyancy forces are taken into account has been considered by Chen (2000). The study carried out in this paper deals only with steady-state flow, but the flow and thermal fields may be unsteady due to either impulsive stretching of the surface or external stream and sudden changes in the surface temperature. Davies (1963) has examined the fact that the boundary layer thickness and drag coefficient diminishes steadily as magnetic force parameter S increases. The rate of heat transfer is decreased with the increasing of magnetic force parameter S and the magnetic field and thermal boundary layer thicknesses increases were discussed by Tan and Wang (1967). The radiation effects on magnetohydrodynamic-free convection flow have been considered by Chen (2008). Dandapat and Gupta (2005) extended the problem to study heat transfer and Datti et al. (2005) analyzed the problem over a non-isothermal stretching sheet. Lok et al. (2007) investigated the non-orthogonal stagnation point for Newtonian and non-Newtonian flows toward a stretching sheet. Numerical solutions of the similarity equation were obtained in detail. They found that the position of stagnation point depends on stretching sheet parameter and angle of incidence. The MHD boundary layer flow over a continuously moving plate for a micropolar fluid has been studied by Rahman and Sattar (2006) and Raptis (1998). The steady flow over a continuous moving surface with a magnetic field has been studied by Vajravelu and Hadjinicolaou (1997) and Ali and Al-Yousef (1998). The unsteady boundary layer flow over a stationary semi-infinite flat plate in the presence of magnetic field has been studied by Pop and Na (1998). Zueco and Ahmed (2010) carried out the numerical solution of the mixed convection MHD flow past a vertical porous plate. Magnetohydrodynamic stagnation point flow toward stretching sheet has been studied by Mahapatra and Gupta (2002).

The case of the MHD stagnation point flow and electrically conducting fluid in the presence of a transverse magnetic field by taking an induced magnetic field has been studied by Mahapatra and Gupta (2004). MHD stagnation point flow and heat transfer toward stretching sheet with induced magnetic field have been studied by Ali et al. (2011). The temperature distribution in the flow over a linearly stretching sheet with uniform surface heat flux has been studied by Dutta et al. (1985). Grubka and Bobba (1985) have analyzed the effects of linear surface stretching and variable surface temperature. Fluid flow and mixed convection transport from a moving plate in rolling and extrusion processes have been studied by Karwe and Jaluria (1988). Chen and Strobel (1980) have dealt with the problem of combined force and free convection in boundary layers adjacent to a continuous horizontal sheet maintained at a constant temperature and moving with a constant velocity. Ingham (1986) has investigated the existence of the solutions for the free convection boundary layer flow near a continuously moving vertical plate with temperature inversely proportional to the distance up the plate. Ali and Al-Yousef (1998) have recently investigated the problem of laminar mixed convection adjacent to a uniformly moving vertical plate with suction or injection. Several studies have been reported on induced magnetic effects on boundary layer flow and heat transfer (Ferdows et al. 2014).

In the present study, we consider the velocity exponent parameter (P), the wall temperature exponent parameter (m), Prandtl number (Pr), the reciprocal magnetic Prandtl number (λ), the thermal radiation (N), and buoyancy force parameters, respectively, to extend the study of Nazar and Chen (2011). Numerical solutions of the problem are found by applying Local Non-Similarity method (LNS).

2 Mathematical Formulation

The induced magnetic field is accounted due to large Reynolds number. The effect of the induced magnetic field component $H = (H_1, H_2)$ considered here is as that of Chen (2008). The geometry of the flow is shown in Fig. 1.

The fundamental equations of 2-D steady incompressible, laminar flow are given below Ferdows et al. (2014), Beg et al. (2009):

Continuity equations:

$$\frac{\partial u}{\partial x} + \frac{\partial v}{\partial y} = 0 \qquad (1)$$

$$\frac{\partial H_1}{\partial x} + \frac{\partial H_2}{\partial y} = 0 \qquad (2)$$

Momentum equation:

$$u\frac{\partial u}{\partial x} + v\frac{\partial u}{\partial y} - \frac{\mu}{4\pi\rho}\left(H_1\frac{\partial H_1}{\partial x} + H_2\frac{\partial H_1}{\partial y}\right) = v\frac{\partial^2 u}{\partial y^2} + \left(u_e\frac{du_e}{dx} - \frac{\mu H_e}{4\pi\rho}\frac{dH_e}{dx}\right) \qquad (3)$$

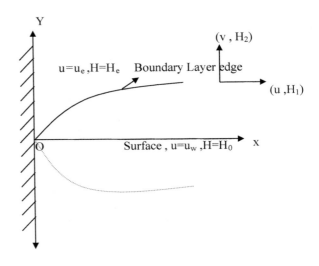

Fig. 1 Schematics of considered geometry

Induction equation:

$$u\frac{\partial H_1}{\partial x} + v\frac{\partial H_1}{\partial y} - H_1\frac{\partial u}{\partial x} - H_2\frac{\partial u}{\partial y} = \mu_e \frac{\partial^2 H_1}{\partial y^2} \qquad (4)$$

Energy equation:

$$u\frac{\partial T}{\partial x} + v\frac{\partial T}{\partial y} = \alpha \frac{\partial^2 T}{\partial y^2} - \frac{\partial q_r}{\partial y} \qquad (5)$$

where x and y are the Cartesian coordinates along the stretching surface and normal to it, respectively. Here u and v are the velocity components along x and y. H_1 and H_2 are the magnetic components along x and y, u_e and H_e are the x-velocity and x-magnetic field at the edge of the boundary layer, v is the kinematic viscosity, ρ is the density and μ dynamic viscosity, $\alpha = \frac{k}{\rho c_p}$ is the thermal diffusivity, k is the thermal conductivity of the fluid, c_p is the specific heat at constant pressure, and q_r is the radiative heat flux. From Rosseland approximation, the following expression can be assigned to radiative heat flux q_r:

$$q_r = -\frac{4\sigma}{3k}\frac{\partial T^4}{\partial y} \qquad (6)$$

where

σ Stefan–Boltzmann constant,
k Rosseland mean absorption coefficient.

Assume that the temperature differences within the flow are sufficiently small such that T^4 can be expressed in a Taylor series about the free steam temperature T_∞ and then neglecting higher order terms as

$$T^4 \approx 4T_\infty^3 T - 3T_\infty^4 \qquad (7)$$

Using (6) and (7) in the last term of Eq. (5), we obtain

$$\frac{\partial q_r}{\partial y} = -\frac{16\sigma_1 T_\infty^3}{3k_1}\frac{\partial^2 T}{\partial y^2} \qquad (8)$$

Introducing q_r in (5), we obtain the following governing boundary layer equations:

Continuity equations:

$$\frac{\partial u}{\partial x} + \frac{\partial v}{\partial y} = 0 \quad (9)$$

$$\frac{\partial H_1}{\partial x} + \frac{\partial H_2}{\partial y} = 0 \quad (10)$$

Momentum equation:

$$u\frac{\partial u}{\partial x} + v\frac{\partial u}{\partial y} - \frac{\mu}{4\pi\rho}\left(H_1\frac{\partial H_1}{\partial x} + H_2\frac{\partial H_1}{\partial y}\right) = v\frac{\partial^2 u}{\partial y^2} + \left(u_e\frac{du_e}{dx} - \frac{\mu H_e}{4\pi\rho}\frac{dH_e}{dx}\right) \quad (11)$$

Induction equation:

$$u\frac{\partial H_1}{\partial x} + v\frac{\partial H_1}{\partial y} - H_1\frac{\partial u}{\partial x} - H_2\frac{\partial u}{\partial y} = \mu_e\frac{\partial^2 H_1}{\partial y^2} \quad (12)$$

Energy equation:

$$u\frac{\partial T}{\partial x} + v\frac{\partial T}{\partial y} = \alpha\frac{\partial^2 T}{\partial y^2} + \frac{16\sigma_1 T_\infty^3}{3k_1}\frac{\partial^2 T}{\partial y^2} \quad (13)$$

The appropriate boundary conditions are

$$\left.\begin{array}{l} u = u_w(x) = Cx^P, v = 0, \frac{\partial H_1}{\partial y} = H_2 = 0, \\ q_w(x) = Bx^m \text{ at } y = 0 \\ u = u_e(x) = Dx^P, H_1 = H_e(x) = H_0 x^P \\ T = T_\infty \text{ as } y \to \infty \end{array}\right\} \quad (14)$$

where C, D, and B are the positive constants, H_0 is the uniform magnetic field at the infinity upstream, P is the velocity exponent parameter, λ is the temperature exponent parameter, T_w is the wall temperature, and T_∞ is the ambient temperature.

3 Mathematical Analysis

To transform the system of governing equations into dimensionless equations, we now introduce the following dimensionless variables (Chen 2000; Ali et al. 2011):

$$\eta = \frac{y}{x}(R_{e_x})^{\frac{1}{2}}$$

$$\xi = \frac{Gr_x \cos\gamma}{(R_{e_x})^{\frac{5}{2}}}$$

$$R_{e_x} = \frac{u_w}{v}x$$

$$\psi(x,y) = v(R_{e_x})^{\frac{1}{2}}f(\xi,\eta) \qquad (*)$$

$$\phi = H_e\left(\frac{vx}{u_w}\right)^{\frac{1}{2}}g(\xi,\eta)$$

$$Gr_x = \frac{g\beta q_w(x)x^4}{kv^2}$$

$$T^4 = 4T_\infty^3 T - 3T_\infty^4$$

$$\theta(\xi,\eta) = \frac{(T-T_\infty)(R_{e_x})^{\frac{1}{2}}}{xq_w(x)/k}$$

where ψ is the stream function, η is the dimensionless distance normal to the sheet, ξ is the buoyancy force parameter, f is the dimensionless stream function, and θ is the dimensionless fluid temperature.

From the above transformations, the non-similar, nonlinear coupled differential Eqs. (9)–(13) become

$$f''' - pf'^2 + \frac{p+1}{2}ff'' + cd^2p + \beta\left[g'^2p - \left(\frac{p+1}{2}\right)gg'' - p\right]$$
$$= \left(m - \frac{5p}{2} + \frac{3}{2}\right)\xi\left[f'\frac{\partial f'}{\partial\xi} - f''\frac{\partial f}{\partial\xi} + \beta\left\{g''\frac{\partial g}{\partial\xi} - g'\frac{\partial g'}{\partial\xi}\right\}\right] \qquad (15)$$

$$\lambda g''' + \frac{p+1}{2}g''f - \frac{p+1}{2}f''g$$
$$= \left(m - \frac{5p}{2} + \frac{3}{2}\right)\xi\left[f'\frac{\partial g'}{\partial\xi} - g'\frac{\partial f'}{\partial\xi} - g''\frac{\partial f}{\partial\xi} + f''\frac{\partial g}{\partial\xi}\right] \qquad (16)$$

$$\frac{1}{P_r}\left(1 + \frac{4}{3}N\right)\theta'' - \left(m - \frac{(p-1)}{2}\right)f'\theta + \frac{p+1}{2}f\theta'$$
$$= \left(m - \frac{5p}{2} + \frac{3}{2}\right)\xi\left[f'\frac{\partial\theta}{\partial\xi} - \theta'\frac{\partial f}{\partial\xi}\right] \qquad (17)$$

The non-similar boundary conditions transform to

$$\left.\begin{array}{l} f(\xi,0) = 0, f'(\xi,0) = 1, g(\xi,0) = 0, \\ g''(\xi,0) = 0, \theta'(\xi,0) = -1 \\ f'(\xi,\infty) = \delta, g'(\xi,\infty) = 1, \theta(\xi,\infty) = 0 \end{array}\right\} \qquad (18)$$

where $\beta = \frac{\mu}{4\pi\rho}\left(\frac{H_o}{C}\right)^2$ is the magnetic force parameter.

$cd = \frac{D}{C}$ is a constant.
$\lambda = \frac{u_e}{v}$ is the reciprocal magnetic Prandtl number.
$P_r = \frac{v}{\alpha}$ is the Prandtl number.
$N = \left(\frac{4\sigma T_\infty^3}{k\alpha}\right)$ is the radiation.

3.1 Local Non-similarity Method

In the local non-similarity method, all the terms in the transformed equations are retained, with the ξ derivatives considering the following transformation:

$$\frac{\partial f}{\partial \xi} = G_1(\xi, \eta);$$

$$\frac{\partial g}{\partial \xi} = G_2(\xi, \eta);$$

$$\frac{\partial \theta}{\partial \xi} = G_3(\xi, \eta);$$

These presents three additional unknown functions, and therefore it is necessary to deduce three further equations to determine $G_1(\xi, \eta)$, $G_2(\xi, \eta)$, and $G_3(\xi, \eta)$. Now, we differentiate the transformed equations with respect to ξ to create the subsidiary equations. The subsidiary equations for $G_1(\xi, \eta)$, $G_2(\xi, \eta)$, $G_3(\xi, \eta)$ contain the terms $\frac{\partial G_1}{\partial \xi}, \frac{\partial G_2}{\partial \xi}, \frac{\partial G_3}{\partial \xi}$, and their η derivatives. When these terms are ignored, the systems of equations for $f(\xi, \eta)$, $g(\xi, \eta)$, $\theta(\xi, \eta)$, $G_1(\xi, \eta)$, $G_2(\xi, \eta)$, and $G_3(\xi, \eta)$ reduce to a system of ordinary differential equations. This form of local non-similarity method is referred to as the second level of truncation, because approximations are by dropping terms in the second-level equation. It is expected that the accuracy of the local non-similarity results will depend upon the truncation level. The equations valid up to second level of truncations are given below.

Now the transformed equations are

$$f''' - pf'^2 + \frac{p+1}{2}ff'' + cd^2 p + \beta\left[g'^2 p - \left(\frac{p+1}{2}\right)gg'' - p\right]$$
$$= \left(m - \frac{5p}{2} + \frac{3}{2}\right)\xi[f'G_1' - f''G_1 + \beta\{g''G_2 - g'G_2'\}]$$

$$\lambda g''' + \frac{p+1}{2}g''f - \frac{p+1}{2}f''g = \left(m - \frac{5p}{2} + \frac{3}{2}\right)\xi[f'G_2' - g'G_1' - g''G_1 + f''G_2]$$

$$\frac{1}{P_r}\left(1+\frac{4}{3}N\right)\theta'' - \left(m - \frac{(p-1)}{2}\right)f'\theta + \frac{p+1}{2}f\theta'$$
$$= \left(m - \frac{5p}{2} + \frac{3}{2}\right)\xi[f'G_3 - \theta'G_1]$$

$$G_1''' - 2pf'G_1' + \frac{p+1}{2}(G_1f'' + fG_1'') + \beta\left[2g'G_2'p - \frac{p+1}{2}(G_2g'' + gG_2'')\right] = \left(m - \frac{5p}{2} + \frac{3}{2}\right)$$
$$\left[\{f'G_1' - f''G_1 - \beta G_2'g' + \beta G_2 g''\} + \xi\left\{(G_1')^2 - G_1''G_1 - \beta(G_2')^2 + \beta G_2 G_2''\right\}\right]$$
(19)

$$\gamma G_2''' + \frac{p+1}{2}(G_1 g'' + fG_2'') - \frac{p+1}{2}(G_2 f'' + gG_1'')$$
$$= \left(m - \frac{5p}{2} + \frac{3}{2}\right)\left\{[G_2'f' - G_1'g' - G_1 g'' + G_2 f''] + \xi\{G_2 G_1'' - G_1 G_2''\}\right]$$
(20)

$$\left(1+\frac{4}{3}N\right)G_3'' + P_r\left[\frac{p+1}{2}(fG_3' + G_1\theta') - \left(m - \frac{(p-1)}{2}\right)(G_3 f' + \theta G_1')\right]$$
$$= P_r\left(m - \frac{5p}{2} + \frac{3}{2}\right)[\{G_3 f' - G_1 \theta'\} + \xi\{G_3 G_1' - G_1 G_3'\}]$$
(21)

The boundary conditions are

$$\left.\begin{array}{llll} G_1(\xi,0) = 0, & G_1'(\xi,0) = 0 & f(\xi,0) = 0, & f'(\xi,0) = 1, \\ G_1'(\xi,\infty) = 0; & G_2'(\xi,0) = 0 & g(\xi,0) = 0, & g''(\xi,0) = 0 \\ G_2'(\xi,\infty) = 0; & G_3(\xi,0) = 0 & \theta'(\xi,0) = -1, & f'(\xi,\infty) = cd \\ G_2''(\xi,0) = 0, & G_3(\xi,\infty) = 0 & g'(\xi,\infty) = 1, & \theta(\xi,\infty) = 0 \end{array}\right\} \quad (22)$$

The non-similarity solutions are obtained by solving together the boundary layer Eqs. (15)–(18) and the auxiliary system (19)–(22). At every grid point, the iteration process continues until the convergence criterion for all the variables, 10^{-6}, is achieved.

4 Results and Discussion

4.1 Flow Profiles

We numerically solve the system of partial differential equations, as mentioned in the previous chapter. Now in order to discuss the results, we solve the problem for different values of the parameters such as Prandtl number Pr, reciprocal of magnetic Prandtl number λ, magnetic force parameter β, radiation parameter N, velocity exponent component p, the wall heat flux exponent parameter m, and carry out the discussion how these parameters do effect the velocity and temperature of the flow field. We also sketch the graph of different profiles to show the effect of mentioned parameter on each profile.

Figure 2a shows the effect of constant cd and also the buoyancy force parameter ξ on the velocity profile. We observe that as the value of the constant cd increases, the corresponding velocity profile increases and when the buoyancy force parameter ξ increases, the velocity profile increases.

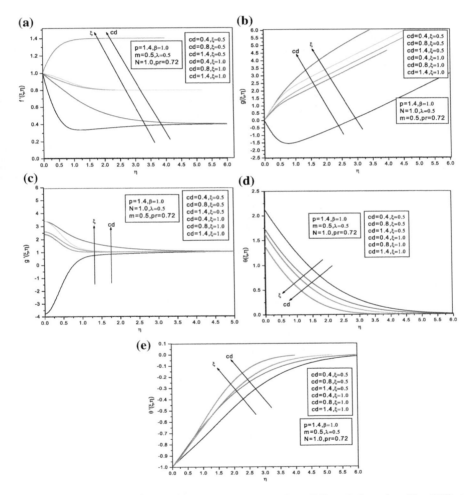

Fig. 2 a Effect of cd on f' profile. **b** Effect of cd on g profile. **c** Effect of cd on g' profile, **d** Effect of cd on θ profile, **e** Effect of cd on θ' profile

Figure 2b–c shows the effect of constant cd and also the effect of buoyancy force parameter ξ on the g and g' profile, respectively. We observe from Fig. 2b that as the values of the constant cd and the buoyancy force parameter ξ increase, the corresponding g' profile increases. Also, we observe from Fig. 2c that the values of the constant cd and the buoyancy force parameter ξ increase, the corresponding g' profile increases.

Figure 2d–e shows the effect of constant cd on the temperature profiles. We observe from Fig. 2d that as the value of the constant cd increases, the corresponding temperature profile increases and when the buoyancy force parameter ξ increases, temperature profile increases. Figure 2e shows that in the increase of the constant cd the θ profile increases and in the increase of buoyancy force parameter ξ, the θ' profile also increases.

Figure 3a–b shows the effect of Prandtl number parameter pr and also the buoyancy force parameter ξ on the velocity profile f' and g profile, respectively. We observe from both figures that as the value of the Prandtl number parameter pr increases, both the f' and g profile increase. When buoyancy force parameter ξ increases, the corresponding f' profile increases, and when it increases, the corresponding g profile increases.

Figure 3c–d shows the effect of Prandtl number parameter pr and also the buoyancy force parameter ξ on the profile g' and the temperature profile θ, respectively. We observe from both figures that as the value of the Prandtl number parameter pr increases, both the g' and θ profiles increase. When buoyancy force parameter ξ increases, there is no change in the corresponding g' profile and the corresponding θ profile.

Figure 3e shows the effect of Prandtl number parameter pr and also the buoyancy force parameter ξ on profile θ' profile. We observe that as the value of the Prandtl number parameter pr increases, the θ' profile increases. When buoyancy force parameter ξ increases, there is no effect on the corresponding the θ' profile.

Figure 4a shows the effect of Prandtl number parameter pr and also the buoyancy force parameter ξ on the velocity profile f'. Also, we observe that there is no effect of Prandtl number Pr and the buoyancy force ξ on the velocity profile f'.

Figure 4b–c shows the effect of radiation parameter N and also the buoyancy force parameter ξ on the profile g and g' profile, respectively. We observe that as the value of the radiation parameter N increases, both the g and g' profiles decrease. As the buoyancy force parameter ξ increases, both the g profile and the g' profiles decrease.

Figure 4d–e shows the effect of radiation parameter N and also the buoyancy force parameter ξ on the profile θ and θ' profile, respectively. We observe from Fig. 4d that as the value of the radiation parameter N increases, the θ profile increases and from Fig. 4e that as the value of the radiation parameter N increases, the θ' profile decreases. As the buoyancy force parameter ξ increases, there is no change in the profile θ and the profile θ'.

Figure 5a–b shows the effect of reciprocal magnetic Prandtl number λ and also the buoyancy force parameter ξ on the profile f' and g profile, respectively. We

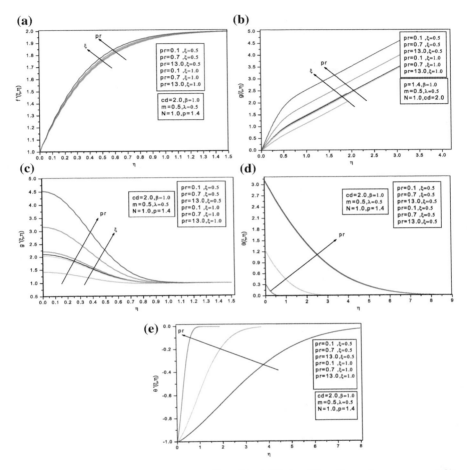

Fig. 3 a Effect of Prandtl number pr on f' profile. **b** Effect of Prandtl number pr on g profile. **c** Effect of Prandtl number pr on g' profile. **d** Effect of Prandtl number pr on θ profile. **e** Effect of Prandtl number pr on θ' profile

observe from Fig. 5a that as the values of reciprocal magnetic Prandtl number λ and the buoyancy force parameter ξ increase, there is no change in the velocity profile f' and from Fig. 5b that as the value f' increases, the g profile decreases. As the buoyancy force parameter ξ increases, the g profile increases.

Figure 5c–d shows the effect of reciprocal magnetic Prandtl number λ and also the buoyancy force parameter ξ on the profile g' and θ profile, respectively. We observe from Fig. 5c that as the value of reciprocal magnetic Prandtl number λ increases, the g' profile decreases and as the buoyancy force parameter ξ increases, the g' profile decreases. We observe from Fig. 5d that as the values of reciprocal

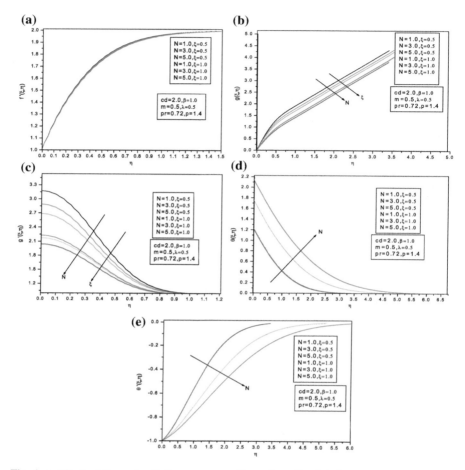

Fig. 4 a Effect of thermal radiation parameter N on f' profile. **b** Effect of thermal radiation parameter N on g profile. **c** Effect of thermal radiation parameter N on θ profile. **d** Effect of thermal radiation parameter N on θ profile. **e** Effect of thermal radiation parameter N on θ' profile

magnetic Prandtl number λ and the buoyancy force parameter ξ increase, there is no change in the profile θ.

Figure 5e shows the effect of reciprocal magnetic Prandtl number λ and also the buoyancy force parameter ξ on the θ' profile. We observe that as the value of reciprocal magnetic Prandtl number λ increases, there is no change in the θ' profile.

Figure 6a shows the effect of magnetic force parameter β and also the buoyancy force parameter ξ on f' profile. We observe that as the values of magnetic force parameter β and the buoyancy force parameter ξ increase, f' profile increases and decreases, respectively.

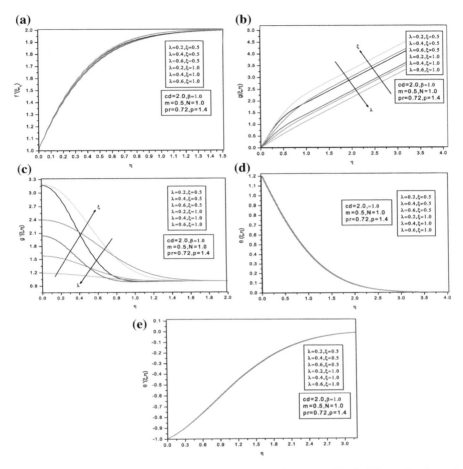

Fig. 5 a Effect of reciprocal magnetic Prandtl number λ on g profile. **b** Effect of reciprocal magnetic Prandtl number λ on g profile. **c** Effect of reciprocal magnetic Prandtl number λ on g' profile. **d** Effect of reciprocal magnetic Prandtl number λ on θ profile. **e** Effect of reciprocal magnetic Prandtl number λ on θ' profile

Figure 6b–c shows the effect of magnetic force parameter β and buoyancy force parameter ξ on the g' and θ profiles, respectively. We observe from Fig. 6b that as the values of the magnetic force parameter β and the buoyancy force parameter ξ increase, the corresponding g' profile decreases and increases, respectively. Also, we observe from Fig. 6c that as the values of the magnetic force parameter β and the buoyancy force parameter ξ increase, the corresponding θ profile increases and decreases, respectively.

Figure 6d shows the effect of reciprocal magnetic Prandtl number λ and also the buoyancy force parameter ξ on the θ' profile. We observe that as the value of

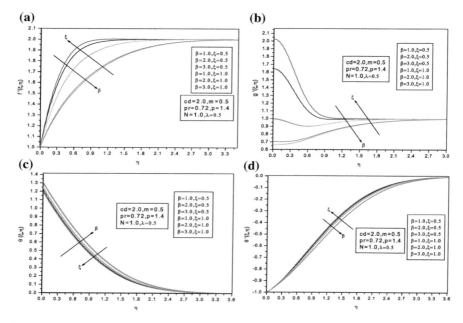

Fig. 6 a Effect of magnetic force parameter β on f' profile. **b** Effect of magnetic force parameter β on g' profile. **c** Effect of magnetic force parameter β on θ profile. **d** Effect of magnetic force parameter β on θ' profile

magnetic force parameter β increases, θ' profile decreases and as the buoyancy force parameter ξ increases, θ' profile increases.

Figure 7a shows the effect of velocity exponent parameter p and also the buoyancy force parameter ξ on f' profile. We observe that as the values of velocity exponent parameter p and the buoyancy force parameter ξ increase, f' profile increases and decreases, respectively.

Figure 7b–c shows the effect of velocity exponent parameter p and also the effect of buoyancy force parameter ξ on the g and g' profiles, respectively. We observe from Fig. 7b that as the values of the velocity exponent parameter p and the buoyancy force parameter ξ increase, the corresponding g profile increases and decreases, respectively. Also, we observe from Fig. 7c that as the values of the velocity exponent parameter p and the buoyancy force parameter ξ increase, the corresponding g' profile increases and decreases, respectively.

Figure 7d–e shows the effect of velocity exponent parameter p and also the effect of buoyancy force parameter ξ on the θ and θ' profiles, respectively. We observe from Fig. 7d that as the value of the velocity exponent parameter p increases, the corresponding θ profile decreases and there is no effect of buoyancy force parameter ξ. Also, we observe from Fig. 7e that as the value of the velocity exponent parameter p increases, the corresponding g' profile decreases when $\eta < 2.0$ and decreases when $\eta > 2.0$, and there is also no effect of buoyancy force parameter ξ.

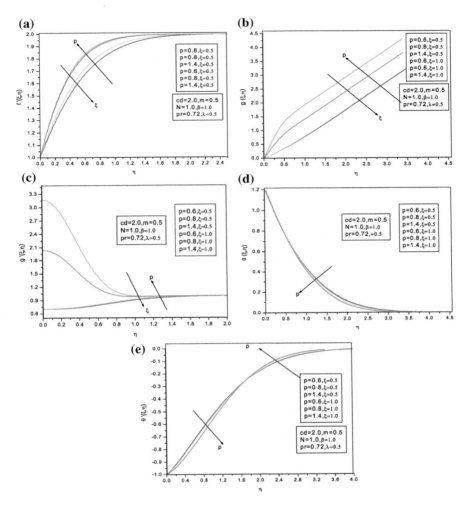

Fig. 7 a Effect of velocity exponent parameter p on f' profile. **b** Effect of velocity exponent parameter p on g profile. **c** Effect of velocity exponent parameter p on g' profile. **d** Effect of velocity exponent parameter p on θ profile. **e** Effect of velocity exponent parameter p on θ' profile

Figure 8a–b shows the effect of wall heat flux exponent parameter m and also the effect of buoyancy force parameter ξ on the f' and g profiles, respectively. We observe from Fig. 8a that there is no effect of wall heat flux exponent parameter m and the buoyancy force parameter ξ on the f' profile. Also, we observe from Fig. 8b that there is no effect of wall heat flux exponent parameter m and the buoyancy force parameter ξ on the g profile.

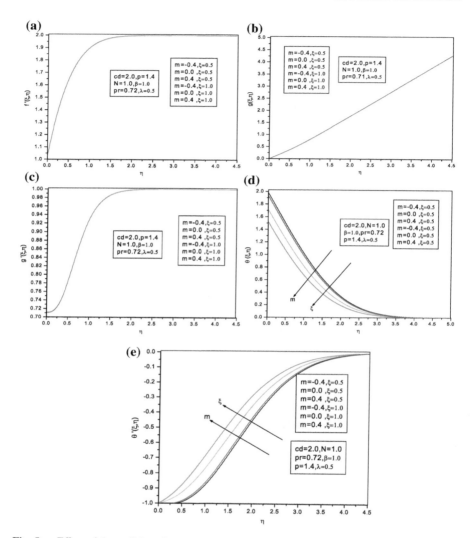

Fig. 8 a Effect of the wall heat flux exponent parameter m on f' profile. **b** Effect of temperature exponent parameter m on g profile. **c** Effect of the wall heat flux exponent parameter m on g' profile. **d** Effect of the wall heat flux exponent parameter m on θ profile. **e** Effect of the wall heat flux exponent parameter m on θ' profile

Figure 8c–d shows the effect of wall heat flux exponent parameter m and also the effect of buoyancy force parameter ξ on the f' and g profiles, respectively. We observe from Fig. 8c that there is no effect of wall heat flux exponent parameter m and the buoyancy force parameter ξ on the g' profile. Also, we observe from Fig. 8d that as the values of the wall heat flux exponent parameter m and the buoyancy force parameter ξ increase, the corresponding θ profile decreases.

Figure 8e shows the effect of wall heat flux exponent parameter m and also the effect of buoyancy force parameter ξ on the θ' profile. We observe that as the value of the wall heat flux exponent parameter m increases, the θ' profile increases and also with the increase of the buoyancy force parameter ξ, the θ' profile increases.

4.2 Physical Parameters

The skin friction coefficient (C_{fx}):
The skin friction coefficient (C_f) and local Nusselt number (Nu_x) are significant in the engineering field. These parameters refer to the wall shear stress and local wall heat transfer rate, respectively.
From equation (*), we have

$$\frac{\partial u}{\partial y} = \frac{v\eta^3 x}{y^3} f''(\xi, \eta)$$

Wall shear stress is given by

$$\tau_x = \mu \left(\frac{\partial u}{\partial y}\right)_{y=0} = \mu \frac{v\eta^3 x}{y^3} f''(\xi, 0) = \mu \frac{v\eta^3 x^3}{x^2 y^3} f''(\xi, 0)$$

$$= \mu \frac{v}{x^2}(Re_x)^{\frac{3}{2}} f''(\xi, 0)$$

$$C_{fx} = \frac{\tau_x}{\rho u_w^2 / 2} = \frac{2\tau_x}{\rho x^2 \frac{(Re_x)^2}{x^2}} f''(\xi, 0) = \frac{2\mu v (Re_x)^{\frac{3}{2}}}{\rho x^2 v^2 \frac{(Re_x)^2}{x^2}} f''(\xi, 0)$$

$$= \frac{\mu (Re_x)^{\frac{3}{2}}}{\rho v (Re_x)^2} 2f''(\xi, 0) = \frac{\rho v (Re_x)^{-\frac{1}{2}}}{\rho v} 2f''(\xi, 0)$$

$$\therefore C_{fx}(Re_x)^{\frac{1}{2}} = 2f''(\xi, \eta)$$

The local Nusselt number (Nu_x):
Again from Newton's law of cooling, we have

$$q_w = h(T - T_\infty)$$
$$h = q_w / (T - T_\infty)$$

where h is the heat transfer coefficient.

Hence, the local Nusselt number Nu_x is given by

$$Nu_x = \frac{hx}{k} = \frac{q_w x}{(T - T_\infty)k} = \frac{q_w x}{\frac{\theta(\xi,0)q_w x}{(R_{e_x})^{\frac{1}{2}}k}} = \frac{(R_{e_x})^{\frac{1}{2}}}{\theta(\xi,0)}$$

$\therefore Nu_x(R_{e_x})^{-\frac{1}{2}} = 1/\theta(\xi,0)$.

We observe the following physical parameters results from our model.

Figure 9a shows the effect of reciprocal of magnetic Prandtl number λ on skin friction coefficient and also the effect of buoyancy force parameter ξ on skin friction coefficient $C_{fx}(R_{e_x})^{\frac{1}{2}}$. Figure 9b is constructed to explain the effects of the reciprocal of magnetic Prandtl number λ and the buoyancy force parameter ξ on the local

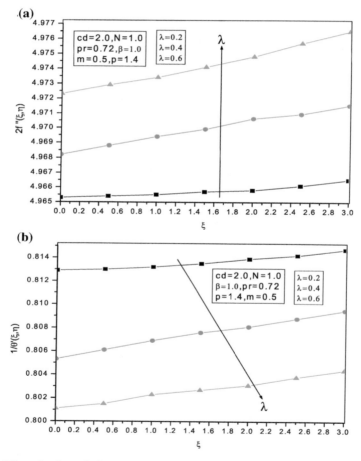

Fig. 9 **a** Effect of reciprocal of magnetic Prandtl number λ on skin friction coefficient. **b** Effect of the reciprocal of magnetic Prandtl number λ on local Nusselt number

Nusselt number. The skin friction coefficient, $C_{fx}(Re_x)^{\frac{1}{2}}$, is presented as a function of ξ at various values of λ. From Fig. 9a, we see that the skin friction coefficient increases as the buoyancy force parameter ξ and the reciprocal of magnetic Prandtl number λ increase, and from Fig. 9b for the increase of the value of λ the local Nusselt number decreases.

Figure 10a shows the effect of magnetic force parameter β and also the effect of buoyancy force parameter ξ on the skin friction coefficient $C_{fx}(Re_x)^{\frac{1}{2}}$. Figure 10b is constructed to explain the effects of the magnetic force parameter β and the

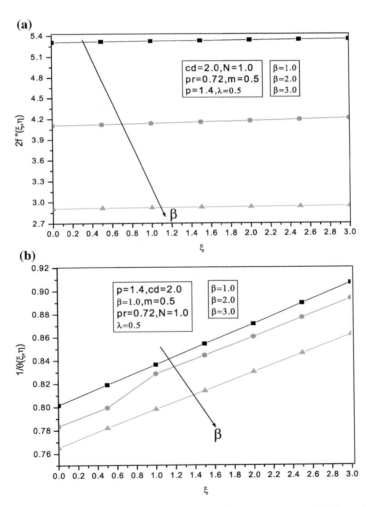

Fig. 10 a Effect of magnetic force parameter β on skin friction coefficient. b Effect of magnetic force parameter β on local Nusselt number

buoyancy force parameter ξ on the local Nusselt number, $Nu_x(Re_x)^{-\frac{1}{2}}$. From Fig. 10a, we see that the skin friction coefficient increases as the buoyancy force parameter ξ increases for a given β and for the increase of the value of β, the skin friction coefficient decreases. From Fig. 10b, we see that the local Nusselt number increases rapidly as the buoyancy force parameter ξ increases for a given β and for the increase of the value of β, the s local Nusselt number increases.

Figure 11a shows the effect of velocity exponent parameter p and also the effect of buoyancy force parameter ξ on the skin friction coefficient $C_{fx}(Re_x)^{\frac{1}{2}}$. Figure 11b

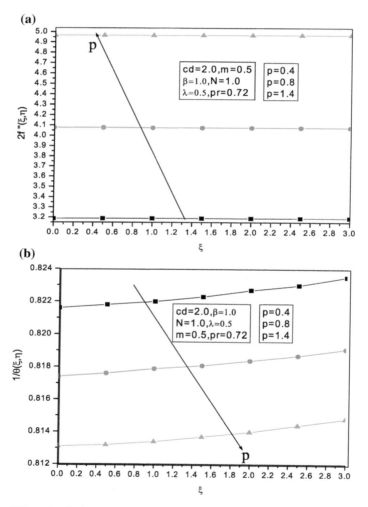

Fig. 11 a Effect of velocity exponent parameter p on the skin friction coefficient. b Effect of velocity exponent parameter p on the local Nusselt number

is constructed to explain the effects of the velocity exponent parameter p and the buoyancy force parameter ξ on the local Nusselt number, $Nu_x(Re_x)^{-\frac{1}{2}}$. From Fig. 11a, we see that the skin friction coefficient increases slowly as the buoyancy force parameter ξ increases for a given p and for the increase of the value of p, the skin friction increases. From Fig. 11b, we see that the local Nusselt number increases as the buoyancy force parameter ξ increases for a given p and for the increase of the value of p, the local Nusselt number decreases.

5 Conclusions

In summing up what has been discussed above, we are remarking the effects of different physical parameters such as radiation parameter N, magnetic force parameter β, Prandtl number Pr, reciprocal of magnetic Prandtl number λ, velocity exponent parameter p, and the wall heat flux exponent parameter m on different flow profiles, coefficients of skin friction, and local Nusselt number which are given as below.

- It is observed that the local Nusselt number decreases as p, β, and λ increase for a given value of buoyancy force parameter.
- The skin friction coefficient increases as p and λ increase for a given value of buoyancy force parameter ξ and it decreases as β increases. Also, the buoyancy force parameter ξ is found to have significant effects on the local Nusselt number and the skin friction coefficient. The local Nusselt number increases with the increasing of buoyancy force parameter (ξ).
- It is also observed that for increasing the value of cd, the values of velocity profile and the magnetic field profile increase but temperature distribution decreases.
- It is noted that the values of the velocity profile, magnetic field profile, and temperature profile increase with the change of Prandtl number Pr.
- The variation in the radiation parameter N leads to decrease the value of magnetic field profile and to increase the temperature profile but there is no change in velocity profile.
- It is also concluded that with the increase of the magnetic force parameter, temperature profile increases but the velocity profile and magnetic field gradient function decrease.
- Therefore, the results obtained explaining the basic thermal behavior of a continuously stretching sheet would be more useful in accomplishing the design for relevant manufacturing processes.

References

Aboeldahab EM, Gendy MSE (2002) Radiation effect on MHD-convection flow of a gas past a semi-infinite vertical plate with variable thermophysical properties for high temperature differences. Can J Phys 80:1609–1619

Ali M, Al-Yousef F (1998) Laminar mixed convection from a continuously moving vertical surface with suction or injection. Heat Mass Transfer 33:301–306

Ali FM, Nazar R, Arfin NM, Pop I (2011) MHD stagnation point flow and heat transfer towards stretching sheet with induced magnetic field. Appl Math Mech 32(4):409–418

Beg OA, Bakier AY, Prasad VR, Zueco J, Ghosh SK (2009) Nonsimilar, laminar, steady, electrically-conducting forced convection liquid metal boundary layer flow with inducedmagnetic field effects. Int J Therm Sci 48(8):1596–1606

Chen CH (2000) Mixed convection cooling of a heated, continuously stretching surface. Heat Mass Transfer 36:79–86

Chen TM (2008) Radiation effects on magnetohydodynamic free convection flow. Technical Note, AIAA. J Thermopy Heat Transfer 22(1):125–128

Chen TS, Strobel FA (1980) Buoyancy effects on boundary layer inclined, continuous, moving horizontal flat plate. ASME J Heat Transfer 102:170–172

Cortell R (2007) Viscous flow and heat transfer over a nonlinearly stretching sheet. Appl Math Comput 184(2):864–873

Dandapat BS, Gupta AS (2005) Flow and heat transfer in a viscoelastic fluid over a stretching sheet. Int J Non-Linear Mech 40:215–219

Datti PS, Prasad KV, Abel MS, Joshi A (2005) MHD viscoelastic fluid flow over a non isothermal stretching sheet. Int J Eng Sci 42:935–946

Davies TV (1963) The magnetohydrodynamic boundary layer in two-dimensional steady flow past a semi-infinite flat plate, Part I, uniform conditions at infinity. Proc R Soc Lond A 273:496–507

Dutta BK, Roy P, Gupta AS (1985) Temperature field in flow over a stretching sheet with uniform heat flux. Int Comm Heat Mass Transfer 12:89–94

Ferdows M, Shamima I, Beg OA (2014) Numerical simulation of Marangoni magnetohydrodynamic bio-nanofluid convection from a non-isothermal surface with magnetic induction effects: a bio-nanomaterial manufacturing transport model. J Mech Med Biol 14(3):1–32

Ghaly AY, Elbarbary EME (2002) Radiation effect on MHD free—convection flow of a gas at a stretching surface with a uniform free stream. J Appl Math 2(2):93–103

Glauert MB (1961) A study of the magnetohydrodynamic boundary layer on a flat plate. J Fluid Mech 10:276–288

Greenspan HP, Carrier GF (1959) The magnetohydrodynamic flow past a flat plate. J Fluid Mech 6:77–96

Gribben RJ (1965) The magnetohydrodynamic boundary layer in the presence of pressure gradient. Proc R Soc A 287:123–141

Grubka LG, Bobba KM (1985) Heat transfer characteristic of a continuous stretching surface with variable temperature. ASME J Heat Transfer 107:248–250

Ibrahim W (2016) The effect of induced magnetic field and convective boundary condition on MHD stagnation point flow and heat transfer of upper-convected Maxwell fluid in the presence of nanoparticle past a stretching sheet. Propul Power Res 5(2):164–175

Ingham DB (1986) Singular and non unique solutions of the boundary layer equations for the flow due to free convection near a continuously moving vertical plate. J Appl Mech Phys (ZAMP) 37:559–572

Karwe MV, Jaluria Y (1988) Fluid flow and mixed convection transport from a moving plate in rolling and extrusion processes. ASME J Heat Transfer 110:655–661

Lok YY, Amin N, Pop I (2007) Comments on: steady two-dimensional oblique stagnation-point flow towards a stretching surface. Reza M, Gupta AS (2005) Fluid Dyn Res 37(2005):334–340. Fluid Dyn Res 39:505–510

Mahapatra TR, Gupta AS (2002) Heat transfer in a stagnation-point flow towards a stretching sheet. Heat Mass Transfer 38:517–521

Mahapatra TR, Gupta AS (2004) Stagnation point-flow of a viscoelastic fluid towards a stretching surface. Int J Non-Linear Mech 39:811–820

Pop I, Na TY (1998) A note on MHD over a stretching permeable surface. Mech Res Commun 25:263–269

Rahman MM, Sattar MA (2006) Magnetohydrodynamic convective flow of a micropolar fluid past a continuously moving vertical porous plate in the presence of heat generation/absorption. ASME J Heat Transfer 128:142–152

Rahman MM, Rahman MA, Samad MA, Alam MS (2009) Heat transfer in a micropolarfluid along a non-linear stretching sheet with a temperature–dependent viscosity and variable surface temperature. Int J Thermophys 30:1649

Raptis A (1998) Flow of a micropolar fluid past a continuously moving plate by the presence of radiation. Int J Heat Mass Transfer 41:2865–2866

Tan CW, Wang CT (1967) Heat transfer in aligned-field magnetohydrodynamic flow past a flat plate. Int J Heat Mass Transfer 11:319–329

Vajravelu K, Hadjinicolaou A (1997) Convective heat transfer in an electrically conducting fluid at a stretching surface with uniform free stream. Int J Eng Sci 35:1237–1244

Zueco J, Ahmed S (2010) Combined heat and mass transfer by convection MHD flow along a porous plate with chemical reaction in presence of heat source. Appl Math Mech 31(10):1217–1230

Printed by Printforce, the Netherlands